INTRODUCTION TO

COMPUTER MATHEMATICS

Girolamo Cardano
1501–1576

Rene Descartes
1596–1650

Abraham De Moivre
1667–1754

Fifteenth Century

1400 In Europe, algorithms using numerals began to replace the abacus for performing arithmetic operations. **(pp. 22, 38–39)**

1478 The first arithmetic book printed with movable type was produced in Italy. **(pp. 38–39)**

Sixteenth Century

1500 Mathematical symbolism began to take its present form. **(p. 50)**

1525 Christoff Rudolff introduced the radical sign in *Die Coss.* **(p. 232)**

1545 Girolamo Cardano published the solution to the general cubic and quartic equations in *Ars Magna.* **(pp. 235, 252)**

1550 Robert Recorde first used the equals sign. **(p. 245)**

1585 Simon Stevin published the first systematic treatment of decimal fractions in *La Disme.* **(p. 125)**

Seventeenth Century

1614 John Napier introduced logarithms as an aid to computation. **(p. 281)**

1622 William Oughtred used sliding logarithmic scales to make a slide rule. **(p. 281)**

1637 Rene Descartes laid the foundations of analytic geometry. **(pp. 207, 260)**

1642 Blaise Pascal built a calculator to add and subtract. **(pp. 71, 138)**

1654 Pierre Fermat and Blaise Pascal founded the theory of probability. **(pp. 71, 138, 382)**

1694 Gottfried Wilhelm von Leibniz completed construction of a calculator that would add, subtract, and multiply. **(pp. 42, 71, 318)**

INTRODUCTION TO

COMPUTER MATHEMATICS

Eleanor H. Ninestein
Fayetteville Technical Institute

Scott, Foresman and Company
Glenview, Illinois   London, England

To Ella Hunter and Dwight

The images used on the cover include a binary code, computer print-out paper, lightning bolts, an artist's rendering of a computer key, a photo micrograph, a digitized 2, a computer keyboard, and mathematical notations on a chalkboard.

Credit lines for photos appearing in the book are in the Acknowledgments section at the end of the book. This section is to be considered an extension of the copyright page. Cover: left column, bottom photo Thomas H. Ives, ©1986; center column, center photo Elia Kairinen and Emil Bernstein, Gillette Research Institute, Rockville, Maryland; all other images produced by Candace Haught.

Library of Congress Cataloging-in-Publication Data

Ninestein, Eleanor H.
   Introduction to computer mathematics.

   Includes bibliographies and index.
   1. Electronic data processing—Mathematics.
I. Title.
QA76.9.M35N56   1987 512′.1 86-13071
ISBN 0-673-18205-3

12345678—RRC—919089888786

PREFACE

This book provides a foundation in mathematics for students who wish to study computer programming or computer-related fields, such as data processing. Because of its flexibility, it could be adapted for a combined class of technical and liberal arts students.

Many students who take a computer mathematics course have had no programming experience, but have been exposed to mathematics. Since the mental processes involved in doing mathematics are often required for computer programming, the book serves as a bridge between these two disciplines. The topics emphasize transferrable skills such as formula manipulation, a top-down approach to problem solving, abstraction from specific cases to general problems, and recognition of common threads in different topics. In a field that changes as rapidly as the computer industry, the ability to think is the most important skill a student can develop.

Recently, there has been an emphasis on algorithms in the curriculum for students of computing. There is, however, a need for a text that stresses *algorithmic thinking* rather than *algorithms*. There is no dichotomy between the traditional approach of problem solving by deductive reasoning and the contemporary approach of problem solving by technique. Reasoning and technique, insight and action, should be inseparable. To meet this goal, this book emphasizes the importance of an intuitive understanding rather than a rigorous development of ideas. The approach throughout the text is to develop general algorithms from specific examples.

Algorithms are introduced in chapter 1 and are stated throughout the book. They are written in as simple and direct a form as possible. Numerous exercises direct the student to put the algorithms in structured form using pseudocode or flowcharts. This approach allows the instructor the choice of stressing either pseudocode or flowcharts and gives the student the opportunity to practice developing algorithms without always having to start from scratch. The student is thus able to test his or her understanding at the level Donald Knuth has suggested; that is, "a person doesn't really understand something until he can teach it to a computer." For students who are familiar with a programming language, most of the algorithms can be developed into short programs.

Pseudocode/flowcharting problems are indicated by the symbol ⮕ to separate them from the routine problems. Problems indicated by an asterisk (✳) are either more difficult (and should be assigned with discretion), or they introduce additional concepts related to, but not covered in, the text. Many of these problems introduce applications to computer programming, but they require no prior knowledge of programming. Sequence and series, parity, and Polish notation are examples of supplementary topics. Answers to the odd-numbered routine problems are given in the answer section at the end of the book. The answers to enrichment problems are not. Many of these problems have simple answers, but perceptiveness is required to find them. If the student looks at the solutions too soon, he or she will derive none of the intended benefits of thinking through the problems.

There is a great deal of diversity in computer mathematics courses. The course may be taught as a data processing course or as a math course. It may be as short as one quarter, with a strong algebra prerequisite, or as long as two semesters, with a review of algebra included. This book is organized to allow the instructor to tailor the course to meet the needs of his or her students. Even if the chapters on algebra are not covered in class, some students (particularly mature and highly motivated individuals returning to school after several years) will find those chapters useful as a reference for review. With the exception of the algebraic topics, each chapter is largely independent of the others. A one-quarter course might cover chapters 1, 2, 3, and 4, with selected topics from 7 and 8 and either 9 or 10. A two-quarter or one-semester course might cover chapters 1 through 6, with selected topics from 7 and 8. A two-semester course could cover the entire book. The following diagram shows the relationship of chapters to each other.

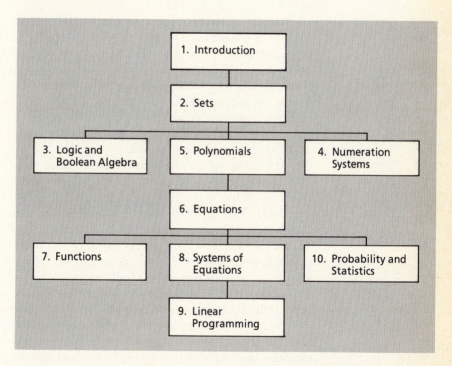

For the instructor, the key features of this text are its flexibility, the emphasis on algorithmic thinking, the large number of problems, the enrichment material, and the references (which are suitable sources for student reports) at the end of each chapter.

For the student, the key features are simplicity, the large number of detailed examples, applications to computer programming, chapter summaries, and chapter tests. Biographical sketches are included for entertainment as well as to remind the student that mathematics, even in the context of electronic computing, is a human pursuit.

Many people have contributed to the completion of this text. I appreciate Michael McLaurin scheduling my classes to accommodate my writing. I also appreciate the attention to detail of the staff at Scott, Foresman and Company. Special thanks go to Steve Quigley, Barbara Maring, and Marisa L. L'Heureux for their patience, insight, and continual optimism. I am also grateful to the following reviewers, who made a number of helpful suggestions:

Kathleen Carlborg
Prairie State College

Norman Carson
Marion Technical College

Caroline Goodman
Parkland College

James Hall
Parkland College

James Hodge
College of Lake County

Neal Howard
Danville Community College

Marilyn Mays
North Lake College

Robert A. McCoy
Southern Oregon State College

John Monroe
University of Akron

Herbert Wright
Marion Technical College

My husband, Dr. V. Dwight House, chairman of the Department of Mathematics and Computer Science at Methodist College, also read the manuscript and has served as critic and consultant throughout the project. His advice and support have been invaluable. There are many people whose influence on the book was less direct, but nonetheless pronounced. Space permits me to thank only Dr. Sara L. Ripy of Agnes Scott College and Mr. John A. Stevenson for the encouragement they have given me through the years.

Eleanor H. Ninestein
Fayetteville Technical Institute

CONTENTS

Chapter One
Introduction

Just as the Jacquard loom used punched cards to control patterns in weaving, Charles Babbage's nineteenth-century design for a programmable calculating machine used punched cards to control the sequence of steps in the solution to a problem. Computer programmers sometimes refer to the steps for solving a problem as an *algorithm*.

1.1   Algorithms

In 1842, Ada Augusta, Countess of Lovelace, made an important observation about Charles Babbage's computing machine. She wrote, "The Analytical Engine has no pretensions whatever to *originate* anything. It can do whatever we *know how to order it* to perform." The modern computer programmer expresses the same idea when he or she says, "The computer did what I *told* it to, not what I *wanted* it to."

To solve a problem, a computer must be told, step by step, what to do. The word *algorithm* is used in both mathematics and computer programming to describe a list of rules for solving a problem.

> **Definition**
> An algorithm is a sequence of unambiguous steps that will produce a solution to a given problem in a finite number of steps.

Algorithms are usually designed to be general. That is, they are stated with variables so that many different problems of the same type can be solved. For computer programming, a variable may be thought of as the name of a storage or memory location. A variable is capable of holding a single value at one time. Each computer language has rules for naming variables. Generally, a variable name need not be a single letter, as it is in algebra. It is a common practice to use words or abbreviations that remind the programmer what a variable stores. For example, RATE and HRS might represent someone's hourly wage and the number of hours worked, respectively.

Developing an algorithm is a task that is largely independent of the computer language that will be used. The technique of "top-down design" is often used to break a problem into several small problems. Each small problem may then be divided into still smaller problems. It is important to know what information is available and what information is sought. Algorithms often begin with input (i.e., getting the data required to solve the problem) and end with output (i.e., a display of the solution).

The first step in learning to write algorithms is learning to read them. A walk-through consists of going through the steps one-by-one, just as a computer would. A common problem in computer programming is to switch the values of two variables. Suppose our two numbers are called FIRST and SECOND. You might be tempted to write:

1. Get the values of FIRST and SECOND.
2. Assign FIRST the value of SECOND.
3. Assign SECOND the value of FIRST.
4. Show the values of FIRST and SECOND.

To walk through this algorithm, we draw a box for each variable, so that we see the contents at each step.

FIRST ☐   SECOND ☐

1. The algorithm is written so that any numbers can be used. Suppose we use FIRST = 2 and SECOND = 3. These numbers are then assigned to the variables.

FIRST ②   SECOND ③

2. The current value of SECOND, 3, is assigned to FIRST.

FIRST ③   SECOND ③

(Since each variable can hold only one value, the old value is replaced by the new.)

3. The current value of FIRST, 3, is then assigned to SECOND.

FIRST ③   SECOND ③

4. FIRST = 3 and SECOND = 3. The algorithm, however, did not accomplish the exchange. A third variable is required for temporary storage of one of the values. The algorithm of Example 1 shows how to solve the problem.

EXAMPLE 1

Walk through the following algorithm for switching the values of two variables. Use A = 2 and B = 3.

1. Get the values of A and B.
2. Assign the value of A to a third variable, C.
3. Assign the value of B to A.
4. Assign the value of C to B.
5. Show the values of A and B.

Solution

1. A = 2 and B = 3. We draw boxes to represent the memory locations A, B, and C and show the current value of each.   A ② B ③ C ☐
2. The current value of A is assigned to C.   A ② B ③ C ②
3. The current value of B is assigned to A.   A ③ B ③ C ②
4. The current value of C is assigned to B.   A ③ B ② C ②
5. The value of A is now 3, and the value of B is 2. ■

EXAMPLE 2

Walk through the following algorithm using M = 3 and N = 5.

1. Get any two whole numbers M and N, with M less than N.
2. Subtract M from N and assign the difference to DIFF1.
3. Multiply N and 10, add M and assign to LARGE.

4. Multiply M and 10, add N and assign to SMALL.
5. Subtract SMALL from LARGE and assign difference to DIFF2.
6. Multiply DIFF1 by 9.
7. Show the values of DIFF1 and DIFF2.

Solution   The values of the variables are shown after each step.

	M	N	DIFF1	LARGE	SMALL	DIFF2
1.	3	5				
2.	3	5	2			
3.	3	5	2	53		
4.	3	5	2	53	35	
5.	3	5	2	53	35	18
6.	3	5	18	53	35	18
7.	DIFF1 = 18 and DIFF2 = 18. ∎

EXAMPLE 3

Walk through the algorithm of Example 2 using M = 2 and N = 7.

Solution   To shorten the process, we will draw a single box for each variable. As the contents change, we cross out the old value and replace it with the new. The steps at which the contents change are indicated in parentheses.

M	N	DIFF1	LARGE	SMALL	DIFF2
2	7	5̶ 45	72	27	45
(step 1)	(step 1)	(steps 2 and 6)	(step 3)	(step 4)	(step 5)

DIFF1 = 45 and DIFF2 = 45. ∎

EXAMPLE 4

Walk through the following algorithm using NUMBER = 5.

1. Get any number, NUMBER.
2. Multiply NUMBER by NUMBER.
3. Subtract 1 from NUMBER.
4. Divide NUMBER by 24 and assign the quotient to QUOTIENT; assign the remainder to REMAINDER.
5. Show the values of QUOTIENT and REMAINDER.

Solution

NUMBER	QUOTIENT	REMAINDER
5̶ 2̶5̶ 24	1	0
(steps 1, 2, and 3)	(step 4)	(step 4)

QUOTIENT = 1 and REMAINDER = 0. ∎

In the next section, you will begin to write algorithms.

1.1 EXERCISES

Problems 1–3 refer to the following algorithm. Walk through it using the suggested values. See Examples 1–4.

1. Get a whole number N between 1 and 1000 (inclusive).
2. Assign A the value of 7.
3. Multiply N by A and assign the product to N.
4. Add 4 to A.
5. Multiply N by A and assign the product to N.
6. Add 2 to A.
7. Multiply N by A and assign the product to N.
8. Show the value of N.

1. Use N = 45.　　　　　　**2.** Use N = 37.　　　　　　**3.** Use N = 23.

Problems 4–6 refer to the following algorithm. Walk through it using the suggested values. See Examples 1–4.

1. Get a whole number N between 1 and 104 (inclusive).
2. Assign N to X. Divide X by 3 and assign the remainder to A.
3. Assign N to X. Divide X by 5 and assign the remainder to B.
4. Assign N to X. Divide X by 7 and assign the remainder to C.
5. Multiply A by 70 and assign the product to A.
6. Multiply B by 21 and assign the product to B.
7. Multiply C by 15 and assign the product to C.
8. Add A, B, and C and assign the sum to D.
9. Divide D by 105 and assign the remainder to D.
10. Show N and D.

4. Use N = 34.　　　　　　**5.** Use N = 73.　　　　　　**6.** Use N = 101.

Problems 7–9 refer to the following algorithm. Walk through it using the suggested values. See Examples 1–4.

1. Get any two whole numbers M and N, with M less than N.
2. Multiply M, N, and 2, and assign the product to X.
3. Multiply M by M.
4. Multiply N by N.
5. Subtract M from N, and assign the difference to Y.
6. Add M and N, and assign the sum to Z.
7. Show X, Y, and Z.

7. Use M = 1 and N = 2.　　**8.** Use M = 2 and N = 3.　　**9.** Use M = 1 and N = 3.

Problems 10–12 refer to the following algorithm. Walk through it using the suggested values. See Examples 1–4.

1. Get any whole number, FIRST.
2. Add 1 to FIRST and assign to SECOND.
3. Add 1 to SECOND and assign to THIRD.
4. Multiply FIRST, SECOND, and THIRD; assign product to PRODUCT.
5. Divide PRODUCT by 6, and assign the quotient to QUOTIENT.
6. Show QUOTIENT.

10. Use FIRST = 3.　　　　**11.** Use FIRST = 8.　　　　**12.** Use FIRST = 10.

Problems 13–15 refer to the following algorithm. Walk through it using the suggested values. See Examples 1–4.

1. Get any three-digit whole number, NUMBER.
2. Divide NUMBER by 100, and assign quotient to LEFT; assign remainder to NUMBER.
3. Divide NUMBER by 10, and assign quotient to MID; assign remainder to RIGHT.
4. Show the values of LEFT, MID, and RIGHT.

13. Use NUMBER = 213.   **14.** Use NUMBER = 596.   **15.** Use NUMBER = 847.

✷16. Is there a relationship between X, Y, and Z in the algorithm of problems 7–9?

✷17. Do you think that the division in the algorithm of problems 10–12 will ever have a remainder? Why or why not?

✷18. What does the algorithm of problems 13–15 accomplish?

1.2   Structured Programming

Computer programmers recognize the importance of making programs easy to read so that the tasks of debugging, maintaining, and modifying a program and subdividing the work among a team of programmers are simplified. Structured programming is used to make programs easier to read. Structured programs use algorithms that are composed of three patterns or structures: sequence, selection, and iteration.

The *sequence structure* is a series of steps to be carried out, or executed, in order. The algorithm for switching the values of two variables uses a sequence structure. If A and B are the variables, the algorithm could be written as follows:

1. Get the values of A and B.
2. Assign the value of A to a third variable, C.
3. Assign the value of B to A.
4. Assign the value of C to B.
5. Show the values of A and B.

This algorithm will work anytime we want to switch the values of two variables. But we may need more flexibility in an algorithm. The *selection structure* allows options. An automatic teller machine must be programmed to add if the transaction is a deposit and to subtract if it is a withdrawal. We can write the algorithm as follows:

1. Get the nature and amount (as a positive number, AMT) of the transaction (from the customer).
2. Get the customer's current balance, BAL (from bank records).
3. If the transaction is a deposit, add AMT to BAL. Otherwise, subtract AMT from BAL.
4. Show the amount of the transaction AMT, and the new balance BAL.

The algorithm specified that the amount AMT is to be a positive number. What would happen if a negative withdrawal were made using this algorithm? In practice, programmers try to anticipate such circumstances and build safeguards into their programs to prevent such errors.

The third structure, *iteration*, uses a sequence of steps that may be executed several times. The word *loop* is often used to describe the repeating steps. A common application of the iteration structure is to keep track of how many items have been processed. A program to compute student grade point averages would process a different number of student records each term. The computer could count the grade reports. We need a variable to store the count. We call it a *counter*, and we assign it a value of 0 at the beginning, just in case another value is already stored there. Assigning a value (often 0, but not always) to a variable before it is used in any calculations is called *initializing* the variable. The algorithm follows:

1. Assign the variable named COUNTER a value of 0.
2. Repeat steps 3 and 4 until the last student record is processed.
3. Get and process a student record.
4. Add 1 to COUNTER.
5. Show the value of COUNTER.

Anytime an iteration structure is used, a decision must be made whether to go through the loop again or whether to exit the loop. We say that a *condition* is tested. If the condition is tested at the beginning of the loop, the structure is said to be a DOWHILE loop. As long as the condition for execution is met at that point, the loop will be executed. Two things must be considered. First, the condition must be initialized before entering the loop. Second, the condition must be changed within the loop so that it eventually terminates the loop.

Often in computer programming it is necessary to add a series of numbers. A running total is kept in a variable called an *accumulator*. The first step is to initialize the accumulator, just as we did with the counter in the previous problem. The condition for execution will be that there are more numbers. The computer must be able to recognize when the last item has been encountered. Sometimes a "dummy" value is used to signal the end of input. We might call it an *end-marker*. The algorithm for this problem follows. The loop is shown in bold type.

1. Assign TOTAL a value of 0 (i.e., initialize the total).
2. Get the first number, NUM.
3. **While NUM is not the end-marker, do steps 4 and 5.**
4. **Add NUM to TOTAL.**
5. **Get the next number, and assign it to NUM.**
6. Show the value of TOTAL.

After the end-marker is encountered at step 5, the condition for execution is tested at the beginning of the loop (step 3). The condition is not met, so the loop is exited, and execution continues at step 6. The value of TOTAL is displayed only once, because step 6 occurs *outside* the loop.

If the algorithm were written in the following manner, the value of TOTAL would be displayed every time a number is added, because step 5 occurs *inside* the loop.

1. Assign TOTAL a value of 0 (i.e., initialize the total).
2. Get the first number, NUM.
3. **While NUM is not the end-marker, do steps 4–6.**
4. **Add NUM to TOTAL.**
5. **Show the value of TOTAL.**
6. **Get the next number, and assign it to NUM.**

One of the reasons for studying mathematics is to practice problem solving and algorithm development. In mathematics, however, you are used to solving *specific* problems. In computer programming, you must solve *general* problems. One technique used to develop algorithms for general problems is to first solve the problem for specific data. Thinking about the steps you used to solve the specific problem will help you to write a general algorithm. This technique is used extensively in the following chapters as well as in the examples of this section.

EXAMPLE 1

Develop an algorithm for finding the average of *n* numbers.

Solution   First we solve a specific problem. Find the average of four numbers: 100, 50, 82, and 76. We would accumulate the total: $100 + 50 + 82 + 76 = 308$. Then we divide by 4: $308 \div 4 = 77$. For the general algorithm, we'll need an accumulator for the sum. Since we don't know the value of *n*, we'll need to count the numbers as we add. Accumulators and counters suggest an iteration structure.

1. Assign TOTAL and COUNTER each a value of 0.
2. Get the first number, NUMBER.
3. While NUMBER is not the end-marker, do steps 4–6.
4. Add NUMBER to TOTAL.
5. Add 1 to COUNTER.
6. Get the next number, NUMBER.
7. Divide TOTAL by COUNTER to get AVERAGE.
8. Show AVERAGE. ■

EXAMPLE 2

Develop an algorithm to calculate an employee's weekly pay, P, if he makes D dollars an hour and works H hours. Time-and-a-half is paid for more than 40 hours per week.

Solution   Suppose the wage is $6.00 per hour and time is 35 hours. The pay would be 35 × $6.00, or $210.00. But for 45 hours, the pay would be 40 × $6.00, or $240 for the first 40 hours and 5 × $9.00, or $45 for the last 5 hours, for a total of $285 dollars. Since there is an option, we need a selection structure. The algorithm follows.

1. Get hourly wage D and time H.
2. If H is less than 40, then multiply D by H to get weekly pay P. Otherwise, do steps 3–6.
3. Multiply 40 by hourly wage D to get regular pay.
4. Calculate overtime rate as 1.5 × hourly rate D.
5. Subtract 40 from time H and multiply by overtime rate to get overtime pay.
6. Add regular pay and overtime pay to get weekly pay P.
7. Show the value of P. ■

1.2 EXERCISES

Solve each problem using the suggested numbers. Then write a general algorithm using the variables. See Examples 1 and 2.

1. An item sells for LIST dollars. If it is on sale at RATE percent off, write an algorithm to determine the amount of discount (DISC) and the sale price (SALE). LIST = $595 and RATE = 15.

2. You fill your gasoline tank with G gallons of gas, drive M miles, spend the night, and drive N miles before you fill the tank again. Write an algorithm to determine the number of miles per gallon your car uses. G = 14 gallons, M = 198 miles, and N = 264 miles.

3. Suppose you borrow B dollars at R percent annual interest, and the monthly payment is D dollars. Write an algorithm to determine the amount of interest owed (I) at the end of one month and the amount that goes toward repayment of principal (P), and assign the remaining principal to B. B = $500, R = 12, and D = $23.54.

4. A person earns SALARY dollars per year and has NUM dependents. If there is an exemption of $1000 for each dependent, write an algorithm to determine the taxable income (INC), and the amount of tax (TAX) owed if the tax rate is RATE percent. SALARY = $15,000, NUM = 3, and RATE = 9.

5. A salesperson is paid S dollars a week plus R percent of all sales over a certain amount A. Write an algorithm to find the pay (P) for a week in which M dollars worth of merchandise is sold. S = $300, R = 5, A = $5000, and M = $7000.

6. Some states charge R percent sales tax on nonfood items and no sales tax on food. Write an algorithm to determine the amount of tax (T) paid on a purchase of P dollars (all in the same category). R = 4 and P = $20 in food.

7. If your bank balance is BAL dollars, write an algorithm to determine the new balance after making a series of transactions, each of amount AMT (but AMT may be different for each transaction). Use a deposit of $0 to signal the end of input. BAL = $320, withdraw $15, deposit $120, and withdraw $88.

8. A polltaker finds that the number of people (X) out of a total of Y on each of several street corners favor a certain political candidate. Write an algorithm to determine the percent (R) of the total polled who favor the candidate. Use 999 to signal the end of input. X = 40 and Y = 60 on the first corner; X = 30 and Y = 50 on the second; and X = 25 and Y = 45 on the third.

9. Suppose that you have a noninterest-bearing account with a balance of BAL dollars. If you withdraw half of the remaining balance at the end of each month, write an algorithm to determine how long it will take to bring the balance down to MIN dollars or less. BAL = $3200 and MIN = $200.

✳**10.** It has been conjectured that for any whole number N, the following algorithm will eventually produce the value N = 1. Walk through it using (a) N = 21, (b) N = 53, and (c) N = 80.

　1. Get a whole number N between 1 and 99 (inclusive).

　2. Assign counter C a value of 0.

　3. While N is greater than 1, do steps 4 and 5.

　4. If N is even, divide by 2 and assign quotient to N. Otherwise, multiply by 3, add 1, and assign result to N.

　5. Add 1 to C.

　6. Show C and N.

✳**11.** If the algorithm of Example 1 is used to calculate a student's average grade, modify it to accommodate an entire class.

✳**12.** Modify the algorithm of Example 2 to process more than one employee.

✳**13.** Modify the algorithm of Example 2 to deliver an error message if an employee works more than 72 hours during the week.

1.3 Pseudocode and Flowcharts

The process of writing an algorithm in a computer language such as COBOL, BASIC, or Pascal is called *coding*. An algorithm should be fully developed before it is coded. Computer programmers have standard ways of writing algorithms, even before they are coded. Two of the most common ways are to use pseudocode or flowcharts.

Pseudocode is written in English using the structures of sequence, selection, and iteration. Pseudocode is not completely standardized. Generally, each step of a sequence structure has the same margin. The steps of a selection or iteration structure are indented to help identify the steps that belong together. The words ENDIF and ENDWHILE are often used to mark the ends of selection and iteration structures, respectively. As you review the notation of algebra in Chapter 2, you may begin to use it to abbreviate some of the English.

EXAMPLE 1

Use pseudocode to write an algorithm for computing the amount W withheld from an employee's annual salary S for social security if the rate is R percent on a maximum of M dollars.

Solution   The algorithm contains a selection structure as part of a sequence structure.

Get the values of S, R, and M.
Replace R by 0.01 × R (i.e., change the percent to a decimal).
If S is less than M, multiply S by R to get W.
  Otherwise, multiply M by R to get W.
  ENDIF.
Show the value of W. ▪

EXAMPLE 2

Use pseudocode to write an algorithm to add the values of all of the cars in a car dealer's inventory.

Solution   After the last car's value is entered, we enter an end-marker. Since no car would actually have a value of $9, we use the number 9.

Assign the TOTAL a value of 0.
Get the first value, VAL.
While VAL is not 9,
  Add VAL to TOTAL.
  Get the next value, VAL.
  ENDWHILE.
Show TOTAL. ▪

Notice that the first value is obtained before entering the loop, and the last step inside the loop is to get a new value (rather than to add). Thus, the "dummy" value of 9 is not added to the total. It is a common practice in data processing to get the first item before the loop begins so that a signal for the end of input can be used in this manner.

Another common method for specifying algorithms for a computer is flow-charting. A flowchart is a pictorial representation of an algorithm. The most frequently used symbols are illustrated in Figure 1.1.

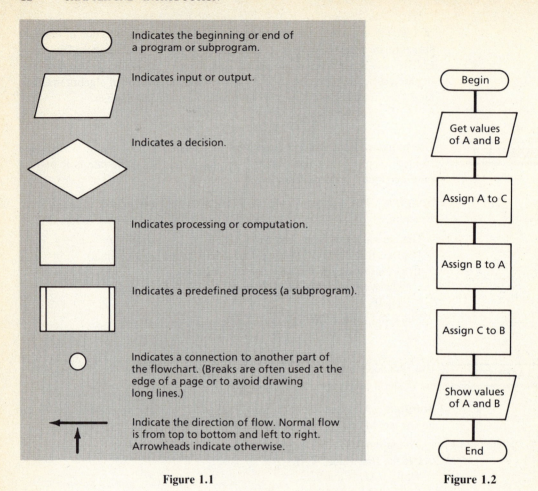

Indicates the beginning or end of a program or subprogram.

Indicates input or output.

Indicates a decision.

Indicates processing or computation.

Indicates a predefined process (a subprogram).

Indicates a connection to another part of the flowchart. (Breaks are often used at the edge of a page or to avoid drawing long lines.)

Indicate the direction of flow. Normal flow is from top to bottom and left to right. Arrowheads indicate otherwise.

Figure 1.1

Begin

Get values of A and B

Assign A to C

Assign B to A

Assign C to B

Show values of A and B

End

Figure 1.2

Structured flowcharts result from putting these symbols together using the sequence, selection, and iteration structures. We illustrate the sequence structure for the problem of switching the values of two variables. See Figure 1.2.

The selection structure is illustrated in Figure 1.3 for the problem of a bank transaction. The small circles are used at the beginning and end, because we are looking at only part of a flowchart. This portion would be joined to the rest of the flowchart where the connectors occur.

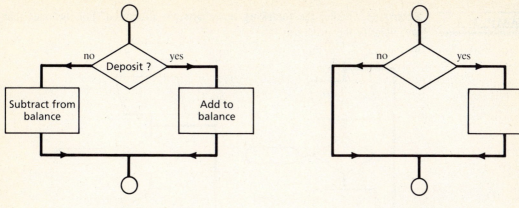

Figure 1.3   Figure 1.4

It is important to realize that one of the actions in a selection structure might be "do nothing." The flowchart would appear as it does in Figure 1.4 or Figure 1.5.

In an iteration structure, the loop begins with a decision whether to execute the loop. If the condition is met, the action is executed. If the condition is not met initially, the loop will not be executed at all. Figure 1.6 shows a flowchart for an iteration structure.

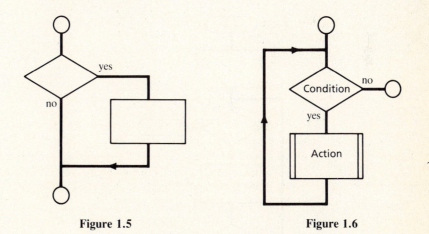

Figure 1.5   Figure 1.6

It is very important for each structure to have only one entry point and one exit point. The action in this example is not a single step, but rather an entire algorithm, which would be shown separately.

EXAMPLE 3

Determine whether the following flowcharts are structured. For the ones that are, identify the structures.

a.   b.

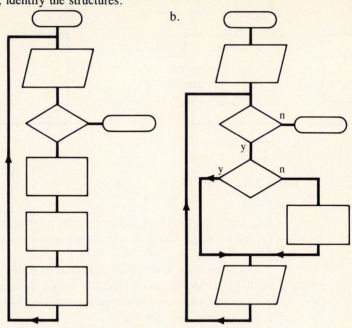

c.

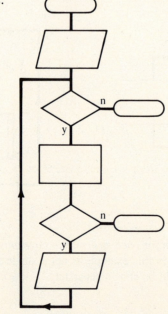

Solution

a. It is not structured. An iteration structure would loop back to a position immediately before the decision.

b. It is structured.

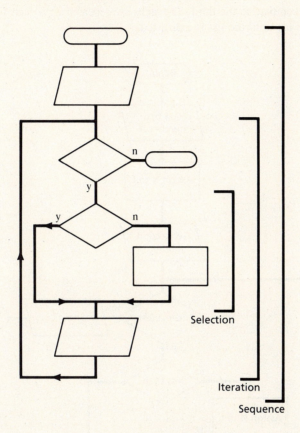

c. It is not structured. There should be only one exit point. ▥

We conclude this section by examining another common problem in computer programming.

EXAMPLE 4

Construct a flowchart for finding the largest number L in a list of numbers.

Solution   We begin by assuming that the first number is the largest. Then we compare the second number to the first. If the second is smaller, then we go on to the third. If the second is larger than the first, then we have a new "largest" to compare to the third. The steps are repeated with each successive number in the list until the list is exhausted.

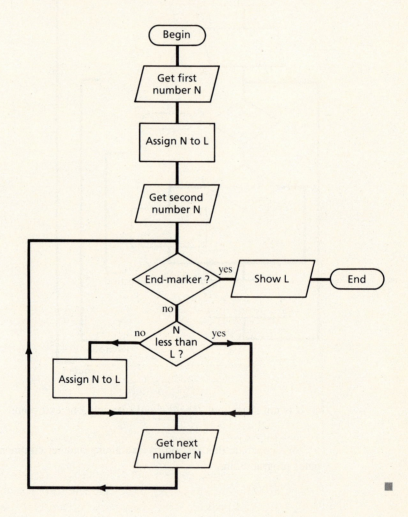

EXAMPLE 5

Construct a flowchart for counting the number of deposits (D) and the number of withdrawals (W) made by a bank customer. Let a deposit of zero serve as an end-marker.

Solution

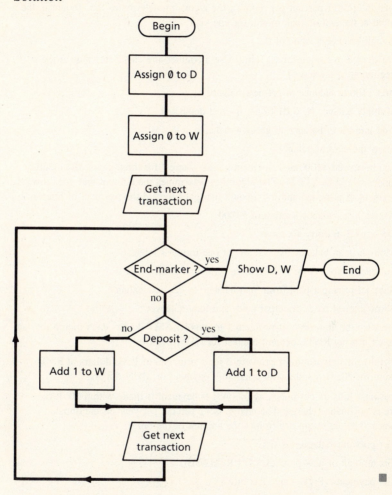

It takes much practice to become proficient at writing algorithms so that they follow the structures of sequence, selection, and iteration. There are many algorithms in this book. You may be asked to write some of them in pseudocode or to construct flowcharts so that you will have the opportunity to practice this skill throughout the course.

1.3 EXERCISES

Use pseudocode to write the following algorithms. See Examples 1 and 2.

1. If $100 is deposited on the first of each month in an account bearing 12 percent annual interest (1 percent per month), the algorithm determines the amount in the account at the end of each month for one year.

   1. Assign balance a value of 0.
   2. Do steps 3–6 twelve times. (Hint: Use a counter and check its value at the beginning of the loop.)
   3. Add $100 to balance to get new balance.
   4. Multiply balance by 0.01 to get interest deposited.
   5. Add interest to balance to get new balance.
   6. Show the current balance.

2. If you borrowed $1000 at 12 percent annual interest for three years, the monthly payment would be $33.21. The algorithm determines the amount that goes toward repayment of principal and the amount that goes toward the payment of interest.

   1. Assign loan balance a value of $1000
   2. Do steps 3–6 thirty-six times.
   3. Multiply loan balance by 0.01 to get one month's interest.
   4. Subtract interest from $33.21 to get principal paid.
   5. Subtract principal paid from loan balance to get new balance.
   6. Show interest paid, principal paid, and loan balance.

3. Generalize the algorithm of problem 1 to deposit AMT dollars each month into an account bearing RATE percent interest for NUM months.

4. Generalize the algorithm of problem 2 for a principal of BAL dollars at RATE percent for NUM months with a monthly payment of MON dollars.

5. A business bills its customers as follows: If payment is made within 30 days, there is no finance charge. The charge is 12 percent per year on the unpaid balance. The algorithm determines the amount of the bill.

   1. Get previous balance (BAL).
   2. Get amount of new purchases (PURCHASE).
   3. Get payments (PAY) since last bill.
   4. Subtract PAY from BAL to get unpaid balance (BAL).
   5. If BAL is greater than zero, then multiply BAL by 0.01 to get charge (CHARGE) on unpaid balance.
      Otherwise assign CHARGE a value of 0.
   6. Add BAL, PURCHASE, and CHARGE to get amount (BAL) of bill.

Which of the following flowcharts are structured? For the ones that are, identify the structures. See Example 3.

6.

7.

8.

9.

10.

11.

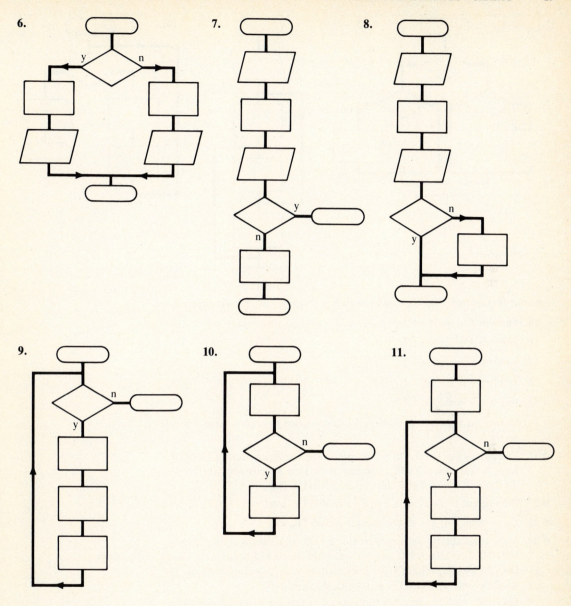

12.   **13.**   **14.**

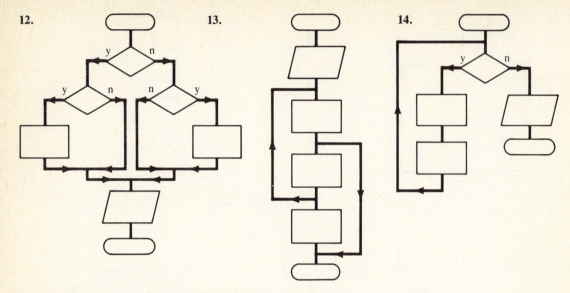

Construct flowcharts for the algorithms in each problem below. See Examples 4 and 5.

15. Problem 1 in section 1.1.

16. Problem 3 in section 1.1.

17. Problem 5 in section 1.1.

18. Problem 7 in section 1.1.

19. Problem 9 in section 1.1.

20. Modify the algorithm in problem 5 of this section to process more than one customer.

✳**21.** There is another version of the iteration structure which uses a DOUNTIL loop. A condition is tested at the *end* of the loop to determine whether to exit the loop. Construct a flowchart for a DOUNTIL loop.

✳**22.** Do problem 1 of this section using a DOUNTIL loop.

✳**23.** Do problem 2 of this section using a DOUNTIL loop.

✳**24.** Many programming languages provide a FOR loop, which is essentially a DOWHILE or DOUNTIL loop for which the number of iterations is known. For example, in problem 1, we knew that there were twelve iterations. Step 2 could be written "For CTR = 1 to 12, do steps 3–6."

   (a) How could the algorithm be modified so that the number of months could be input as part of the data?

   (b) Under what conditions would it be better to use a DOWHILE loop and an end-marker rather than a FOR loop?

CHAPTER REVIEW

An _____ is a sequence of unambiguous steps that will produce a solution to a given problem in a finite number of steps.

A _____ consists of going through the steps of an algorithm one-by-one just as a computer would.

The _____ structure is a series of steps to be executed in order.

The _____ structure allows options.

The _____ structure uses a sequence of steps that may be executed several times.

The word _____ is often used to describe repeating steps.

Assigning a value to a variable before it is used in any computations is called _____ the variable.

Programmers often write their algorithms in English in a form called _____.

A _____ is a pictorial representation of an algorithm.

FOR FURTHER READING

Arganbright, Deane E. "The Electronic Spreadsheet and Mathematical Algorithms," *The College Mathematics Journal,* Vol. 15 No. 2 (March 1984), 148–157.

Arndt, A. B. "Al-Khowarizmi," *The Mathematics Teacher,* Vol. 76 No. 9 (December 1983), 668–670.

Knuth, Donald E. "Algorithmic Thinking and Mathematical Thinking," *The American Mathematical Monthly,* Vol. 92 No. 3 (March 1985), 170–181.

Maurer, Stephen B. "Two Meanings of Algorithmic Mathematics," *The Mathematics Teacher,* Vol. 77 No. 6 (September 1984), 430–435. (The difference between creating and performing algorithms is discussed.)

Biography   Leonardo Fibonacci

Leonardo Fibonacci was born in the latter part of the twelfth century. The name "Fibonacci" is short for "filius Bonacci," meaning "the son of Bonaccio." His father, Guilielmo Bonacci, was a prominent Italian merchant who was appointed to the position of "public scribe" in what is now Algeria.

Leonardo travelled to Egypt, Syria, Greece, and Sicily. As he travelled, he studied methods of calculating. He found the Hindu-Arabic numerals to be superior for computation. When he returned to Italy, he wrote a book called *Liber Abbaci*, which means "book of the abacus." The name is misleading, for the book is really about computing *without* an abacus. He introduced the Hindu-Arabic numerals, explained the concept of place value, and gave examples of arithmetical computations using the system. He had studied the work of Mohammed ibn Musa al-Khowarizmi and was influenced by his style. At about the same time, other European scholars were beginning to translate the works of al-Khowarizmi into Latin. The books of al-Khowarizmi and Fibonacci were influential in converting Europe from Roman numerals to our present system. The arithmetical procedures for computing came to be known as *algorithms*, a corruption of the name al-Khowarizmi.

One of the problems in the *Liber Abbaci* gives rise to a sequence of numbers that bears Fibonacci's name. A man has one pair of rabbits confined by a wall. The problem is to find how many pairs can be bred from it in one year if each pair begins to breed in the second month after birth and breeds a pair every month thereafter.

Little else is known about Fibonacci's life. He continued to write about mathematics and its commercial applications. He probably died about 1240.

Chapter Two
Sets

Just as office workers use filing cabinets to store records, computer programmers use *files* for storing data. The programming concepts of records and files are related to mathematical sets. A file may be thought of as a set of records.

2.1   Set Notation

Many concepts in mathematics can be explained using set theory. The ideas introduced here will be useful in subsequent chapters.

The idea of a set is so basic that rather than define the word, we will develop the idea intuitively by considering several examples. A set is a collection of objects. We might speak of a set of people taking a math course, a set of golf clubs, or a set of pots and pans. In mathematics, the sets are often sets of numbers.

Sets may be specified in any of three ways: by name, by roster, or by description.

In Section 2.4 we will discuss some of the sets that have special names, such as the integers and the real numbers.

When the roster method is used, the members of the set are separated by commas and are listed between braces { }. If there are many members in the set, it may not be convenient, or even possible, to list them all. Three dots, called an ellipsis, indicate that some of the objects have been omitted from the list, but they are included in the set, nevertheless. When you use the ellipsis, be sure to specify enough members that the pattern is clearly established. For example, $\{1, 2, 3, 4, 5\}$ specifies a set with only five members. But $\{1, 3, 5, 7, 9, \ldots\}$ specifies a set with infinitely many members. The ellipsis indicates that odd numbers beyond 9 are included in the set. The set $\{1, 2, 3, 4, \ldots, 90\}$ contains the numbers 1 through 90. The set $\{1, 4, \ldots, 49\}$ is not clearly specified. This set could be $\{1, 4, 9, 16, 25, 36, 49\}$, obtained by squaring 1, 2, 3, 4, 5, 6, and 7 respectively, or it could be the set $\{1, 4, 7, 10, 13, \ldots, 49\}$, in which each element is three more than the previous one. There were not enough objects given to determine the pattern.

Rather than list individual members of a set, we might use set-builder notation to describe the objects in the set. This method is often used for sets of numbers. Thus, $\{x \mid x$ is an even number and $x > 6\}$ is read "The set of all x such that x is an even number and x is greater than 6." The braces indicate a set. The $\mid$ symbol is read "such that," and the symbol $>$ means "is greater than." From the description we know that the set could also be written $\{8, 10, 12, 14, 16, \ldots\}$.

Sometimes it is helpful to assign a name to a set. We use upper case letters for set names. We could write $A = \{8, 10, 12, 14, 16, \ldots\}$ or $B = \{x \mid x$ is a positive multiple of 3$\}$.

The symbol $\in$ is read "belongs to" or "is an element of." We could say $8 \in A$, $3 \in B$, $6 \in B$, and so on. The symbol $\notin$ is read "does not belong to" or "is not an element of." Thus, $9 \notin A$.

There are three set operations that we will consider: intersection, union, and complementation.

Intersection is indicated by the symbol $\cap$.

Definition

$$A \cap B = \{x \mid x \in A \text{ and } x \in B\}.$$

Notice that the intersection of two sets is a set.

EXAMPLE 1

If $A = \{2, 3, 5, 6\}$ and $B = \{1, 2, 3, 4\}$, find $A \cap B$.

Solution We look for the elements that are in both sets.
$A = \{\mathbf{2}, \mathbf{3}, 5, 6\}$ and $B = \{1, \mathbf{2}, \mathbf{3}, 4\}$. Therefore $A \cap B = \{2, 3\}$. ∎

EXAMPLE 2

If $C = \{1, 3, 5\}$ and $D = \{2, 4, 6\}$, find $C \cap D$.

Solution There are no elements that are in both sets. Therefore $C \cap D = \{\ \}$. ∎

The set $\{\ \}$ is called the "empty" or "null" set. It is a set that contains no objects. Sometimes the symbol $\varnothing$ is used instead of $\{\ \}$.

The elements of a set may themselves be sets. Thus $\{\{1, 2\}, \{3\}\}$ is a set containing two sets. The sets $\{\{1\}\}$ and $\{\varnothing\}$ each contain one set. Do not confuse the empty set $\varnothing$ with the set $\{\varnothing\}$ containing one object.

The second set operation is *union*. The symbol for union is $\cup$.

Definition

$$A \cup B = \{x \mid x \in A \text{ or } x \in B\}.$$

One way to find the union of two sets is to put all of the elements from the sets into one set. Rather than list all the elements of A followed by all the elements of B, we usually list them in their "natural order" (if there is one), and each element is listed only once.

EXAMPLE 3

If $A = \{2, 3, 5, 6\}$ and $B = \{1, 2, 3, 4\}$, find $A \cup B$.

Solution Putting the elements of A and B into one set, we have $A \cup B = \{1, 2, 3, 4, 5, 6\}$. ∎

EXAMPLE 4

If $C = \{1, 3, 5, 7, 9\}$ and $D = \{2, 4, 6\}$, find $C \cup D$.

Solution Putting the elements of C and D into one set, we have $C \cup D = \{1, 2, 3, 4, 5, 6, 7, 9\}$. ∎

To belong to $A \cap B$, an object must belong to both A and B. To belong to $A \cup B$, an object may belong to only one of the two sets, or it may belong to both. The "or" in the definition of union is used in the "inclusive sense." That is, $x \in A$ or $x \in B$ means x belongs to either A or B, or both.

The third set operation is *complementation*. Before the complement of a set is defined, it is necessary to introduce the universal set (U). The *universal set* is the largest set under consideration for a given problem. If $C = \{x|\ x < 6\}$, we need to know the largest set from which we are allowed to choose x. If the universal set is $\{1, 2, 3, 4, 5, 6\}$, then $C = \{1, 2, 3, 4, 5\}$. But if the universal set is $\{-2, 0, 2, 4, 6\}$, then $C = \{-2, 0, 2, 4\}$. The complement of set A is denoted by A'.

Definition

$$A' = \{x|\ x \in U, x \text{ is } \mathbf{not} \text{ in } A\}.$$

EXAMPLE 5

If $U = \{1, 3, 5, 7, 9\}$ and $A = \{1, 5, 9\}$, find A'.

Solution $U = \{\mathbf{1}, 3, \mathbf{5}, 7, \mathbf{9}\}$. Since 1, 5, and 9 belong to A, we can see which elements of U do *not* belong to A. Therefore $A' = \{3, 7\}$. ∎

EXAMPLE 6

If $U = \{1, 3, 5, 7, 9\}$ and $B = \{1, 3, 5\}$, find B'.

Solution $U = \{\mathbf{1}, \mathbf{3}, \mathbf{5}, 7, 9\}$. We see that 7 and 9 do *not* belong to B. Therefore $B' = \{7, 9\}$. ∎

When parentheses appear in an expression, the operation in parentheses is done first.

EXAMPLE 7

If $A = \{2, 3, 5, 6\}$, $B = \{1, 2, 3, 4\}$, and $C = \{1, 3, 5\}$, find $(A \cap B) \cup C$.

Solution   Since $A \cap B$ is in parentheses, we consider it first.
$A \cap B = \{2, 3\}$, so $(A \cap B) \cup C = \{2, 3\} \cup \{1, 3, 5\} = \{1, 2, 3, 5\}$. ∎

EXAMPLE 8

If $A = \{2, 3, 5, 6\}$, $B = \{1, 2, 3, 4\}$, and $C = \{1, 3, 5\}$, find $A \cap (B \cup C)$.

Solution $B \cup C = \{1, 2, 3, 4, 5\}$, so
$A \cap (B \cup C) = \{2, 3, 5, 6\} \cap \{1, 2, 3, 4, 5\} = \{2, 3, 5\}$. ∎

The words *and, or,* and *not* used in these three definitions are very important. You will find it helpful in translating verbal problems into mathematical problems if you associate the word "and" with intersection, "or" with union, and "not" with complementation.

$$A \cap B = \{x \mid x \in A \text{ and } x \in B\}$$

$$A \cup B = \{x \mid x \in A \text{ or } x \in B\}$$

$$A' = \{x \mid x \in U, x \text{ is \textbf{not} in } A\}$$

Examples 9–14 use the sets named below:

U = The set of students at your school.

M = The set of students taking math.

E = The set of students taking English.

A = The set of students with an A average.

EXAMPLE 9

Use set notation to specify each set.

(a) The set of students at your school who are taking math and English.

(b) The set of students at your school who are taking math or English.

Solution

(a) $M \cap E$

(b) $M \cup E$ ∎

EXAMPLE 10

Use set notation to specify each set.

(a) The set of students at your school who are not taking math.

(b) The set of students at your school who have an A average and who are not taking math.

Solution

(a) M'

(b) $A \cap M'$ ∎

Parentheses are used to punctuate math expressions in the way that commas punctuate English sentences. Commas indicate which phrases are grouped together in a sentence. Parentheses indicate which sets are grouped together in a mathematical expression.

EXAMPLE 11

Use set notation to specify each set.

(a) The set of students who are taking math or English, but do not have an A average. ("But" means "and" in this sentence.)

(b) The set of students who are taking math, or who are taking English but do not have an A average.

Solution

(a) $(M \cup E) \cap A'$

(b) $M \cup (E \cap A')$ ∎

2.1 EXERCISES

For problems 1–6, use the following sets:

$U = \{1, 2, 3, \ldots 10\}$

$A = \{2, 3, 5, 7\}$

$B = \{3, 6, 9\}$

Find each set. See Examples 1–6.

1. (a) $A \cap B$

(b) $A \cup B$

2. (a) A'

(b) B'

3. (a) $(A \cup B)'$

(b) $A' \cap B'$

4. (a) $(A \cap B)'$

(b) $A' \cup B'$

5. (a) $A' \cap B$

(b) $A \cap B'$

6. (a) $A' \cup B$

(b) $A \cup B'$

For problems 7–12, use the following sets:

$U = \{1, 3, 5, 7, 9, 11\}$

$P = \{1, 3\}$

$Q = \{5, 7\}$

$R = \{9, 11\}$

Find each set. See Examples 1–6.

7. (a) $P \cap Q$

(b) $Q \cup R$

8. (a) $P \cup R$

(b) $P \cap R$

9. (a) $P' \cap Q$

(b) $Q' \cup R$

10. (a) $P \cap R'$

(b) $(P \cap Q)'$

11. (a) $P' \cup Q'$

(b) $(Q \cup R)'$

12. (a) $Q' \cup R'$

(b) $Q' \cap R'$

For problems 13–25, use the following sets:

$U = \{2, 4, 6, 8, 10\}$

$X = \{2, 4, 6\}$

$Y = \{4, 8\}$

$Z = \{4, 10\}$

Find each set. See Examples 7 and 8.

13. $(X \cap Y) \cap Z$

14. $X \cap (Y \cap Z)$

15. $X \cap (Y \cup Z)$

16. $(X \cap Y) \cup Z$

17. $(X \cap Y) \cup (X \cap Z)$

18. $(X \cup Z) \cap (Y \cup Z)$

19. $X \cup (Y \cap Z)$

20. $(X \cup Y) \cap (X \cup Z)$

21. $X \cap (Y \cup Z)$

22. $(X \cap Z) \cup (Y \cap Z)$

23. $X' \cap (Y \cup Z)$

24. $X \cap (Y \cup Z)'$

25. $X' \cup (Y \cap Z)$

For problems 26–39, use the following sets:

$U = $ The set of all people insured by a certain company.

$W = $ The set of women insured by the company.

$S = $ The set of smokers insured by the company.

$O = $ The set of overweight people insured by the company.

Write a description for each set using ∩, ∪, and ′. See Examples 9–11.

26. The set of women who smoke.

27. The set of men who smoke.

28. The set of people who smoke or are overweight.

29. The set of people who do not smoke.

30. The set of women who do not smoke.

31. The set of men who do not smoke.

32. The set of men who are not overweight.

33. The set of women who are not overweight.

34. The set of people who do not smoke or are not overweight.

35. The set of people who do not smoke and are overweight.

36. The set of people who do not smoke and are not overweight.

37. The set of women who do not smoke and are not overweight.

38. The set of women who smoke or are overweight.

39. The set of men who smoke or are overweight.

For problems 40–48, use the sets below:

   U = The set of students in your class.
   F = The set of people 40 years of age or older.
   T = The set of people 20 years of age or younger.
   M = The set of people born east of the Mississippi River.
   R = The set of people born west of the Rocky Mountains.

Write a description for each set using ∩, ∪, and ′. See Examples 9–11.

40. The set of people 40 or over who were born east of the Mississippi.

41. The set of people 40 or over who were not born east of the Mississippi.

42. The set of people 20 years of age or younger who were not born west of the Rockies.

43. The set of people 20 years of age or younger who were born west of the Rockies.

44. The set of people born east of the Rockies.

45. The set of people born west of the Mississippi.

46. The set of people who are 20 years of age or younger or who are 40 years of age or older.

47. The set of people over 20 but under 40.

48. The set of people who were born between the Mississippi River and the Rocky Mountains.

✳**49.** When is $A \cup B$ identical to $A \cap B$?

✳**50.** Fill in the next four elements: {2, 3, 5, 7, 11, . . .}. A program, called the "Sieve of Eratosthenes," which generates this set of numbers, is often used as a *benchmark*. A benchmark is a program used to test and compare computers.

✳**51.** Fill in the next four elements: {1, 2, 3, 5, 8 . . .}. This set of numbers is called the "Fibonacci sequence." A program to generate this set of numbers is sometimes used as a benchmark.

➠**52.** Use pseudocode or a flowchart to show an algorithm for computing the first 20 Fibonacci numbers.

✳**53.** The number of elements in a set A is called the *cardinal number* of the set and is denoted $n(A)$.

(a) When does $n(A \cup B)$ equal $n(A) + n(B)$?

(b) Determine a general formula for $n(A \cup B)$.

2.2   Venn Diagrams

It is often helpful to visualize the relationship between sets by drawing a diagram. Set diagrams are called *Venn diagrams* after a nineteenth-century logician, John Venn. The universal set is usually shown as a rectangle (see Figure 2.1). Sets within the universal set are often depicted as circular regions (see Figure 2.2).

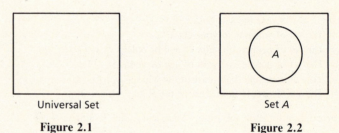

Universal Set

Figure 2.1

Set *A*

Figure 2.2

The basic set operations are illustrated in Figures 2.3, 2.4, and 2.5.

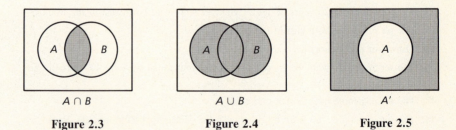

$A \cap B$

Figure 2.3

$A \cup B$

Figure 2.4

A'

Figure 2.5

Even if $A \cap B = \varnothing$, the diagram shown in Figure 2.3 is adequate. For example, if $A = \{1, 2, 3\}$ and $B = \{4, 5, 6\}$, then $A \cap B = \varnothing$. But the diagram could be drawn as shown in Figure 2.6.

Sometimes it is necessary to draw a Venn diagram for an expression containing more than one set operation. For example, to show $(A \cap B)'$, we first visualize $A \cap B$. Since $(A \cap B)'$ is that part of U that is *not* in $A \cap B$, we shade all of U, *except* $A \cap B$. See Figure 2.7.

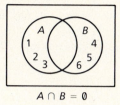

$$A \cap B = \emptyset$$

Figure 2.6

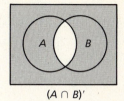

$$(A \cap B)'$$

Figure 2.7

For some problems, you may want to visualize each set individually before using the intersection or union. Consider $A' \cup B'$. We shade A' with horizontal lines and B' with vertical lines. See Figures 2.8 and 2.9.

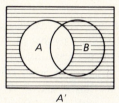

$$A'$$

Figure 2.8

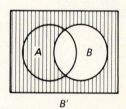

$$B'$$

Figure 2.9

When these two sets are shaded on the same diagram, it is easy to see the union of the two sets. See Figure 2.10.

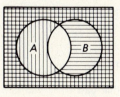

Figure 2.10

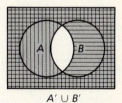

$$A' \cup B'$$

Figure 2.11

A' is shaded horizontally and B' is shaded vertically. Since $A' \cup B' = \{x \mid x \in A'$ **or** $x \in B'\}$, $A' \cup B'$ includes everything that is shaded horizontally *or* vertically. We shade in gray that area of the diagram in which the lines run horizontally, or vertically, or both. See Figure 2.11. We write this procedure as an algorithm.

> **To illustrate the intersection or union of sets on a Venn diagram:**
> 1. Shade the first set with horizontal lines.
> 2. Shade the second set with vertical lines.
> 3. (a) The intersection of the two sets has lines that run horizontally **and** vertically.
>   (b) The union of the two sets has lines that run horizontally **or** vertically (or both).

EXAMPLE 1

Illustrate $A' \cap B'$ on a Venn diagram.

Solution A' is shaded horizontally and B' is shaded vertically. Since $A' \cap B' = \{x | x \in A' \textbf{ and } x \in B'\}$, $A' \cap B'$ includes everything that is shaded horizontally *and* vertically. The area of the diagram with lines running both horizontally and vertically is shaded in gray to indicate the final answer. See Figure 2.12. ∎

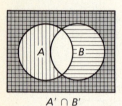

$A' \cap B'$

Figure 2.12

Two sets that contain the same elements are said to be *equal*. In exercises 4(a) and (b) of Section 2.1, you saw that $(A \cap B)' = A' \cup B'$ for $A = \{2, 3, 5, 7\}$ and $B = \{3, 6, 9\}$. It is impossible to verify this relationship for every pair of sets A and B by listing members. Therefore, we verify it by using a Venn Diagram for arbitrary sets A and B. We show $(A \cap B)'$ on one Venn Diagram and $A' \cup B'$ on another. See Figures 2.7 and 2.11. Since the two diagrams are identical regardless of what elements are in A and B, we know that the sets are equal.

> **To show that two sets are equal:**
> 1. Illustrate each set on a Venn diagram.
> 2. Compare the two diagrams. If they are identical, the two sets are equal.

EXAMPLE 2

Show that $(A \cup B)' = A' \cap B'$.

Solution $(A \cup B)'$ is that part of U that is *not* in $(A \cup B)$. See Figure 2.13. $A' \cap B'$ was shown in Figure 2.12. Since the two diagrams are identical, the two sets are equal. ∎

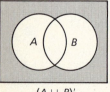

$(A \cup B)'$

Figure 2.13

The statements we have just examined are known as *De Morgan's laws*, after Augustus De Morgan, an English mathematician who lived in the nineteenth century.

De Morgan's Laws

$$(A \cap B)' = A' \cup B'$$

$$(A \cup B)' = A' \cap B'$$

2.2 EXERCISES

Draw a Venn diagram to illustrate each set. See Example 1.

1. $A \cap B$

2. $A \cup B$

3. A'

4. $(A')'$

5. $A \cap B'$

6. $A' \cap B$

7. $A \cup B'$

8. $A' \cup B$

9. $A \cap A$

10. $A \cup A$

11. $A \cup A'$

12. $A \cap A'$

13. $A \cap U$

14. $A \cup U$

15. $A \cup \varnothing$

16. $A \cap \varnothing$

17. $A' \cap B'$

18. $A' \cup B'$

Show that each statement is true. See Example 2.

19. $(A')' = A$

20. $(A \cap B) \cup A = A$

21. $(A \cap B) \cap A = A \cap B$

22. $(A \cup B) \cup A = A \cup B$

23. $(A \cup B) \cap A = A$

24. $A \cup A' = U$

25. $A \cap A' = \varnothing$

26. $(A' \cap B)' = A \cup B'$ (Hint: See problem 6 above for $A' \cap B$)

27. $(A \cup B')' = A' \cap B$ (Hint: See problem 7 above for $A \cup B'$)

28. $(A' \cup B)' = A \cap B'$

29. $(A \cap B')' = A' \cup B$

30. $(A' \cap B')' = A \cup B$

✳**31.** When does $(A \cup B)'$ equal $A' \cup B'$?

✳**32.** When does $(A \cap B)'$ equal $A' \cap B'$?

✳**33.** A is a subset of B, written $A \subseteq B$, if every element of A is an element of B. Furthermore $\varnothing \subseteq A$ for every set A. If $A = \{a, b\}$, its subsets are $\varnothing$, $\{a\}$, $\{b\}$, and $\{a, b\}$.

(a) List the subsets of $\{a, b, c\}$.

(b) Determine a general formula for the number of subsets of a set with n elements.

(c) Explain why adding one element to a set doubles the number of subsets.

2.3 Venn Diagrams with Three Sets

When three sets are illustrated on a Venn diagram, they are drawn as shown in Figure 2.14 or Figure 2.15. This type of diagram is adequate to depict any relationship involving three sets.

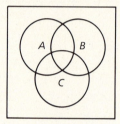

Figure 2.14

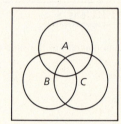

Figure 2.15

Often an expression that relates three sets will contain parentheses. An expression in parentheses usually involves intersection, union, or complementation. Figures 2.16, 2.17, and 2.18 are provided as a review of these operations.

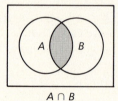

$A \cap B$

Figure 2.16

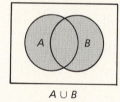

$A \cup B$

Figure 2.17

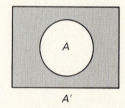

A'

Figure 2.18

We can use the technique of horizontal and vertical shading if we consider the part of the expression in parentheses as a single set.

To illustrate an expression involving three sets:
1. Consider the part of the expression in parentheses as a single set. Shade it with horizontal lines.
2. Consider the other part of the expression as a single set. Shade it with vertical lines.
3. (a) The intersection of two sets has lines that run horizontally **and** vertically.
   (b) The union of two sets has lines that run horizontally **or** vertically (or both).

EXAMPLE 1

Illustrate $A \cap (B \cup C)$ using a Venn diagram.

Solution   We shade $(B \cup C)$ with horizontal lines. We shade A with vertical lines. $A \cap (B \cup C)$ has both horizontal *and* vertical lines, so we shade that part of the diagram in gray. See Figure 2.19.

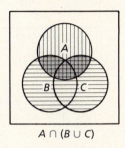

$A \cap (B \cup C)$

Figure 2.19   ▪

EXAMPLE 2

Illustrate $(A \cap B) \cup (A \cap C)$ on a Venn diagram.

Solution   We shade $(A \cap B)$ with horizontal lines. We shade $(A \cap C)$ with vertical lines. Since $(A \cap B) \cup (A \cap C)$ has horizontal *or* vertical lines, we shade that area in gray. See Figure 2.20.

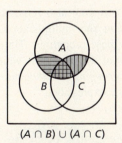

$(A \cap B) \cup (A \cap C)$

Figure 2.20

Because the shaded parts of Figures 2.19 and 2.20 are identical, we know that $A \cap (B \cup C) = (A \cap B) \cup (A \cap C)$.   ▪

EXAMPLE 3

Show that $A \cup (B \cap C) = (A \cup B) \cap (A \cup C)$.

Solution   Illustrate the two sets on separate Venn diagrams and compare. See Figures 2.21 and 2.22.

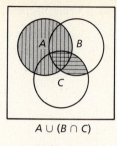

$A \cup (B \cap C)$

Figure 2.21

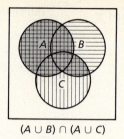

$(A \cup B) \cap (A \cup C)$

Figure 2.22

The two statements we have just examined are known as distributive properties because they are similar to the distributive property of algebra: $a \times (b + c) = (a \times b) + (a \times c)$. We write $A \cap (B \cup C) = (A \cap B) \cup (A \cap C)$ and say that intersection distributes over union. We write $A \cup (B \cap C) = (A \cup B) \cap (A \cup C)$ and say that union distributes over intersection.

Distributive Properties
1. $A \cap (B \cup C) = (A \cap B) \cup (A \cap C)$
2. $A \cup (B \cap C) = (A \cup B) \cap (A \cup C)$

You may also remember the commutative and associative properties for addition and multiplication. The set operations of intersection and union also have these properties. The commutative property says that we can change the *order* of the operands, while the associative property says that we can change their *grouping* (or association) as long as the same operation is involved throughout the expression.

Commutative Properties
1. $A \cap B = B \cap A$
2. $A \cup B = B \cup A$

Associative Properties
1. $(A \cap B) \cap C = A \cap (B \cap C)$
2. $(A \cup B) \cup C = A \cup (B \cup C)$

EXAMPLE 4

Determine which properties are used in each statement.

(a) $(A \cap B) \cap C = (B \cap A) \cap C$

(b) $(A \cup B) \cup C = C \cup (A \cup B)$

(c) $A \cup (B \cap C) = (A \cup B) \cap (A \cup C)$

Solution

(a) $(A \cap B) \cap C = (B \cap A) \cap C$. The commutative property was used to change the order of A and B.

(b) $(A \cup B) \cup C = C \cup (A \cup B)$. The commutative property was used to change the order of $(A \cup B)$ and C.

(c) $A \cup (B \cap C) = (A \cup B) \cap (A \cup C)$. The distributive property was used to distribute union over intersection. ■

EXAMPLE 5

Determine which properties are used in each statement.

(a) $(A \cap B) \cap C = A \cap (B \cap C)$

(b) $A \cap (B \cup C) = (A \cap B) \cup (A \cap C)$

(c) $A \cup (B \cup C) = (A \cup C) \cup B$

Solution

(a) $(A \cap B) \cap C = A \cap (B \cap C)$. The associative property was used to change the grouping from B with A on the left to B with C on the right.

(b) $A \cap (B \cup C) = (A \cap B) \cup (A \cap C)$. The distributive property was used to distribute intersection over union.

(c) $A \cup (B \cup C) = A \cup (C \cup B) = (A \cup C) \cup B$. Both the commutative and associative properties were used in this example. The commutative property allowed the order to change. The associative property allowed the grouping to change from C with B to C with A. ■

2.3 EXERCISES

Draw a Venn diagram to illustrate each set. See Examples 1 and 2.

1. $A' \cap (B \cup C)$ 2. $A \cap (B \cap C)'$ 3. $A \cup (B \cap C)'$ 4. $A' \cup (B \cap C)$

5. $A \cup (B \cap C')$ 6. $A \cap (B \cap C')$ 7. $(A \cup B) \cap C'$ 8. $(A \cup B) \cup C'$

9. $(A \cap B) \cup C'$ 10. $(A \cap B) \cap C'$ 11. $A' \cap (B \cup C)'$ 12. $A' \cup (B \cap C)'$

13. $A' \cap (B \cap C)'$ 14. $A' \cup (B \cup C)'$ 15. $(A \cap B') \cap C'$ 16. $(A' \cap B) \cup C'$

Draw Venn diagrams to show that each statement is true. See Example 3.

17. $A \cap (B \cap C) = (A \cap B) \cap C$ 18. $A \cup (B \cup C) = (A \cup B) \cup C$

19. $A \cap (B \cup C)' = (A \cap B') \cap C'$ 20. $A \cap (B \cap C)' = (A \cap B') \cup (A \cap C')$

Determine which properties are used in each statement. See Examples 4 and 5.

21. $B \cup (C \cap D) = (B \cup C) \cap (B \cup D)$
22. $(B \cap C) \cap D = D \cap (B \cap C)$
23. $(B \cup C) \cup D = (C \cup B) \cup D$
24. $B \cap (C \cap D) = (B \cap D) \cap C$
25. $(B \cup C) \cup D = B \cup (C \cup D)$
26. $Q \cap (R \cup P) = (Q \cap R) \cup (Q \cap P)$
27. $(Q \cup R) \cup P = P \cup (Q \cup R)$
28. $Q \cup (R \cup P) = (Q \cup P) \cup R$
29. $(Q \cup R) \cup P = Q \cup (R \cup P)$
30. $(Q \cup R) \cup P = (R \cup Q) \cup P$

✳**31.** An instructor asked a class of thirty-five students how many knew BASIC. Twenty students raised their hands. When asked how many knew FORTRAN, ten raised their hands. When asked how many knew Pascal, fifteen raised their hands. Nine knew both BASIC and FORTRAN, four knew both FORTRAN and Pascal, while eight knew both BASIC and Pascal. Only three knew all three languages. Determine how many knew only one of the languages, and how many knew none of them.

✳**32.** A Venn diagram illustrating two sets has four nonoverlapping regions. A Venn diagram illustrating three sets has eight nonoverlapping regions. Explain why including an additional set doubles the number of nonoverlapping regions.

✳**33.** Construct a Venn diagram to illustrate four sets. (Hint: It is impossible to do with four circles.)

2.4   Sets of Numbers

$N = \{1, 2, 3, 4, \ldots\}$ is the set of *natural numbers*. When the number 0 is included, the set is called the set of *whole numbers*. When you first began to study numbers, you learned certain "number facts" about this set. You may have displayed these facts in addition and multiplication tables like Tables 2.1 and 2.2. To find the sum of two numbers, locate the first number on the left side of the addition table. Draw an imaginary line across that row. Locate the second number at the top of the table. Draw an imaginary line down that column. The two imaginary lines intersect at the entry of the table that is the sum. The multiplication table is used in the same way to find the product of two numbers.

Table 2.1 **Addition Table**

+	0	1	2	3	4	5	6	7	8	9
0	0	1	2	3	4	5	6	7	8	9
1	1	2	3	4	5	6	7	8	9	10
2	2	3	4	5	6	7	8	9	10	11
3	3	4	5	6	7	8	9	10	11	12
4	4	5	6	7	8	9	10	11	12	13
5	5	6	7	8	9	10	11	12	13	14
6	6	7	8	9	10	11	12	13	14	15
7	7	8	9	10	11	12	13	14	15	16
8	8	9	10	11	12	13	14	15	16	17
9	9	10	11	12	13	14	15	16	17	18

Table 2.2 **Multiplication Table**

×	0	1	2	3	4	5	6	7	8	9
0	0	0	0	0	0	0	0	0	0	0
1	0	1	2	3	4	5	6	7	8	9
2	0	2	4	6	8	10	12	14	16	18
3	0	3	6	9	12	15	18	21	24	27
4	0	4	8	12	16	20	24	28	32	36
5	0	5	10	15	20	25	30	35	40	45
6	0	6	12	18	24	30	36	42	48	54
7	0	7	14	21	28	35	42	49	56	63
8	0	8	16	24	32	40	48	56	64	72
9	0	9	18	27	36	45	54	63	72	81

$I = \{ \ldots, -2, -1, 0, 1, 2, 3, \ldots \}$ is the set of *integers*. The set of integers can be regarded as an extension of the set of natural numbers. Rather than discard the natural numbers and their "number facts," we continue to use the old rules and develop a few new ones for negative numbers. (You may want to review the rules for signed numbers in the appendix at this time.)

$Q = \{x \mid x = a/b, \text{ where } a, b \in I, b \neq 0\}$ is the set of *rational numbers*. (We use the letter Q for "quotient.") That is, a rational number is any number that can be written as a ratio of two integers, as long as the denominator is not zero. Thus 4/5, −9/7, and 7/8 are rational numbers. The number 2 is rational since it can be written as 2/1, 4/2, and so on. Since $\sqrt{4}$ can be written as 2, it is also rational. The rational numbers form an extension of the integers. The old rules are still valid, but it is necessary to develop some new ones to handle denominators, so we can do problems with fractions.

Any number that can be written as a fraction can be written as a decimal by performing long division. An important characteristic of these decimals is that they can be considered infinite repeating decimals. That is, a finite number of digits form a pattern that is repeated infinitely. For example, 1/3 = 0.333 . . ., where the "3" is repeated; 1/2 = 0.5000 . . ., where the "0" is repeated; and 3/11 = 0.272727 . . ., where the "27" is repeated. We could say that $Q = \{x \mid x$ **can be written as an infinite repeating decimal**$\}$.

Any number that cannot be written as an infinite repeating decimal is said to be *irrational*. For example, the number 0.1010010001 . . . (where the number of zeroes is increased by one between each pair of successive ones), has no repeating pattern. It is harder to see that numbers like $\sqrt{2}$ and $\sqrt{3}$ are irrational, but mathematicians have shown that they are. In fact, $\sqrt{a}$, where a is a positive integer, will either be an integer, or it will be an irrational number. The number π was shown to be irrational by Johann Heinrich Lambert in 1761.

$R = \{x \mid x$ **is a rational number or** x **is an irrational number**$\}$ is the set of *real numbers*. We could say that R is formed by taking the union of the rational and irrational numbers.

The number sets can be shown on a Venn diagram. Here it is clear that each set is an extension of the previous one. See Figure 2.23.

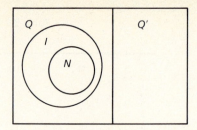

Figure 2.23

Sets of Numbers
Natural numbers: $N = \{1, 2, 3, 4, \ldots\}$
Integers: $I = \{ \ldots, -2, -1, 0, 1, 2, 3, \ldots \}$
Rational numbers: $Q = \{x | \; x = a/b$ where $a, b \in I, b \neq 0\}$ or
   $Q = \{x | \; x$ can be written as an infinite repeating decimal$\}$
Real numbers: $R =$ rational numbers $\cup$ irrational numbers

EXAMPLE 1

For each number in the table, check the sets to which it belongs.

Solution

Number	Natural no.	Integers	Rational no.	Real no.
2	x	x	x	x
-3		x	x	x
1/3			x	x
$\sqrt{3}$				x

We usually classify a number according to the first set to which it belongs. Thus, we would say $2 \in N$, $-3 \in I$, $1/3 \in Q$, and $\sqrt{3} \in R$. ■

EXAMPLE 2

Write in roster form: $\{x | \; x \in N, x \geq 6\}$.

Solution $\{6, 7, 8, 9, 10, \ldots\}$ ■

EXAMPLE 3

Write in roster form: $\{x | x \in I, x$ is a multiple of 4$\}$.

Solution $\{\ldots, -8, -4, 0, 4, 8, 12, 16, \ldots\}$ ■

Since N, I, Q, and R represent sets of numbers, it is possible to perform set operations using them.

EXAMPLE 4

Find $Q \cap N$.

Solution   Since every element of N is also an element of Q, $Q \cap N = N$. ▦

EXAMPLE 5

Find $I \cup N$.

Solution   Since every element of N is also an element of I, $I \cup N = I$. ▦

Some computer languages support the use of set concepts. One application of sets is to make a program user-friendly. The instruction to enter an integer between 1 and 10 might be given within a DOWHILE loop. The algorithm is shown in pseudocode:

Show message: ''Enter an integer between 1 and 10.''

Get number from keyboard.

While number is not in the set {1, 2, 3, 4, 5, 6, 7, 8, 9, 10},

    Show message: ''Try again—Enter an integer between 1 and 10.''

    Get number from keyboard.

ENDWHILE.

The program would not ''crash'' if incorrect data, such as 2.5 or 400, were entered.

2.4 EXERCISES

For each number in the table, check the sets to which it belongs. See Example 1.

	Number	Natural no.	Integers	Rational no.	Real no.
1.	0				
2.	2				
3.	−3				
4.	2/3				
5.	$\sqrt{5}$				
6.	$\sqrt{9}$				
7.	1 4/5				
8.	3.14				
9.	5				
10.	−1/2				
11.	−12				
12.	$\sqrt{7}$				
13.	π				
14.	5.1				
15.	7/5				

Write each set in roster form. See Examples 2 and 3.

16. $\{x \mid x \in N, x < 5\}$

17. $\{x \mid x \in I, x < 5\}$

18. $\{x \mid x \in N, x \text{ is a multiple of } 3\}$

19. $\{x \mid x \in I, x \text{ is a multiple of } 3\}$

20. $\{x \mid x \in I, x \text{ is divisible by } 5\}$

21. $\{x \mid x \in N, x \text{ is divisible by } 5\}$

22. $\{x \mid x \text{ is an even integer}\}$

23. $\{x \mid x \text{ is an odd integer}\}$

24. $\{x \mid x \text{ is an odd natural number}\}$

25. $\{x \mid x \text{ is a non-negative integer}\}$

26. $\{x \mid x \text{ is a rational number with numerator } 2\}$

27. $\{x \mid x \text{ is a rational number with numerator } 3\}$

28. $\{x \mid x \text{ is a rational number with denominator } 5\}$

29. $\{x \mid x \text{ is a rational number with denominator } 7\}$

Find each set. See Examples 4 and 5.

30. $N \cup I$

31. $N \cap I$

32. $Q \cup I$

33. $Q \cap I$

34. $N \cup Q$

35. $N \cap Q$

36. $R \cap Q$

37. $R \cup Q$

38. $N \cap R$

39. $N \cup R$

40. $I \cap R$

41. $I \cup R$

42. $Q' \cap Q$

➡43. In 1675, Gottfried Wilhelm von Leibniz showed that $\pi = 4 - 4/3 + 4/5 - 4/7 + 4/9 - 4/11 + \ldots$. It requires 152 terms of this series to produce π correct to the nearest hundredth (3.14). Use pseudocode or a flowchart to show an algorithm to determine the sum of 152 terms of this series.

✳44. Explain why *every* fraction can be written as a repeating decimal.

✳45. The following argument shows that 0.363636 . . . (where 36 repeats) is equal to the fraction 4/11.

$$\text{Let } x = 0.363636 \ldots$$
$$\text{Then } 100x = 36.363636 \ldots$$

If we subtract, we have
$$\underline{-x = -0.363636 \ldots}$$
$$\text{Thus } 99x = 36, \text{ and } x = 36/99 \text{ or } 4/11.$$

(a) What fraction is equal to 0.9999999 . . .?

(b) What fraction is equal to 4.31275275275 . . .?

✳46. The repeating digits of an infinite decimal may be indicated by a bar drawn above them. For example $113/900 = 0.12\overline{5}$, $113/990 = 0.1\overline{14}$, and $113/999 = 0.\overline{113}$. It is possible to convert an infinite repeating decimal to a fraction using the fact that the number of nines in the denominator is the same as the number of barred digits, and the number of zeroes is the same as the number of unbarred digits after the decimal point. How can the barred and unbarred digits be used to determine the numerator?

✳47. It is possible to draw a line exactly $\sqrt{2}$ units in length by using the Pythagorean theorem. The theorem states that for a right triangle having sides of lengths a, b, and c, with side of length c opposite the right angle, $c^2 = a^2 + b^2$. If we draw a right triangle having two perpendicular sides each one unit in length, the third side must be $\sqrt{2}$ units in length. Explain how to draw a line exactly $\sqrt{3}$ units in length.

2.5 Scientific Notation and Significant Digits

If we divided the length of a football field into three unequal segments and asked each of three people to measure one segment, we might get the following results: first length = 30 meters, second length = 28.3 meters, and third length = 32.69 meters. But to say that the length of the field is 90.99 meters is misleading. It implies a degree of precision that was not obtained for this problem. Measurement, by its nature, involves approximations. The first person measured to the nearest meter. The second person measured to the nearest tenth of a meter, while the third measured to the nearest hundredth. If the exact measurements were 30.41 meters, 28.34 meters, and 32.69 meters, the exact sum would be 91.44 meters. Our original sum of 90.99 is off by almost half a meter.

This example illustrates an important concept: *The result of a computation involving approximate numbers is no more precise than the least precise data that is used in the computation.* In this case, we could say that the length of the field is 91 meters to the nearest meter.

Because computers often display six or eight decimal places, the result of a computation may appear to be more precise than it actually is. As a general rule, when adding approximate numbers, the sum should be rounded to the same number of decimal places as the addend with the fewest decimal places.

To add approximate numbers:

1. Line up the decimal points and add.
2. Round off the sum to the same number of decimal places as the addend with the fewest decimal places.

EXAMPLE 1

Add the approximate numbers below, and round off appropriately.

(a) $43.6 + 4.36 + 0.436$

(b) $1.42 + 72.3 + 0.51$

(c) $632 + 85.4 + 2.31$

Solution

(a) $43.6 + 4.36 + 0.436 = 48.396$, which is rounded off to one decimal place (43.6 has only one place) to become 48.4

(b) $1.42 + 72.3 + 0.51 = 74.23$, which is rounded off to one decimal place (72.3 has only one place) to become 74.2

(c) $632 + 85.4 + 2.31 = 719.71$, which is rounded off to the nearest whole number (632 is a whole number) to become 720. ■

For multiplication, we use the concept of significant digits. Digits are the symbols 0, 1, 2, 3, 4, 5, 6, 7, 8, and 9 that we use in various combinations to write numbers.

> **Definition**
> A digit is *significant* if
> (a) it is a *nonzero* digit,
> (b) it is a zero *between* significant digits, or
> (c) it is a *terminal* zero *after* a decimal point. (A zero is terminal if no nonzero digits follow it.)

It might help to see what this definition leaves out—that is, which digits are *not* significant. A zero is *not significant* if it is a *beginning* zero (i.e., no nonzero digits precede it) or it is a *terminal* zero *before* a decimal point (unless specified otherwise).

Suppose that two people have electric bills of \$198.39 and \$200.00. The first might say, "My electric bill was \$200!" The two zeroes in 200 are not significant, because they indicate rounding off to the nearest hundred. The second person might say, "My electric bill was \$200 to the dollar!" The phrase "to the dollar" indicates that the two zeroes are significant in this number. The number is correct to the nearest whole number. We say that terminal zeroes before the decimal point are not significant unless there is some additional information to indicate that they are.

EXAMPLE 2

How many significant digits are in each group of numbers?

(a) 493, 4.93, 49.3
(b) 507, 5.07, 50.7
(c) 26.0, 2.60, 0.260
(d) 160, 1600, 16,000
(e) 0.18, 0.018, 0.0018

Solution
(a) There are three: each digit is a nonzero number.
(b) There are three: the zero is *between* significant digits.
(c) There are three: the zeroes are *terminal* zeroes *after* a decimal point.
(d) There are two: the zeroes are *terminal* zeroes *before* a decimal point.
(e) There are two: the zeroes are *beginning* zeroes. ■

EXAMPLE 3

How many significant digits are in the number 0.0020?

Solution There are two: the first zeroes are *beginning* zeroes; the last zero is a *terminal* zero *after* the decimal point. ■

EXAMPLE 4

How many significant digits are there in the number 10.00?

Solution   There are four: the last two zeroes are *terminal* zeroes *after* the decimal point; the first occurs *between* significant digits. ■

EXAMPLE 5

How many significant digits are there in the number 10?

Solution   There is only one: the zero is a *terminal* zero *before* the decimal point. ■

> **To multiply approximate numbers:**
> 1. Line up the rightmost digits, multiply, and place the decimal point.
> 2. Round off the answer to the same number of significant digits as the factor with the fewest significant digits.

EXAMPLE 6

Perform the multiplication using the approximate numbers below, and round off the answer appropriately:

(a) 493×160
(b) 5.07×10.00
(c) 5.07×10

Solution
(a)   493 has three significant digits, and 160 has only two. Since $493 \times 160 = 78{,}880$ and the answer is rounded to two significant digits, we have 79,000.
(b)   5.07 has three significant digits, and 10.00 has four. Since $5.07 \times 10.00 = 50.70$ and the answer is rounded to three significant digits, we have 50.7.
(c)   5.07 has three significant digits, and 10 has only one. Since $5.07 \times 10 = 50.7$ and the answer is rounded to one significant digit, we have 50. ■

It is important to realize that these rules for rounding apply only when using approximate numbers, such as measurements. They do *not* apply when the numbers are known to be exact values. For instance, if one ball-point pen costs \$.79, we would say that three cost $3 \times \$.79 = \$2.37$, because there are exactly three pens and each one costs exactly \$.79.

Scientific notation is often used to write very large or very small numbers. Rubik's cube is said to have 43,252,003,274,489,856,000 or about 43 quintillion different configurations. The expression 4.3×10^{19} is scientific notation for 43,000,000,000,000,000,000. That is, when you multiply 4.3 by 10^{19}, the decimal point is moved over 19 places to the right. This result is reasonable if you remember, for example, that $10^2 = 100$, and multiplication by 100 moves the decimal point two places to the right. To write a number in scientific notation, express it in the form $n \times 10^p$ where $1 \leq |n| < 10$ (read "the absolute value of n is greater than or equal to 1 and less than 10") and p is an integer.

> **To express a number in scientific notation:**
> 1. Place the decimal point so that there is exactly one nonzero digit before it.
> 2. Let $p =$ the number of decimal places the decimal point should be moved to put it back in its original position.
> 3. (a) Multiply by 10^p if the decimal point should be moved to the right.
>   (b) Multiply by 10^{-p} if the decimal point should be moved to the left.

EXAMPLE 7

Express each number in scientific notation:

(a)   49

(b)   490

(c)   58.23

Solution
(a) $49 = 4.9 \times 10^1$.
(b) $490 = 4.9 \times 10^2$.
(c) $58.23 = 5.823 \times 10^1$. ∎

EXAMPLE 8

Express each number in scientific notation:

(a)   0.231

(b)   0.0056

(c)   5823

Solution
(a) $0.231 = 2.31 \times 10^{-1}$
(b) $0.0056 = 5.6 \times 10^{-3}$
(c) $5823 = 5.823 \times 10^3$ (It is true that $5823 = 58.23 \times 10^2$, but this expression is not scientific notation, because 58.23 is not between 1 and 10.) ∎

Scientific notation is sometimes used to clarify the number of significant digits in a number. For example, 2×10^2 has only one significant digit, but 2.00×10^2 has three significant digits. Writing "2.00" indicates that the two zeroes are significant.

EXAMPLE 9

Express each number in decimal form:

(a) 6.3×10^4

(b) 2.13×10^{-1}

(c) 4.3×10^{-2}

Solution

(a) 63,000

(b) 0.213

(c) 0.043 ∎

All computers must use numbers within a limited range. Most can display about six or eight significant digits. When a number is very large or very small, it will be shown in scientific notation. Computers usually replace "$\times 10$" by the letter E. The number 6.3×10^4 might be written as 6.3E04 or 6.3 E4. It is not uncommon for a machine to work with and display numbers in the range of 10^{-38} to 10^{38} in scientific notation. Without scientific notation, a six or eight significant digit display would place a severe limitation on computation.

2.5 EXERCISES

Assume that the following additions involve approximate numbers. Add and round off appropriately. See Example 1.

1. $6.92 + 14.348$

2. $25.1 + 6.32 + 0.547$

3. $7.53 + 12.214$

4. $34.9 + 7.05 + 0.268$

5. $82 + 53.4 + 6.93$

State the number of significant digits in each number. See Examples 2–5.

6. 25.6	**7.** 3.14	**8.** 1.732
9. 7083	**10.** 80	**11.** 9000
12. 200	**13.** 71.30	**14.** 2.500
15. 6.0	**16.** 0.046	**17.** 0.0032
18. 0.067	**19.** 0.0300	**20.** 0.0070

Assume that the following multiplications involve approximate numbers. Multiply and round off appropriately. See Example 6.

21. 6.92×1.4

22. $0.213 \times 25 \times 4000$

23. $0.005 \times 12 \times 2.400$

24. 5.32×2.7

25. $31.2 \times 5 \times 400$

Write each number below in scientific notation. See Examples 7 and 8.

26. 45   **27.** 58   **28.** 231   **29.** 409

30. 6000   **31.** 8732   **32.** 0.027   **33.** 0.006

34. 0.0345   **35.** 0.4278

Write each number below in decimal form. See Example 9.

36. 6.7×10^3 **37.** 2.34×10^2

38. 8.35×10^{-2} **39.** 5.08×10^{-1}

40. 9.1×10^{-3} **41.** 8.2×10^2

42. 6.204×10^1 **43.** 3.452×10^3

44. 7.89×10^{-2} **45.** 6.57×10^{-1}

✸46. Edward Kasner, a mathematician, once asked his nephew to propose a name for the number 1×10^{100}. The child called it a ''googol,'' and the name is now commonly used.

   (a) How many zeroes are there when one googol is expressed in decimal notation?

   (b) A googolplex is expressed by writing a ''1'' followed by a googol zeroes. Write one googolplex in scientific notation.

✸47. Which is larger: A googol raised to the power 1 googolplex or 1 googolplex raised to the power 1 googol?

✸48. A certain microcomputer displays real numbers whose absolute values are 1 or larger, but less than 10, in decimal form rounded to six significant digits. All other numbers are shown in scientific notation with six significant digits. Show the output as it would be given for each number:

   (a) 32

   (b) -4.32

   (c) 7.0021

   (d) $-3,762,142$

   (e) 28.27143

   (f) 89,412,631

   (g) -0.000021579

✸49. A number is in normalized exponential form when it is written as $n \times 10^p$ where $0.1 \le |n| < 1$. Write the following numbers in normalized exponential form.

   (a) 26.31

   (b) 0.00251

   (c) 5863

   (d) 4.536

2.6   Order of Operations

Consider the problem $2 + 3 \times 5$. Some people might say the answer is 25, because $2 + 3 = 5$, and $5 \times 5 = 25$. Others might say the answer is 17, because $3 \times 5 = 15$, and $2 + 15 = 17$. So that there will not be two or more different answers to every problem, mathematicians have agreed that 17 is the correct answer to this problem. Knowing the order in which the operations are performed is just as important as knowing the addition and multiplication tables. Many programming languages follow the mathematical convention for the order of operations.

Order of Operations
1. Operations within grouping symbols are done first. Within grouping symbols, the order listed below is followed.
2. Exponents.
3. Multiplication and division (from left to right).
4. Addition and subtraction (from left to right).

Grouping symbols include parentheses ( ), brackets [ ], braces { }, and the fraction bar. Evaluate the expression within grouping symbols by using the standard order for exponents, multiplication, division, addition, and subtraction. If there are nested grouping symbols (one inside the other), work from the inside out. For example, in the expression $1 + \{2 - [3 + 2(4 + 5)]\}$, $(4 + 5)$ would be done first, leaving $1 + \{2 - [3 + 2(9)]\}$. The next step would be $[3 + 2(9)] = 21$, leaving $1 + \{2 - 21\}$. Since $2 - 21 = -19$, we have $1 + \{-19\} = -18$.

Multiplication and division are listed at the same level in the order of operations because they are performed from left to right as they occur in the expression. For example, $8 \div 4 \times 2 = 2 \times 2 = 4$, but $8 \div (4 \times 2) = 8 \div 8 = 1$, because the parentheses call for the multiplication to be done first. Addition and subtraction are listed at the same level in the order of operations, so they, too, are performed from left to right as they occur in the expression.

EXAMPLE 1

Evaluate $2(4 + 3)^2$.

Solution $2(\mathbf{4 + 3})^2$ The parentheses say that $4 + 3$ comes first.

$2(\mathbf{7})^2$ The exponent applies only to 7.

$2(\mathbf{49})$ Multiplication is done next.

98 ∎

EXAMPLE 2

Evaluate $12 \div 4 \, (2)$.

Solution   There are no grouping symbols, and there are no exponents.

$\qquad$ **(12 ÷ 4)**(2)   The division is done first.

$\qquad\qquad$ **3 (2)**   The multiplication is done next.

$\qquad\qquad\quad$ 6 ■

EXAMPLE 3

Evaluate $-1 \, [2 \, (3 + 1) - 5]$.

Solution $-1 \, [2 \, \mathbf{(3 + 1)} - 5]$ (inner grouping)

$\qquad\quad -1 \, [\mathbf{2(4)} - 5]$ (outer grouping)

$\qquad\quad -1 \, [\mathbf{8 - 5}]$

$\qquad\quad -1 \, [\mathbf{3}]$ (multiplication)

$\qquad\quad -3$ ■

EXAMPLE 4

Evaluate $3[2^2 - 3(2)]$.

Solution $\quad 3[\mathbf{4 - 6}]$

$\qquad\quad 3\mathbf{(-2)}$

$\qquad\quad -6$ ■

The rules for the order of operations are ambiguous for a problem like -3^2. Mathematicians treat the negative sign as the sign of the answer. That is $-3^2 = -9$. Many computers, however, treat the negative sign as part of the number to be squared. Thus $-3^2 = (-3)(-3) = 9$. Two things are important about this difference in interpretation. First, you should know how the computer on which you are working handles this problem, and second, you should use parentheses to avoid confusion. $(-3)^2$ clearly means $+9$ while $-(3^2)$ clearly means -9.

In BASIC, Pascal, and most other computer languages, the plus $(+)$ and minus $(-)$ signs are used to indicate addition and subtraction, but the asterisk (*) is used for multiplication and the slash (/) for division. In writing arithmetic expressions for the computer, every multiplication must be indicated. In algebra, *AB* means *A* times *B*, but a computer would interpret AB as a single variable. To say A times B, you would write A*B. Also, fractions cannot be written in vertical form. We write $\dfrac{3 - 2}{4 - 1}$ as $(3 - 2)/(4 - 1)$. Without parentheses, $3 - 2/4 - 1$ would give us $3 - 0.5 - 1$. (The order of operations requires that division be done before subtraction.) In programming, parentheses are the only grouping symbol. You should use parentheses anytime they help to clarify the intent, whether they are necessary or not.

115855

To write an arithmetic expression for the computer:
1. Use + for addition.
2. Use − for subtraction.
3. Use * for multiplication.
   Every multiplication must be indicated.
4. Use / for division.
5. Use ( ) for grouping.

EXAMPLE 5

Write each expression below as you would for a programming language such as BASIC or Pascal:

(a) 2 + 5

(b) 7A

(c) 3(X + 5)

Solution

(a) 2 + 5 appears the same way in programming.

(b) 7A is written as 7*A since every multiplication has to be indicated.

(c) 3(X + 5) is written as 3*(X + 5). ▪

EXAMPLE 6

Write each expression below as you would for a programming language such as BASIC or Pascal:

(a) $\dfrac{7}{XY}$

(b) $\dfrac{7Y}{X}$

(c) 7[2 + 3(A − 1)]

Solution

(a) 7/(X*Y)   Use parentheses so the multiplication will be done before the division.

(b) (7*Y)/X or 7*Y/X   The parentheses are used for clarity.

(c) 7*(2 + 3*(A − 1))   Parentheses are the only grouping symbol. ▪

In most versions of BASIC, the caret (^) is the exponentiation symbol. We write X^3 for *x* cubed. It is better programming practice to write X*X*X, though. On one popular microcomputer, for example, 163^3 is given as

4330745.7, an error of 1.3, while 163*163*163 is 43307747, the exact answer. Pascal does not have an exponentiation symbol, so it is natural to use repeated multiplication instead of exponents when the exponents are small.

It is also good programming practice to write arithmetic expressions so that division is done last whenever possible. Consider the following problem: "A utility company has 200,000 customers whose average bill over a 31-day period is $86. What is the daily income to the utility company?" If you first find the average daily bill of one customer ($86 ÷ 31) rounded off to $2.77 and multiply by 200,000, you would get $554,000.00 If you first find the monthly income of the company (200,000 × $86 = $17,200,000), and then find the daily income ($17,200,000 ÷ 31), you would get $554,838.71. The difference of $838.71 is due solely to round-off error introduced by the division 86 ÷ 31. That is, the round-off error introduced by division is compounded or magnified when the multiplication is done.

2.6 EXERCISES

Evaluate each expression. See Examples 1–4.

1. $4 + 2(5)$

2. $(4 + 2)(5)$

3. $8 - 2(3)$

4. $(8 - 2)(3)$

5. $(5 + 3^2)(7)$

6. $(5 + 3)^2(7)$

7. $-8 \div 4 \div 2$

8. $-8 \div (4 \div 2)$

9. $-8 \div 4(2)$

10. $-8 \div (4 \cdot 2)$

11. $4[(5 + 3) - 2^2]$

12. $4[(5 + 3) - 2]^2$

13. $6[7 - (1 - 3)]^2$

14. $6[7 - (1 - 3^2)]$

15. $2[5 + 3(7 - 4^2) - 6]$

16. $2[5 + 3(7 - 4)^2 - 6]$

17. $2[(5 + 3)(7 - 4) - 6]$

18. $3(4 + 1)^2 + 2(3 - 5) - 6(7)$

19. $3\{1 + 2[4 + 5(6 - 7) + 8] - 9\}$

20. $2\{2[2(2 + 1) + 1] + 1\}$

Write each expression as you would for a programming language such as BASIC or Pascal. See examples 5 and 6.

21. $4Y + 2X$

22. $4(Y - 2X)$

23. $(X + 2Y)(2X - Y)$

24. $(X - 3)(X - 4)$

25. $X - 2Y + 3$

26. $7[X - 3(2Y + 4) + 2]$

27. $3[X + 2(Y - 1)]$

28. $X[X + 2(X - 1)]$

29. $\dfrac{4AB}{5CD}$

30. $4 + \dfrac{X}{Y} - 5$

31. $\dfrac{4 + X}{Y - 5}$

32. $A + \dfrac{3}{B} + 2$

33. $\dfrac{A + 3}{B + 2}$

34. $X\{X\,[X\,(X + 1) + 1] + 1\} + 1$

35. $3\{X\,[Y\,(Z - 2) - 3] - 4\}$

36. $2\{\,X[\,2(Y - 1) + 2] - 3\}$

37. 2^3

38. $X^2 + Y^2$

39. XY^2

40. $\dfrac{2^2}{Y^2}$

✳**41.** In the early 1950s, the Polish logician Jan Lukasiewicz invented a notation that does not require parentheses. This notation, called Polish notation, is used in some computer languages and by some calculators. In postfix Polish notation, the operators are written after the operands. If the operations are done from left to right, the operands of each operator are the two that have most recently been evaluated. Several examples follow.

Polish notation	Algebraic meaning	
XY −	X − Y	The 2 operands preceding " − " are X and Y.
ABC + *	A(B + C)	The " + " is preceded by B and C. Thus, we have (B + C). This expression becomes one operand for the next operation. Thus "*" applies to A and (B + C).
BC + A*	(B + C)A	The " + " is preceded by B and C. Thus, we have (B + C). This expression becomes one operand for the next operation. Thus "*" applies to (B + C) and A.
YZ/X +	Y/Z + X	The "/" is preceded by Y and Z. Thus, we have Y/Z. This expression becomes one operand for the next operation.
		The " + " applies to (Y/Z) and X.
ZXY* −	Z − XY	The "*" is preceded by X and Y. Thus, we have XY. This expression becomes one operand for the next operation. The " − " applies to Z and XY.

Write the following expressions in Polish notation:

(a) ABC

(b) AB + C

(c) X − YZ

(d) YZ − X

(e) Y(Z − X)

(f) (A + B)/(C + D)

2.7 Properties of Real Numbers

Many students of elementary algebra take the properties of real numbers for granted. But we must pay special attention to them in programming. These properties are called "field properties."

Field Properties
1. **Closure properties:**
    If $a, b \in R$, then $a + b \in R$.
    If $a, b \in R$, then $ab \in R$.
2. **Commutative properties:**
    If $a, b \in R$, then $a + b = b + a$
    If $a, b \in R$, then $ab = ba$
3. **Associative properties:**
    If $a, b, c \in R$, then $(a + b) + c = a + (b + c)$
    If $a, b, c \in R$, then $(ab)c = a(bc)$
4. **Distributive property:**
    If $a, b, c \in R$, then $a(b + c) = ab + ac$
5. **Identity properties:**
    If $a \in R$, then $a + 0 = 0 + a = a$
    If $a \in R$, then $a(1) = 1(a) = a$
6. **Inverse properties:**
    If $a \in R$, then there is a number $b \in R$ such that $a + b = b + a = 0$
    If $a \in R$, $a \neq 0$, then there is a number $c \in R$ such that $ac = ca = 1$

It is not enough to memorize these properties. You must understand them thoroughly.

   1. **Closure property:** The closure property says that when we perform an operation on any two elements from a set, the result is also a member of the set.

EXAMPLE 1

Consider the natural numbers. For which of the four arithmetic operations is this set closed?

Solution

Closed for addition: When we add two natural numbers, the sum is a natural number.

Not closed for subtraction: If we subtract two natural numbers, the result is not always a natural number. For example, $2 - 3 = -1$.

Closed for multiplication: When we multiply two natural numbers, the product is a natural number.

Not closed for division: If we divide two natural numbers, the result is not always a natural number. For example $2 \div 3 = 2/3$. ∎

   2. **Commutative property:** The commutative property says that the *order* of the operands may be changed without affecting the answer.

EXAMPLE 2

Consider the natural numbers. Which of the four arithmetic operations are commutative?

Solution

Addition is commutative: $2 + 3 = 3 + 2$, and so on.

Subtraction is not commutative: For example, $2 - 3 \neq 3 - 2$.

Multiplication is commutative: $2(3) = 3(2)$, and so on.

Division is not commutative: For example, $2 \div 3 \neq 3 \div 2$
$$2/3 \neq 3/2. \quad \blacksquare$$

3. Associative property: The associative property says that when the same operation is used more than once, the *grouping* of the operands may be changed without affecting the answer.

EXAMPLE 3

Consider the natural numbers. Which of the four arithmetic operations are associative?

Solution

Addition is associative: $(2 + 3) + 4 = 2 + (3 + 4)$
$$5 \quad + 4 = 2 + \quad 7.$$

Subtraction is not associative: $(2 - 3) - 4 \neq 2 - (3 - 4)$
$$-1 \; - 4 \neq 2 - (-1)$$
$$- 5 \neq 3.$$

Multiplication is associative: $(2 \cdot 3)4 = 2(3 \cdot 4)$
$$(6)4 \quad = \quad 2(12).$$

Division is not associative. $(2 \div 3) \div 4 \neq 2 \div (3 \div 4)$
$$(2/3) \quad \div 4 \neq 2 \div \quad (3/4)$$
$$1/6 \neq 8/3. \quad \blacksquare$$

4. Distributive property: You saw the two distributive properties for set intersection and union in Section 2.3. A distributive property is used when two *different* operations appear in the same expression. There is only one distributive property for real numbers. We say that multiplication distributes over addition.

EXAMPLE 4

Consider the natural numbers.

(a) Is multiplication distributive over addition?

(b) Is addition distributive over multiplication?

Solution

(a) Multiplication is distributive over addition:
$$2(3 + 4) = 2(3) + 2(4)$$
$$2 \quad (7) \quad = \quad 6 \; + \; 8.$$

(b) Addition is not distributive over multiplication:

$$2 + (3 \times 4) \neq (2 + 3) \times (2 + 4)$$
$$2 + \quad 12 \quad \neq \quad 5 \quad \times \quad 6$$
$$14 \neq 30. \quad \blacksquare$$

5. **Identity property:**   We could describe an identity for an operation as an element of the set that has no effect on an expression when it is one of the operands. We say that 0 is the identity for addition in the set of real numbers, and 1 is the identity for multiplication in the set of real numbers. Notice the requirement for commutativity in the definition.

EXAMPLE 5

Consider the set of natural numbers. Which of the four arithmetic operations have an identity?

Solution

Addition: There is no identity, since $0 \notin N$.

Subtraction: There is no identity.

Multiplication: The identity is 1.

Division: There is no identity. (You might think that 1 is the identity, since $3 \div 1 = 3$, but the definition would require that $1 \div 3 = 3$ also.) $\blacksquare$

6. **Inverse property:**   Two numbers are said to be inverses under an operation if they yield the identity for that operation. Notice the requirement for commutativity in the definition.

EXAMPLE 6

Consider the set of rational numbers. For which of the four arithmetic operations are there inverses for each nonzero number?

Solution

Addition: Every number has an inverse. The identity is 0 and $2/3 + (-2/3) = 0$, and so on.

Subtraction: There are no inverses, because there is no identity.

Multiplication: Every nonzero number has an inverse. The identity is 1 and $(2/3)(3/2) = 1$, and so on.

Division: There are no inverses, because there is no identity. $\blacksquare$

EXAMPLE 7

Determine which of the six field properties is illustrated by each statement below:

(a) $3 + (4 + 5) = 3 + (5 + 4)$

(b) $3 + (4 + 5) = (4 + 5) + 3$

(c) $3 + (4 + 5) = (3 + 4) + 5$

Solution

(a) The commutative property was used to change the order of the addends from 3 + **(4 + 5)** to 3 + **(5 + 4)**.

(b) The commutative property was used. The quantity (4 + 5) is the same on both sides of the equation. The order was changed from **3** + (4 + 5) to (4 + 5) + **3**.

(c) The associative property was used. The grouping was changed from 4 with 5 to 4 with 3. ∎

EXAMPLE 8

Determine which of the six field properties is illustrated in each statement below:

(a) 3(4 + 5) = 3(4) + 3(5)

(b) 3 + 0 = 0 + 3 = 3

(c) (7/8)(8/7) = 1

Solution

(a) The distributive property.

(b) The identity property for addition. Adding 0 has no effect on 3.

(c) The inverse property for multiplication. The product of the two numbers 7/8 and 8/7 is 1, which is the identity for multiplication. ∎

Although we use the field properties automatically in doing arithmetic with real numbers, they do not necessarily hold for computer arithmetic. For example, consider (1/39 + 1/18) + 1/157. One popular microcomputer gives 0.875660079. But when the associative property is used to regroup the terms as 1/39 + (1/18 + 1/157), the answer is given as 0.87566008. (Not every computer will show a discrepancy for this particular problem, but similar discrepancies will occur in other problems.) The discrepancy is not large, but that it exists at all is a violation of the associative property. The distributive property also fails for certain problems. For example (1/131)(1/86 + 1/112) is given as 0.000156919683, but when the distributive property is used, (1/131)(1/86) + (1/131)(1/112) is given as 0.000156919682. The commutative property, however, holds for computer arithmetic as well as for the arithmetic of real numbers.

2.7 EXERCISES

Which of the following sets are closed under addition? Under multiplication? See Example 1.

1. Integers

2. Positive integers

3. Negative integers

4. Even integers

5. Odd integers

6. Rational numbers

7. Irrational numbers

8. {0, 1}

9. {1, −1}

10. {x| x is a positive three-digit integer}

Which of the following sets have an identity for addition? For multiplication? See Example 5.

11. Integers

12. Positive integers

13. Negative integers

14. Even integers

15. Odd integers

16. Rational numbers

17. Irrational numbers

18. {0, 1}

19. {1, −1}

20. In Section 2.3, you saw that intersection and union have the commutative, associative, and distributive properties. Is there an identity for set intersection? For set union?

Which of the following sets have an inverse for each number under addition? For each nonzero number under multiplication? See Example 6.

21. Integers

22. Positive integers

23. Negative integers

24. Even integers

25. Odd integers

26. Rational numbers

27. Irrational numbers

28. {0, 1}

29. {1, −1}

30. {x| x is a positive three-digit integer}

Determine which of the six field properties are illustrated by each statement below. See Examples 7 and 8.

31. $(3 + 4) + 5 = 3 + (4 + 5)$

32. $(3 \cdot 4) \cdot 5 = 5 \cdot (3 \cdot 4)$

33. $(3 + 4) + 5 = (4 + 3) + 5$

34. $3(4 + 5) = 3(4) + 3(5)$

35. $7 + 0 = 7$

36. $(5/6)(6/5) = 1$

37. $8(1) = 8$

38. $5 + (−5) = 0$

39. $(3 \cdot 4) \cdot 5 = 4 \cdot (3 \cdot 5)$

40. $(3 + 4) + 5 = (5 + 4) + 3$

41. $(2/3) + (−2/3) = 0$

42. $(1/9)(9) = 1$

✳**43.** Let ˆ denote the operation of exponentiation (that is, $a \char94 b = a^b$) defined for integers.

   (a) Give an example to show that ˆ is not, in general, commutative.

   (b) For what values of a and b does $a \char94 b$ equal $b \char94 a$?

✳**44.** Is the operation ˆ (see problem 43, above) associative?

✳**45.** Let ° denote an operation defined on the integers such that $a \circ b = a + b − 2$.

   (a) Is ° a commutative operation?

   (b) Is there an identity element in I for the operation °?

   (c) For each integer, is there an inverse in I under the operation °?

✳**46.** Explain why the following sets are not closed under division.

   (a) Integers

   (b) Rational numbers

   (c) Real numbers

2.8 Integer Arithmetic

The computer representation of a real number is often a rounded value. You have seen some of the problems that can arise because of round-off error. For many purposes, such errors are insignificant. For accounting, however, results must be accurate to the penny. Financial calculations involving addition, subtraction, and multiplication are often done by using integers for the number of pennies involved. Integers are represented within the computer without round-off error. In many programming languages, it is possible to do "integer arithmetic."

In integer arithmetic, we treat the integers as though the set were closed under division as well as addition, subtraction, and multiplication. That is, the quotient of two integers is an integer. (Division by zero is still undefined.) The usual order of operations is followed, but any fractional part of a quotient is discarded. We say that the number is *truncated*. Truncation is not the same as rounding. For example, 9.9 is truncated to 9, but 9.9 is rounded to 10.

To perform an operation in integer arithmetic:

1. Follow the usual order of operations.
2. Drop any fractions from the quotient of a division problem.

EXAMPLE 1

Perform the following operations using real number arithmetic; then perform the same operations using integer arithmetic.

(a) 7/3

(b) 2/5 + 3/5

(c) (2 + 3)/5

Solution

(a) Real: 7/3 = 2 1/3
Integer: 7/3 = 2 (the fraction 1/3 is dropped)

(b) Real: 2/5 + 3/5 = 5/5 = 1
Integer: 2/5 + 3/5 = 0 + 0 = 0

(c) Real: (2 + 3)/5 = 5/5 = 1
Integer: (2 + 3)/5 = 5/5 = 1 ■

EXAMPLE 2

Perform the following operations using real number arithmetic; then perform the same operations using integer arithmetic.

(a) 5/3

(b) 1/4 + 3/4

(c) (5 + 5)/4

Solution

(a)   Real: 5/3 = 1 2/3

Integer: 5/3 = 1 (2/3 is dropped)

(b)   Real: 1/4 + 3/4 = 4/4 = 1

Integer: 1/4 + 3/4 = 0 + 0 = 0

(c)   Real: (5 + 5)/4 = 10/4 = 5/2 = 2 1/2

Integer (5 + 5)/4 = 10/4 = 2 (1/2 is dropped) ∎

It may be useful to know the remainder in such an integer division. For example, if you are printing a long list of data, you may want to skip every fifth line for readability. You could number the lines and print a blank line for those line numbers that have a remainder of zero when divided by five. Pascal and some versions of BASIC allow you to calculate the remainder by using MOD. The word MOD is from the word *modulo*. The set {0, 1, 2, 3} is an example of a modulo system if we define addition and multiplication by Tables 2.3 and 2.4.

Table 2.3						Table 2.4			
+	0	1	2	3		×	0	1	2	3
0	0	1	2	3		0	0	0	0	0
1	1	2	3	0		1	0	1	2	3
2	2	3	0	1		2	0	2	0	2
3	3	0	1	2		3	0	3	2	1

There is a physical interpretation for addition defined in this way. The numbers 0, 1, 2, and 3 are placed around the face of a dial (see Figure 2.24). Addition is done by moving a center hand. For example, if the hand were at 3, adding 3 would move it through 3 spaces: 3 + 3 = 2.

This concept can be generalized.

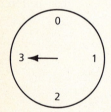

Figure 2.24

> **Definition**
>
> A modulo *n* system is a set {0, 1, 2, 3 . . . *n* − 1} with addition and multiplication of two numbers *a* and *b* from the set defined and written as follows:
>
> (a) $(a + b)$ MOD *n* is the remainder for the division $(a + b)/n$.
>
> (b) (ab) MOD *n* is the remainder for the division $(ab)/n$.

Thus (5 + 3) MOD 4 = 0, because 5 + 3 = 8, and 8 ÷ 4 has a remainder of 0. Likewise, (5 × 3) MOD 4 = 3, because 5 × 3 = 15, and 15 ÷ 4 has a remainder of 3. We could abbreviate "the remainder is 3" by writing R 3. Thus 15 ÷ 4 = 3 R 3.

EXAMPLE 3

Evaluate each expression:

(a) 9 MOD 5

(b) (4 + 3) MOD 5

(c) (2 + 3 × 4) MOD 6

Solution
(a) 9/5 = 1 R 4; therefore 9 MOD 5 = 4
(b) (4 + 3)/5 = 7/5 = 1 R 2; therefore (4 + 3) MOD 5 = 2
(c) (2 + 3 × 4)/6 = 14/6 = 2 R 2; therefore (2 + 3 × 4) MOD 6 = 2 ■

EXAMPLE 4

Evaluate each expression:

(a) (7 + 8) MOD 12

(b) (13 + 14) MOD 26

(c) (6 × 6) MOD 12

Solution
(a) (7 + 8)/12 = 15/12 = 1 R 3; therefore (7 + 8) MOD 12 = 3
(b) (13 + 14)/26 = 27/26 = 1 R 1; therefore (13 + 14) MOD 26 = 1
(c) (6 × 6)/12 = 36/12 = 3 R 0; therefore (6 × 6) MOD 12 = 0 ■

2.8 EXERCISES

Perform the indicated operations using real number arithmetic; then perform the same operations using integer arithmetic. See Examples 1 and 2.

1. 12/5 **2.** 14/3 **3.** 21/6 **4.** 3/7 + 4/7
5. 7/11 + 9/11 + 6/11 **6.** 22/7 + 13/7 **7.** 4/5 · 20 **8.** 4 · 20/5
9. 21 · 3/7 **10.** 21 · (3/7) **11.** (3/4)(8/9)(6) **12.** 3/4 · 8/9 · 6
13. 48/5 · 2/3 **14.** 48/(5 · 2)/3 **15.** (48/5)(2/3) **16.** (1/2)(2/3) + (1/4)(8/3)
17. (1/3)(3/5) + (1/10)(8) **18.** (2/3)(3/5) + (3/4) **19.** 15/4 · 2/3 **20.** 7 · 9/2/3

Evaluate each expression below. See Examples 3 and 4.

21. (2 + 1) MOD 3 **22.** (3 + 2) MOD 4
23. (4 + 4) MOD 5 **24.** (3 × 3 + 4) MOD 5
25. (2)(2 + 1) MOD 3 **26.** (1 + 2)(3) MOD 4
27. (4 + 5)(5) MOD 7 **28.** (5)(4)(3) MOD 7
29. (3)(3)(3) MOD 4 **30.** (2)(2)(2) MOD 3
31. (4)(3)(2) MOD 5 **32.** [5(2) + 3(4)] MOD 6
33. [3(5) + 4(2)] MOD 6 **34.** [6(3) + 2(5)] MOD 7
35. [3(5) + 4(2)] MOD 7 **36.** [3(5) + 4(2)] MOD 8

A set S with an operation ° defined for any two members of S is a group if:

(a) S is closed under the operation °.

(b) The operation ° is associative.

(c) There is an identity element in S for the operation °.

(d) For each element of S there is in S an inverse under the operation °.

Group theory has applications in chemistry and physics, as well as in solutions to puzzles such as Rubik's cube. In computer science, it has applications to error detection and correction of codes. For each set and operation, determine whether the system is a group.

✳**37.** Integers, addition

✳**38.** {0, 1, 2}, MOD 3 addition

✳**39.** {0, 1, 2}, MOD 3 multiplication

✳**40.** {0, 1} multiplication

✳**41.** Rational numbers, ° defined as $a \circ b = ab - 2$

➠**42.** Use pseudocode or a flowchart to show an algorithm that will list the squares of the numbers from 1 to 50, skipping every fifth line.

CHAPTER REVIEW

$\{x | \ x \in A \textbf{ and } x \in B\}$ = _____.

$\{x | \ x \in A \textbf{ or } x \in B\}$ = _____.

The largest set under consideration for a given problem is known as the _____ set.

$\{x | \ x \in U, \ x \notin A\}$ = _____.

A diagram that illustrates set relationships is called a _____ diagram.

$\{1, 2, 3, \ldots\}$ is the set of _____ numbers.

$\{\ldots, -2, -1, 0, 1, 2, \ldots\}$ is the set of _____.

$\{x | \ x = a/b \text{ where } a, b, \in I, \ b \neq 0\}$ is the set of _____ numbers.

Any number that can be written as an infinite repeating decimal is a _____ number.

Any number that cannot be written as an infinite repeating decimal is an _____ number.

The set of real numbers is the union of the set of _____ numbers and the set of _____ numbers.

A number in scientific notation is expressed in the form $n \times 10^p$ where _____ $\leq |n| <$ _____ and p is an _____.

$(a + b)$ MOD n is the remainder for the division _____.

CHAPTER TEST

Let $U = \{1, 2, 3, 4, 5, 6, 7\}$
    $A = \{1, 3, 5, 7\}$
    $B = \{3, 4, 5, 6\}$
    $C = \{1, 4, 5, 7\}$

Find each set.

1. (a) $A \cap B$
   (b) $B \cup C$

2. (a) $A' \cap C$
   (b) $B' \cup C$

3. (a) $A' \cap (B \cup C)$
   (b) $A \cup (B \cap C)'$

Let U = The set of applicants for a data processing job.
    C = The set of applicants who know COBOL.
    F = The set of applicants who know FORTRAN.
    B = The set of applicants who know BASIC.

Write a description for each set using $\cap$, $\cup$, and $'$.

4. (a) The set of applicants who know COBOL or FORTRAN.
   (b) The set of applicants who know FORTRAN and BASIC.

5. (a) The set of applicants who know BASIC but not FORTRAN.
   (b) The set of applicants who know COBOL or FORTRAN, but not BASIC.

6. (a) The set of applicants who don't know BASIC or COBOL.
   (b) The set of applicants who know BASIC and COBOL, but not FORTRAN.

Draw Venn diagrams to illustrate each set.

7. $A' \cap B$

8. $A \cup B'$

Show that each statement is true.

9. $(A \cap B)' = A' \cup B'$

10. $(A \cup B)' = A' \cap B'$

Draw Venn diagrams to illustrate each set.

11. $A \cap (B \cup C)$

12. $(A \cap B)' \cup C$

Draw Venn diagrams to show that each statement is true.

13. $(A \cup B) \cup C' = A \cup (B \cup C')$

14. $A \cap (B \cap C)' = (A \cap B') \cup (A \cap C')$

Determine which properties are illustrated by each statement.

15. (a) $A \cap (B \cup C) = (A \cap B) \cup (A \cap C)$
    (b) $A \cup (B \cup C) = (A \cup B) \cup C$
    (c) $A \cup (B \cup C) = (B \cup C) \cup A$

For each number in the table, check the set(s) to which it belongs:

	Number	Natural numbers	Integers	Rational numbers	Real numbers
16(a)	3				
(b)	$-2/3$				
(c)	$\sqrt{11}$				
(d)	-5				
(e)	1.3				
(f)	$\sqrt{9}$				

Write each set in roster form.

17. $\{x \mid x \in N, x \geq 5\}$

18. $\{x \mid x \in I, x < 2\}$

Find each set:

19. $N \cup R$

20. $Q \cap I$

21. Assume that the following additions involve approximate numbers, and round off the sum appropriately.

(a) $41.23 + 6.2$

(b) $7.93 + 0.147$

22. State the number of significant digits in each number.

(a) 0.0023

(b) 0.230

(c) 230

23. Assume that the following multiplications involve approximate numbers, and round off the product appropriately.

(a) 0.41×0.003

(b) 0.413×25

24. Write each number in scientific notation.

(a) 835

(b) 0.0051

25. Write each number in decimal form.

(a) 7.93×10^{-2}

(b) 6.4×10^{3}

Evaluate each expression.

26. $(1 + 2)^{3}(3)$

27. $100 \div 20 \div 5$

28. $5[2^{3} + 2(4 - 1)^{2} - 1]$

Write each expression as you would for a programming language such as BASIC or Pascal.

29. $3[X - 3(Y - 1)]$

30. $\dfrac{3A + B}{4CD}$

31. Is the set of even positive integers closed under addition? Multiplication?

32. Does the set of odd positive integers have an identity under addition? Multiplication?

33. Does each element in the set of even positive integers have an inverse for addition? For multiplication?

34. Determine which of the field properties are illustrated by each statement.

(a) $(2 \cdot 4) \cdot 6 = 2 \cdot (4 \cdot 6)$

(b) $(2 + 4) + 6 = 6 + (2 + 4)$

(c) $4/7 + (-4/7) = 0$

35. Determine which of the field properties are illustrated by each statement:

(a) $2(4 + 6) = 2(4) + 2(6)$ (b) $5(1) = 5$ (c) $4(1/4) = 1$

Evaluate each expression using real number arithmetic; then evaluate the same expression using integer arithmetic.

36. $2(2/9) + 5/9$ **37.** $2/3(3/4) + 1/2$

Evaluate each expression:

38. $(3 \times 2 + 5)$ MOD 6   **39.** $3(2 + 5)$ MOD 7   **40.** $[2(3) + 4(4)]$ MOD 5

41. The following flowchart shows an algorithm that generates prime numbers less than 100. Modify it so that pairs of "twin primes," (that is, primes that differ by 2) are found.

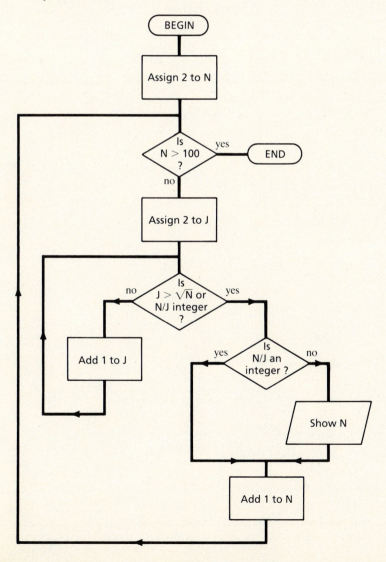

FOR FURTHER READING

Cooper, Curtis. "An Application of a Generalized Fibonacci Sequence," *The College Mathematics Journal*, Vol. 15 No. 2 (March 1984), 145–157. (A generalized Fibonacci sequence is used to solve a coin-tossing problem.)

DeSanto, Carmine. "A Classroom Note on Venn diagrams for Four and Five Sets," *Mathematics and Computer Education*, Vol. 18 No. 2 (Spring 1984), 107–110. (Venn diagrams for four and five sets are shown.)

Kung, Mou-Liang, and George C. Harrison. "Is the Venn diagram good enough?" *The College Mathematics Journal*, Vol. 15 No. 1 (January 1984), 48–50. (It is shown that circles cannot be used to construct a general Venn diagram for four sets.)

Leavitt, W. G. "Repeating Decimals," *The College Mathematics Journal*, Vol 15 No. 4 (September 1984), 299–308.

Biography   Pythagoras

Pythagoras was born on the Greek island of Samos about 570 B.C. He probably travelled to Egypt and India, where he acquired the sense of mysticism that influenced his philosophy. When he returned to Samos, the island was ruled by the tyrant Polycrates, so he moved to Crotona in the southern part of Italy. There he founded a school for the study of mathematics, philosophy, music, and astronomy.

His students were mostly upper-class men, but people of all classes attended his lectures. Women also came to hear him, breaking a law that forbade their attendance at public meetings. A young woman named Theano was among them, and despite the difference in their ages, she and Pythagoras were married.

An elite group of students, called the Pythagoreans, formed a close-knit society with secret rites and symbols. They ate neither meat nor beans, and they believed in keeping their discoveries secret from the public. They left no written records, and they credited discoveries to the founder of the school, so it is impossible to know exactly what contributions to mathematics Pythagoras himself made. The Pythagorean theorem was known to the ancient Babylonians, but the Pythagoreans are believed to be the first to have proved it.

The Pythagoreans also discovered that there are numbers that cannot be written as ratios of two integers. There is a legend that a Pythagorean named Hippasus revealed this secret. His drowning (whether accidental or intentional) was considered just punishment by the sect.

The Pythagoreans became politically powerful, and forces opposed to their aristocratic ways broke up the society. Pythagoras and Theano went into exile, and he died at about age 75, perhaps murdered by his enemies.

Chapter Three
Logic and Boolean Algebra

A microprocessor chip is an integrated circuit fabricated on a piece of silicon less than 5 mm square and 0.5 mm thick. It contains most of the circuitry necessary for interpreting the instructions and performing the operations of a computer program. The logic involved is similar to mathematical logic.

3.1 Symbolic Logic

Mathematics and computer programming depend on reasoning clearly. Language is so complex that sometimes reasoning is difficult to follow. Logic allows us to study the methods of distinguishing correct reasoning from incorrect reasoning.

> **Definition**
> A *proposition* is a declarative sentence that is either true or false, but not both.

The following sentences are propositions:

Two plus three equals five.	(true)
Four is less than four.	(false)
Three is a square root of nine.	(true)

The following sentences are not propositions:

Four plus five.	(not a sentence)
Is three equal to one?	(not declarative)
This statement is false.	(If the statement is true, then it must be false; if it is false, then it must be true.)

A proposition that expresses a single affirmative thought is called a *simple proposition*. When "and" or "or" is used to join two or more propositions, it is called a *logical connective*. Although it does not join two thoughts, the word "not" is usually classified as a logical connective. A sentence using a logical connective is called a *compound proposition*. The *components* of a compound proposition are the propositions that compose it.

In symbolic logic, we use symbols so we can examine the *structure* of propositions without regard to the meaning of the words. Usually a single letter is used to represent a simple proposition. To write a compound proposition symbolically, we use a single letter for each simple proposition and appropriate symbols for the logical connectives.

A sentence of the form *p **and** q* is called a *conjunction*. A sentence of the form *p **or** q* is called a disjunction. A sentence of the form ***not** p* is called a *negation*.

> **Logical Connectives**
> | | | |
> |---|---|---|
> | Conjunction | $\wedge$ | means "and" |
> | Disjunction | $\vee$ | means "or" |
> | Negation | $\sim$ | means "not" |

Consider the proposition, "x is a rational number and x is less than four." We let r represent the proposition "x is a rational number," and we let f represent the proposition "x is less than four." Then $r \wedge f$ would represent the conjunction.

EXAMPLE 1

Write as a proposition in symbolic form: "x is a negative integer or x is a positive rational number."

Solution

n: x is a negative integer

p: x is a positive rational number

"*x is a negative integer or x is a positive rational number*" is written $n \vee p$. ▨

The letters used in symbolic logic are often p, q, and r. But they may be chosen, instead, to remind you of the proposition itself. In Example 1, we used n for *n*egative and p for *p*ositive.

It is a good idea to let the letter represent a proposition in its affirmative form. For example, if we want to say "x is not equal to five," we let p represent "x is equal to five" and use $\sim p$ for "x is *not* equal to five."

EXAMPLE 2

Write as a proposition in symbolic form: "There are two solutions, and they are not rational."

Solution

t: there are two solutions

r: they are rational

"*There are two solutions*, and *they are not rational*" is written $t \wedge \sim r$. ▨

Translating English sentences into symbolic form calls for special care. It is not always possible to treat a declarative sentence as a proposition without some modification. You should understand the *intent* of the sentence. For example, the word "and" is not always stated. Consider the proposition, "two is a natural number, an integer, and a rational number." This statement really means $n \wedge i \wedge r$ where n, i, and r represent the propositions:

n: two is a natural number

i: two is an integer

r: two is a rational number.

The word "but" may mean "and," as it does in the statement "x is greater than two, but x is less than five." This proposition is written as $t \wedge f$ where we have the propositions:

t: *x* is greater than two (or $x > 2$)

f: *x* is less than five (or $x < 5$).

The same proposition (or thought) can be expressed by different sentences. Consider the two sentences: 1. Many people say that Ada Lovelace was the first computer programmer. 2. Ada Lovelace is often cited as the first computer programmer. The *sentences* are different, but they could be represented by the same letter, because the *proposition* is the same.

Sentences written in different tenses often express the same proposition. For example, suppose the local meteorologist says, "It will be hot and humid on Tuesday." On Tuesday, you remark, "It is hot, but not humid." The sentences "It will be hot" and "It is hot" would be represented by the same letter in symbolic logic, because they both refer to the local temperature on Tuesday. That is, they express the same *proposition,* even though the *sentences* are different.

Still another complication is that the same sentence may express two different propositions. For example, consider the statements: 1. In 1642, Pascal set out to build a calculating machine, and he was successful. 2. Later Leibniz set out to build a calculating machine, and he was successful. The phrase "he was successful" means Pascal was successful in the first sentence, but it means Leibniz was successful in the second. The simple *sentence* "He was successful" is the same, but the *propositions* are different and would require different letters.

Parentheses should be used with symbolic logic. A comma in the English sentence often gives a clue to the placement of parentheses.

EXAMPLE 3

Write as a proposition in symbolic form: "*x* is a negative integer, or it is an integer greater than three and is a perfect square."

Solution   *n*: *x* is a negative integer

   t: *x* is an integer greater than three

   s: *x* is a perfect square

The proposition is written $n \lor (t \land s)$. ▪

EXAMPLE 4

Write as a proposition in symbolic form: "*x* is a negative integer or it is an integer greater than three, and is a perfect square.

Solution   *n*: *x* is a negative integer

   t: *x* is an integer greater than three

   s: *x* is a perfect square

The proposition is $(n \lor t) \land s$. ▪

The only difference between the propositions in Examples 3 and 4 was the placement of the comma, yet the meaning is quite different. In Example 3, $x \in \{ \ldots -3, -2, -1 \}$ or $x \in \{ 4, 9, 16, 25, \ldots \}$. That is, *x* belongs to the union of the two sets, so $x \in \{\ldots -3, -2, -1, 4, 9, 16, 25, \ldots\}$. In Example 4, $x \in \{\ldots -3, -2, -1, 4, 5, 6, 7, 8, \ldots\}$ and $x \in \{4, 9, 16, 25, \ldots\}$. That is, x belongs to the intersection of the two sets, so $x \in \{4, 9, 16, 25, \ldots\}$.

The words "neither . . . nor" are often used to express the negation of an "either . . . or" sentence.

EXAMPLE 5

Write as a proposition in symbolic form: "zero is neither positive nor negative."

Solution p: zero is positive

n: zero is negative

To say it is neither says it is not either. That is, we have $\sim(p \vee n)$. ■

EXAMPLE 6

Write as a proposition in symbolic form: "Seven is neither less than five nor greater than ten."

Solution l: seven is less than five

g: seven is greater than ten

If it is not either, then $\sim(l \vee g)$. ■

3.1 EXERCISES

Write the symbolic form of each sentence (about a data processing curriculum) using the given letters. See Examples 1–6.

  c: You take COBOL.
  b: You take BASIC.
  p: You take Pascal.
  f: You take FORTRAN.

1. You must take COBOL or FORTRAN.

2. You may take BASIC and Pascal.

3. You may take COBOL and FORTRAN.

4. You must take COBOL, and BASIC or Pascal.

5. You must take COBOL and BASIC, or Pascal.

6. You may not take BASIC and Pascal.

Problems 7–23 are typical of compound sentences that might be incorporated into computer programs for different businesses.

Write the symbolic form of each sentence (about an applicant for automobile insurance) using the given letters. See Examples 1–6.

  a: The applicant is 25 or over.
  m: The applicant is male.
  v: The applicant has had a traffic violation within the last three years.
  t: The applicant will have two or more cars insured.

7. The applicant is female and is under 25.

8. The applicant has had no violations or will have two or more cars insured.

9. The applicant is male and is over 25, and has had no violations in the last three years.

10. The applicant is female and will have two or more cars insured, and she is 25 or over.

11. The applicant is female and is 25 or over, but has only one car insured.

12. The applicant is neither male nor 25 or over.

Write the symbolic form of each sentence (about the inventory of a business) using the given letters. See Examples 1–6.

- *s*: The item is on the shelf.
- *o*: The item is out of stock.
- *r*: The item has been reordered.
- *d*: The item has been discontinued.

13. The item is out of stock but has been reordered.

14. The item has been discontinued or is out of stock.

15. The item is out of stock and has been reordered, but has been discontinued.

16. The item is on the shelf or is out of stock, but has not been discontinued.

17. The item is on the shelf, or it is out of stock and has been discontinued.

18. The item is neither out of stock nor discontinued.

Write the symbolic form of each sentence (about airline reservations) using the given letters. See Examples 1–6.

- *w*: The passenger desires a window seat.
- *f*: The passenger desires the first-class section.
- *s*: The passenger desires the smoking section.
- *a*: The passenger will travel alone.
- *r*: The passenger desires a round-trip ticket.

19. The passenger will travel alone and desires a round-trip ticket.

20. The passenger desires a window seat or the smoking section.

21. The passenger wants the first class section, but neither a window seat nor the smoking section.

22. The passenger wants a window seat or the smoking section, but not in the first-class section.

23. The passenger wants neither a window seat nor the smoking section.

Write the symbolic form of each sentence using the given letters. See Examples 1–6.

$$t : x > -2$$
$$f : x < 5$$
$$z : x = 0$$

24. $x > -2$ and $x < 5$

25. $x > -2$ but $x \neq 0$

26. $x > -2$ or $x = 0$

27. $x < 5$ or $x \neq 0$

28. $x > -2$ and $x < 5$, but $x \neq 0$

29. $x < 5$ and $x \neq 0$

30. $x \leq -2$ or $x \geq 5$

✳**31.** Ann, Bill, and Carlos are eating their lunch. One of them has an apple, one has a banana, and one has a carrot. Ann has neither the banana nor the carrot. Bill does not have the banana. Carlos does not have the apple, and he does not have the carrot. What does each one have?

✳**32.** Four programmers, Mr. Alvarez, Ms. Bellowski, Mr. Chin, and Ms. Dodd, each know one computer language. Alvarez doesn't know Pascal, and he doesn't know BASIC. Bellowski knows BASIC or FORTRAN. Alvarez knows COBOL, or Chin does not know Pascal. Dodd knows COBOL or FORTRAN. Chin knows neither BASIC nor COBOL. Which language does each programmer know?

✳**33.** Four students, Juan, Kurt, Lee, and Maria, each have one home computer. The computers are an Apple, a Commodore, an IBM, and a TRS-80. Juan plays softball with the owners of the Apple and the Commodore. Kurt got the owners of the IBM and the Commodore interested in programming. Lee sold the Apple to its present owner, and Maria has dated both the owner of the Apple and the owner of the Commodore. Kurt has the IBM or Juan does not have the IBM. Which computer belongs to each student?

3.2 Truth Tables

In 1854, the self-taught English mathematician George Boole published a book called *An Investigation of the Laws of Thought*. In this book Boole explained that there is a relationship between sets and logic.

Recall from Chapter 2 that

$A \cap B = \{ x \mid x \in A \text{ and } x \in B \}$,

$A \cup B = \{ x \mid x \in A \text{ or } x \in B \}$ where "or" is used inclusively, and

$A' = \{ x \mid x \in U, x \notin A \}$.

That is, $x \in A \cap B$ if x belongs to *both* sets A and B.

$x \in A \cup B$ if x belongs to *at least one* of the two sets A and B.

$x \in A'$ if x does *not* belong to A.

Truth Values of Compound Propositions

$p \wedge q$ is true if *both* propositions p and q are true.

$p \vee q$ is true if *at least one* of the propositions p or q is true.

$\sim p$ is true if p is *not* true.

EXAMPLE 1

Consider the four compound propositions and determine whether each one is true or false.

(a) Candy is sweet, and potato chips are salty.
(b) Candy is sweet, and potato chips are bitter.
(c) Candy is bitter, and potato chips are salty.
(d) Candy is bitter, and potato chips are sweet.

Solution
(a) True, because both components are true.
(b) False, because the first component is true and the second is false.
(c) False, because the first component is false and the second is true.
(d) False, because both components are false. ■

If a proposition is true, we say that its *truth value* is "true." If it is false, its truth value is "false." Whether a compound proposition is true or false depends on the truth values of the components. Therefore we can examine the structure of a compound proposition without regard to the actual meaning of the sentence.

EXAMPLE 2

Consider the following four compound sentences, which have the same structure as the four in Example 1. Determine whether each is true or false.

(a) Five is less than seven, and seven is less than nine.
(b) Five is less than seven, and nine is greater than ten.
(c) Three is greater than four, and seven is less than nine.
(d) Three is greater than four, and nine is greater than ten.

Solution
(a) True, because both components are true.
(b) False, because the first component is true and the second is false.
(c) False, because the first component is false and the second is true.
(d) False, because both components are false. ■

There are four possible combinations of truth values for the two components of a conjunction. These are listed in a chart, called a *truth table*.

p	q	$p \wedge q$
T	T	T
T	F	F
F	T	F
F	F	F

EXAMPLE 3

Consider the disjunctions below, and determine the truth value of each.

(a) Five is less than seven, or seven is less than nine.

(b) Five is less than seven, or nine is greater than ten.

(c) Three is greater than four, or seven is less than nine.

(d) Three is greater than four, or nine is greater than ten.

Solution

(a) True, because both components are true.

(b) True, because the first component is true.

(c) True, because the second component is true.

(d) False, because neither component is true. ■

The truth table for a disjunction follows.

p	q	$p \vee q$
T	T	T
T	F	T
F	T	T
F	F	F

Since a negation applies to one proposition, there are only two rows in its truth table. They are shown in the truth table that follows.

p	$\sim p$
T	F
F	T

The three truth tables you have seen are the "building blocks" from which we construct truth tables for compound propositions. Consider $\sim p \vee q$. The truth value of the compound proposition depends on the truth values of p and q. We list the four possible combinations of truth values for p and q.

p	q	$\sim p \vee q$
T	T	
T	F	
F	T	
F	F	

We number these two columns as "1" and "2." Under the $\sim$ symbol, we list the truth values of $\sim p$. They are the opposites of the truth values of p. We number this column "3." Under q, we list its truth values and number this column "4." Finally, under the $\vee$ symbol, we list the truth values of the disjunction using the values in columns 3 and 4 to represent the truth values of the propositions $\sim p$ and q respectively.

p	q	$\sim$	p	$\vee$	q
T	T	F		T	T
T	F	F		F	F
F	T	T		T	T
F	F	T		T	F
1	2	3		5	4

> **To construct the truth table for a compound proposition with two letters:**
> 1. List the letters used and the compound proposition as column headings.
> 2. List all possible combinations of truth values.
>   a. Under the first letter, list two T's followed by two F's.†
>   b. Under the second letter, alternate T's and F's, starting with T.
> 3. Write the truth value of each compound proposition under its connective and number the column.
> 4. The truth values in the final (highest numbered) column are the truth values for the compound proposition.

EXAMPLE 4

Construct a truth table for $\sim(p \wedge q)$.

Solution   We list the possible truth values of p and q.

p	q	$\sim(p \wedge q)$
T	T	
T	F	
F	T	
F	F	

We know the truth value of $(p \wedge q)$, so we list these values under the $\wedge$ symbol.

p	q	$\sim(p \wedge q)$
T	T	T
T	F	F
F	T	F
F	F	F

†The order in which the possible truth values are listed is not important, but a systematic procedure is helpful when truth tables are to be compared. The answers in the back of the book assume this order.

Finally, we know the truth values for a negation. Under the $\sim$ symbol, we list the truth values that are opposites of the truth values of $(p \wedge q)$.

p	q	$\sim(p \wedge q)$	
T	T	**F**	T
T	F	**T**	F
F	T	**T**	F
F	F	**T**	F
1	2	**4**	3

Sometimes a statement is easier to understand if it is rephrased. If we want to replace one compound proposition with another, however, we must be sure that the truth values are the same for every combination of truth values of its components.

Definition
Two propositions are said to be *logically equivalent* if they have the same truth values for every combination of truth values for the simple propositions used as components. The symbol for logical equivalence is $\equiv$.

EXAMPLE 5

Show that $\sim p \vee \sim q \equiv \sim(p \wedge q)$.

Solution   The truth table for $\sim(p \wedge q)$ was shown in Example 4. We construct the truth table for $\sim p \vee \sim q$.

p	q	$\sim p$	$\vee$	$\sim q$
T	T	F	**F**	F
T	F	F	**T**	T
F	T	T	**T**	F
F	F	T	**T**	T
1	2	3	**5**	4

We conclude that $\sim(p \wedge q)$ is logically equivalent to $\sim p \vee \sim q$, since the truth values of the final column are identical on every line.

The statement we examined in Example 5 is known as De Morgan's law. In Chapter 2, you learned De Morgan's laws for set theory:

$(A \cap B)' = A' \cup B'$ and $(A \cup B)' = A' \cap B'$. It is also true that $\sim(p \vee q)$ is logically equivalent to $\sim p \wedge \sim q$. An examination of the truth tables shows that it is. See problem 11.

EXAMPLE 6

Use De Morgan's law and the fact that $\sim(\sim p) \equiv p$ to rewrite each statement below.

(a) $\sim(\sim p \wedge q)$

(b) $\sim(r \vee \sim s)$

Solution

(a) $\sim(\sim p \wedge q) \equiv \sim(\sim p) \vee \sim q \equiv p \vee \sim q$

(b) $\sim(r \vee \sim s) \equiv \sim r \wedge \sim(\sim s) \equiv \sim r \wedge s$ ▮

We now use the idea of logical equivalence with English sentences.

EXAMPLE 7

Determine whether the two statements are logically equivalent.

(a) Six is not divisible by both 2 and 5.

(b) Six is not divisible by 2 or it is not divisible by 5.

Solution t: six is divisible by 2.
     f: six is divisible by 5.

(a) The first proposition is written $\sim(t \wedge f)$.

(b) The second proposition is written $\sim t \vee \sim f$.

It is not necessary to construct truth tables, because these two statements are logically equivalent by De Morgan's law. ▮

EXAMPLE 8

Determine whether the two statements are logically equivalent:

(a) The programmer does not know both COBOL and FORTRAN.

(b) The programmer does not know COBOL and does not know FORTRAN.

Solution c: The programmer knows COBOL.
     f: The programmer knows FORTRAN.

(a) $\sim(c \wedge f)$.

(b) $\sim c \wedge \sim f$.

We construct truth tables.

c	f	$\sim(c \wedge f)$		c	f	$\sim c \wedge \sim f$	
T	T	**F** T		T	T	**F** F **F**	
T	F	**T** F		T	F	**F** F **T**	
F	T	**T** F		F	T	**T** F **F**	
F	F	**T** F		F	F	**T** T **T**	
1	2	4 3		1	2	3 **5** 4	

Comparing the truth values of $\sim(c \wedge f)$ in column 4 of the first table with those of $\sim c \wedge \sim f$ in column 5 of the second table, we see that the two propositions are not logically equivalent. ■

3.2 EXERCISES

Construct truth tables for each proposition. See Example 4.

1. $r \wedge s$

2. $r \vee s$

3. $\sim r$

4. $\sim(p \vee q)$

5. $\sim p \wedge \sim q$

6. $\sim p \vee q$

7. $u \vee \sim w$

8. $\sim u \vee w$

9. $u \wedge \sim w$

10. Show that $\sim(\sim p)$ is logically equivalent to p. See Example 5.

11. Show that $\sim p \wedge \sim q$ is logically equivalent to $\sim(p \vee q)$. See Example 5.

For problems 12–20, use De Morgan's laws to rewrite the proposition. See Example 6.

12. $\sim(r \vee s)$

13. $\sim(r \wedge s)$

14. $\sim(p \wedge \sim q)$

15. $\sim(\sim p \vee q)$

16. $\sim(\sim p \wedge \sim q)$

17. $\sim(\sim r \vee \sim s)$

18. $\sim(u \vee \sim w)$

19. $\sim p \wedge \sim q$

20. $\sim p \vee \sim q$

Determine whether the sentences are logically equivalent. See Examples 7 and 8.

21. (a) The customer has neither a savings account nor a $200 minimum balance.

  (b) The customer does not have a savings account and does not have a $200 minimum balance.

22. (a) The customer does not have a $200 minimum balance and does not have a $500 average daily balance.

  (b) The customer does not have a $200 minimum balance and a $500 average daily balance.

23. (a) The customer does not have a savings account or a checking account.

  (b) The customer does not have a savings account or does not have a checking account.

24. (a) The customer does not have a savings account and a checking account.

  (b) The customer does not have a savings account or does not have a checking account.

The following logic puzzles have been repeated in various forms for many years.

✳25. Suppose you are on a tropical island where the members of one tribe (which we will call Tribe A) always tell the truth while members of the other tribe (which we will call Tribe B) always lie. You come to a fork in the road and must ask a native which way to go, but you don't know to which tribe the native belongs. What one yes-or-no question could you ask to determine immediately which branch to take?

✳**26.** You are confronted by three natives (see problem 25 above). You ask the first which tribe he belongs to. You are unable to hear the response, so the second one says, "He said he is a member of Tribe B." The third one says, "One of them is a member of Tribe A." To which tribe does each belong?

➠**27.** Use pseudocode or a flowchart to show an algorithm that accepts a list of numbers as input and counts the values which are neither greater than five nor less than two.

3.3 Compound Propositions

To avoid ambiguity in a compound proposition, we follow the standard order for connectives. Any expression within parentheses takes priority over other expressions. Parentheses may be used for clarity, even if they are not required. In the absence of parentheses, or for an expression within parentheses, negations are done first, followed by conjunctions, then disjunctions. If an expression contains more than one of any connective, they are considered from left to right.

> **Order of Connectives**
> 1. The truth value of any expression within parentheses is evaluated first. Within parentheses, the connectives are considered in the following order:
> 2. Negations (from left to right).
> 3. Conjunctions (from left to right).
> 4. Disjunctions (from left to right).

EXAMPLE 1

Place parentheses in each expression to show the correct order of connectives.

(a) $q \vee r \wedge p$

(b) $\sim p \wedge q \vee r$

(c) $p \wedge q \vee \sim r$

(d) $p \wedge q \wedge r$

Solution

(a) $q \vee (r \wedge p)$

(b) $((\sim p) \wedge q) \vee r$

(c) $(p \wedge q) \vee (\sim r)$

(d) $(p \wedge q) \wedge r$ ▪

To construct a truth table for a proposition composed of three simple propositions, we have to consider eight possibilities. As an example, we examine $(p \wedge q) \vee \sim r$. To include all eight possibilities, we are systematic. Under p, we list four T's followed by four F's. Under q, we alternately list two T's and two F's. Under r, we list alternating T's and F's. These columns are numbered 1, 2 and 3 respectively.

The parentheses around $(p \wedge q)$ tell us that expression takes priority. We examine the individual truth values of p and q to determine the truth value of $(p \wedge q)$ for each line and fill in column 4 under the $\wedge$ symbol.

Under the $\sim$ symbol in column 5, we list the truth values of $\sim r$. Finally, in column 6 under the $\vee$ symbol, we list the truth values of the disjunction using the values in columns 4 and 5 as the truth values of the components.

p	q	r	$(p$	$\wedge$	$q) \vee \sim r$	
T	T	T		T	**T**	F
T	T	F		T	**T**	T
T	F	T		F	**F**	F
T	F	F		F	**T**	T
F	T	T		F	**F**	F
F	T	F		F	**T**	T
F	F	T		F	**F**	F
F	F	F		F	**T**	T
1	2	3		4	**6**	5

The procedure can be generalized as an algorithm:

To construct a truth table for a compound proposition with three letters:

1. List each letter and the compound proposition as column headings.
2. List all eight possible combinations of truth values.
   a. List four T's followed by four F's in the first column.
   b. Alternately list two T's and two F's in the second column.
   c. Alternately list T's and F's in the third column.
3. Following the conventional order of connectives, list the truth value for each compound proposition under its connective and number the column.
4. The truth values listed in the final (highest numbered) column are the truth values of the compound proposition.

EXAMPLE 2

Construct a truth table for $p \wedge (q \vee \sim r)$.

Solution

p	q	r	$p \wedge (q \vee \sim r)$				
T	T	T	T	T	T	T	F
T	T	F	T	T	T	T	T
T	F	T	T	F	F	F	F
T	F	F	T	T	F	T	T
F	T	T	F	F	T	T	F
F	T	F	F	F	T	T	T
F	F	T	F	F	F	F	F
F	F	F	F	F	F	T	T
1	2	3	7	8	5	6	4

~r has the opposite truth values from r
truth values of disjunction, columns 4 and 5
truth values of conjunction, columns 6 and 7 ∎

To decide whether two propositions are logically equivalent, we examine their truth tables. If their truth values are the same for the final (highest numbered) column, they are logically equivalent.

EXAMPLE 3

Determine whether $(p \wedge q) \vee \sim r$ is logically equivalent to $p \wedge (q \vee \sim r)$.

Solution The truth table for $(p \wedge q) \vee \sim r$ was shown just before Example 2. The truth table for $p \wedge (q \vee \sim r)$ was shown in Example 2. The last columns (column 6 in one and column 8 in the other) are different in the sixth and eighth rows of the truth tables, so the propositions are not logically equivalent. ∎

EXAMPLE 4

Determine whether $(p \wedge q) \vee r$ is logically equivalent to $(p \vee r) \wedge (q \vee r)$.

Solution The truth tables follow.

p	q	r	$(p \wedge q) \vee r$		
T	T	T	T	T	T
T	T	F	T	T	F
T	F	T	F	T	T
T	F	F	F	F	F
F	T	T	F	T	T
F	T	F	F	F	F
F	F	T	F	T	T
F	F	F	F	F	F
1	2	3	4	6	5

p	q	r	$(p \vee r) \wedge (q \vee r)$		
T	T	T	T	T	T
T	T	F	T	T	T
T	F	T	T	T	T
T	F	F	T	F	F
F	T	T	T	T	T
F	T	F	F	F	T
F	F	T	T	T	T
F	F	F	F	F	F
1	2	3	4	6	5

Since the last columns (column 6 in each) are identical in the two tables, the two propositions are logically equivalent. ■

Sometimes a proposition will be true for every possible combination of truth values.

Definition
A *tautology* is a proposition that is true for every combination of truth values of the simple propositions that compose it.

EXAMPLE 5

Determine whether the proposition $(p \wedge q) \vee (\sim p \vee \sim q)$ is a tautology.

Solution   We construct the truth table.

p	q	$(p \wedge q) \vee (\sim p \vee \sim q)$				
T	T	T	**T**	F	F	F
T	F	F	**T**	F	T	T
F	T	F	**T**	T	T	F
F	F	F	**T**	T	T	T
1	2	3	**7**	4	6	5

We conclude that the proposition is a tautology because it is true in every row of the truth table. See column 7. ■

3.3 EXERCISES

Insert parentheses in each expression to show the proper order of connectives. See Example 1.

1. $p \wedge \sim q \vee r$ **2.** $\sim p \vee q \wedge \sim r$ **3.** $p \vee \sim q \wedge r$ **4.** $p \vee \sim q \vee r$

5. $\sim p \wedge q \wedge \sim r$ **6.** $\sim p \vee \sim q \wedge r$ **7.** $p \wedge q \vee r \wedge s$ **8.** $p \vee q \wedge r \vee s$

9. $\sim p \vee q \vee r \wedge \sim s$ **10.** $\sim p \wedge q \wedge r \vee \sim s$ **11.** $p \wedge \sim q \wedge r \vee \sim s$ **12.** $p \vee \sim q \vee r \wedge s$

Construct a truth table for each proposition below. See Example 2.

13. $(p \vee \sim q) \vee r$ **14.** $p \wedge (\sim q \vee r)$ **15.** $p \vee (\sim q \wedge r)$ **16.** $(p \vee \sim q) \wedge r$

17. $\sim (p \wedge q) \vee r$ **18.** $\sim p \vee (\sim q \vee r)$ **19.** $\sim (p \vee q) \wedge r$ **20.** $\sim p \wedge (\sim q \wedge r)$

21. $\sim [p \wedge (q \vee r)]$ **22.** $\sim p \vee (\sim q \wedge \sim r)$

Problems 23–27 refer to problems 13–22 above. For each pair of propositions, determine whether they are logically equivalent. See Examples 3 and 4.

23. Are the propositions in problems 13 and 14 logically equivalent?

24. Are the propositions in problems 15 and 16 logically equivalent?

25. Are the propositions in problems 17 and 18 logically equivalent?

26. Are the propositions in problems 19 and 20 logically equivalent?

27. Are the propositions in problems 21 and 22 logically equivalent?

Determine whether each proposition is a tautology. See Example 5.

28. $(p \vee q) \wedge (\sim p \wedge \sim q)$ **29.** $(p \wedge \sim q) \vee (q \wedge \sim p)$ **30.** $(p \vee \sim q) \wedge (q \vee \sim p)$

✳**31.** How many rows would be required on a truth table for a compound proposition with n letters?

3.4   The Conditional

A proposition of the form "if p then q" is called a *conditional*. If . . . then is the logical connective. The proposition p is called the *antecedent* and q is called the *consequent*.

Logical Connective

Conditional: $\rightarrow$ means "if . . . then"

Let z be "the divisor is zero," and let d be "the division is defined." The statement "If the divisor is zero, then the division is undefined" would be written in symbolic form as $z \rightarrow \sim d$.

EXAMPLE 1

Write in symbolic form:

(a)   If the factors have the same sign, then the product is positive.

(b)   If the factors do not have the same sign, then the product is not positive.

Solution

s: the factors have the same sign

p: the product is positive

(a) $s \rightarrow p$

(b) $\sim s \rightarrow \sim p$ ▪

In the conditional, the word "then" may be omitted. "If p, q" means "if p, then q." It may also be written as "q if p." It is important to realize that p is still the antecedent (since it follows "if") and q the consequent.

EXAMPLE 2	Write in symbolic form:

(a) If it rains, the graduation ceremony will be held in the auditorium.

(b) The graduation ceremony will be held in the auditorium if it rains.

Solution

   r: it rains

   a: the graduation ceremony will be held in the auditorium.

(a) "it rains" follows "if," so it is the antecedent: $r \to a$

(b) "it rains" follows "if," so it is the antecedent: $r \to a$ ■

Many sentences can be stated in conditional form even though "if . . . then" does not appear in them. For example, the sentence "Where there is smoke, there is fire" could be written "If there is smoke, then there is fire."

The phrase "only if" is often misinterpreted. "Your program will work only if the computer is plugged in" says that if your program works, then the computer is plugged in. It does *not* say that if the computer is plugged in, your program will work. The component following "only if" is the consequent. The statement "only the good die young" could be written "Only if you are good will you die young" and therefore conveys the thought that "if you die young, then you are good."

Alternate Forms of the Conditional

If p, then q

If p, q

p only if q

q if p

q whenever p

EXAMPLE 3	Write in symbolic form:

(a) You can log onto the computer if you know the password.

(b) You can log onto the computer only if you know the password.

Solution

   l: you can log onto the computer

   p: you know the password

(a) "you know the password" follows "if," so it is the antecedent: $p \to l$

(b) "you know the password" follows "only if," so it is the consequent: $l \to p$ ■

EXAMPLE 4

Write in symbolic form:

(a)   You will lose the contents of memory whenever power is interrupted.

(b)   You will lose the contents of memory only if power is interrupted.

Solution

   m: you will lose the contents of memory

   p: power is interrupted

(a)   "whenever" means "if," so "the power is interrupted" is the antecedent: $p \rightarrow m$

(b)   "the power is interrupted" follows "only if," so it is the consequent: $m \rightarrow p$ ■

In the absence of parentheses, the negation, conjunction, and disjunction are considered before the conditional. Thus $s \rightarrow p \vee q$ means $s \rightarrow (p \vee q)$ rather than $(s \rightarrow p) \vee q$.

Order of Connectives

1. The truth value of any expression in parentheses is evaluated first. Within parentheses the connectives are considered in the following order:
2. Negations (from left to right).
3. Conjunctions (from left to right).
4. Disjunctions (from left to right).
5. Conditionals (from left to right).

EXAMPLE 5

Insert parentheses in the following expressions to show the proper order.

(a) $p \wedge q \rightarrow r \vee s$

(b) $\sim p \rightarrow q$

Solution

(a) $p \wedge q \rightarrow r \vee s$ means $(p \wedge q) \rightarrow (r \vee s)$

(b) $\sim p \rightarrow q$ means $(\sim p) \rightarrow q$ ■

If the components of $p \rightarrow q$ are reversed, the proposition $q \rightarrow p$ is called the *converse* of $p \rightarrow q$. If each component of $p \rightarrow q$ is negated, the proposition $\sim p \rightarrow \sim q$ is called the *inverse* of $p \rightarrow q$. If the components of $p \rightarrow q$ are both reversed and negated, the proposition $\sim q \rightarrow \sim p$ is called the *contrapositive* of $p \rightarrow q$.

The proposition $\sim (p \rightarrow q)$ is called the *negation* of the proposition $p \rightarrow q$. Often, we state the negation of a proposition $p \rightarrow q$ by prefacing the proposition $p \rightarrow q$ with the phrase "It is not the case that."

EXAMPLE 6

Consider the proposition "If you can learn Pascal, then you can learn BASIC." Write the

(a)   converse.

(b)   inverse.

(c)   contrapositive.

(d)   negation.

Solution

Let p = you can learn Pascal

Let b = you can learn BASIC

(a)   converse: If you can learn BASIC, then you can learn Pascal.

(b)   inverse: If you cannot learn Pascal, then you cannot learn BASIC.

(c)   contrapositive: If you cannot learn BASIC, then you cannot learn Pascal.

(d)   negation: It is not the case that if you can learn Pascal, then you can learn BASIC.   ■

3.4 EXERCISES

Write the symbolic form of each statement using the given letters. See Examples 1 and 2.

f: you can draw a flowchart
p: you can write a computer program

1. If you can draw a flowchart, then you can write a computer program.

2. If you can write a computer program, then you can draw a flowchart.

s: your program is structured
e: your program is easy to follow

3. If your program is structured, then it is easy to follow.

4. If your program is not structured, then it will not be easy to follow.

l: your logic is clear
e: your program is easy to debug

5. If your logic is clear, then your program will be easy to debug.

6. If your program is easy to debug, then your logic must be clear.

d: you document your programs
e: your programs will be easy to maintain

7. If you document your programs, they will be easy to maintain.

8. If you do not document your programs, they will not be easy to maintain.

h: you expose a floppy disk to extreme heat
m: you expose a floppy disk to a magnetic field
d: you may damage your floppy disk

9. If you expose a floppy disk to extreme heat or to a magnetic field, you may damage it.

10. If you do not expose a floppy disk to extreme heat, you will not damage it.

    d: you want to damage a floppy disk

    b: you bend a floppy disk

    h: you handle it where the holes are

11. If you do not want to damage a floppy disk, you should not bend it or handle it where the holes are.

12. If you bend a floppy disk or handle it where the holes are, you must want to damage it.

Write the symbolic form of each statement. See Examples 3 and 4.

13. You can write a computer program only if you can draw a flowchart.

14. A number is divisible by six only if it is divisible by two and three.

15. It hurts only when I laugh.

16. When it rains, it pours.

17. When the cat is away, the mice will play.

18. A rolling stone gathers no moss.

19. A stitch in time saves nine.

20. A cluttered desk is a sign of genius.

Insert parentheses in each expression to show the correct order for the connectives.
See Example 5.

21. $p \vee \sim q \rightarrow r$

22. $p \vee q \rightarrow r$

23. $p \rightarrow q \wedge r$

24. $\sim p \rightarrow q \vee r$

25. $\sim p \wedge q \rightarrow r$

26. $p \rightarrow q \wedge r \rightarrow s$

27. $p \wedge r \rightarrow q \vee s$

28. $p \vee q \wedge s \rightarrow r$

29. $p \wedge q \vee s \rightarrow r$

Write the converse, inverse, contrapositive, and negation of each statement. See
Example 6.

30. If you can draw a flowchart, then you can write a computer program.

31. If your program is structured, then it is easy to follow.

32. If your logic is clear, then your program will be easy to debug.

33. If you document your programs, they will be easy to maintain.

34. If you expose a floppy disk to extreme heat, you will damage it.

35. If you know the password, you can log onto the computer.

36. If there is a power interruption, then you will lose the contents of memory.

✳**37.** The following logic puzzle is a popular one that has appeared in various forms. Johnny Carson has even discussed it on the Tonight Show!

    "Three intelligent people are applying for a job. They are told to close their eyes while each one has either a black hat or a white hat placed on his or her head. The one who deduces the color of the hat on his or her head first gets the job. Each one is given a white hat. They are then told to open their eyes and to raise a hand if they see a white hat on someone else. After a short time, one of them stands up and says, 'My hat is white.' How did he or she know?"

3.5 Truth Tables with the Conditional

In order to construct a truth table for the conditional, we examine an example in detail. Let p be "the temperature stays below 10° F for three days." Let q be "the lake freezes." On a televised weather report, the meteorologist says, "If the temperature stays below 10° F for three days, the lake will freeze." We write this proposition as $p \rightarrow q$.

If the meteorologist is telling the truth, then $p \rightarrow q$ is true. If the meteorologist is lying, then $p \rightarrow q$ is false. There are four possibilities.

1. The temperature stays below 10° F for three days (thus, p is true), and the lake freezes (q is true). The meteorologist told the truth.
2. The temperature stays below 10° F for three days (p is true), but the lake does not freeze (q is false). The meteorologist lied.
3. The temperature does not stay below 10° F for three days (p is false), but the lake freezes anyway (q is true). We conclude that the meteorologist told the truth. He or she did not say the lake would freeze *only if* the temperature stayed below 10° F for three days. Maybe two days would be sufficient. If the temperature were below 10° on the third day, the lake would still be frozen. Since the meteorologist did not address the question of what happens if the temperature does *not* stay below 10° F for three days, we give him or her the benefit of the doubt, and assume he or she is telling the truth.
4. The temperature does not stay below 10° F for three days (p is false), and the lake does not freeze (q is false). Again the meteorologist did not address the question of what happens if the temperature does not stay below 10° F for three days. We must assume that he or she told the truth.

We can generalize the results of this analysis to construct a truth table for any conditional.

p	q	$p \rightarrow q$
T	T	T
T	F	F
F	T	T
F	F	T

The conditional is false only when the antecedent is true and the consequent false. We can use this information to construct truth tables for other propositions.

EXAMPLE 1

Construct a truth table for $\sim q \rightarrow \sim p$.

Solution

p	q	$\sim q \rightarrow \sim p$		
T	T	F	**T**	F
T	F	T	**F**	F
F	T	F	**T**	T
F	F	T	**T**	T
1	2	3	**5**	4

    Example 1 shows that the truth values for $\sim q \rightarrow \sim p$ are identical to those for $p \rightarrow q$, so we conclude that the two propositions are logically equivalent. In the last section, you learned that $\sim q \rightarrow \sim p$ is the contrapositive of $p \rightarrow q$. If a proposition is difficult to work with, it may be easier to work with the contrapositive.

EXAMPLE 2

Construct a truth table for $q \rightarrow p$.

Solution

p	q	$q \rightarrow p$		
T	T	T	**T**	T
T	F	F	**T**	T
F	T	T	**F**	F
F	F	F	**T**	F
1	2	3	**5**	4

    From the truth table in Example 2 and the truth table for $p \rightarrow q$, we conclude that a statement $p \rightarrow q$ and its converse $q \rightarrow p$ are *not* logically equivalent.

EXAMPLE 3

Construct a truth table for $\sim p \rightarrow \sim q$.

Solution

p	q	$\sim p \rightarrow \sim q$		
T	T	F	**T**	F
T	F	F	**T**	T
F	T	T	**F**	F
F	F	T	**T**	T
1	2	3	**5**	4

From Example 3 and the truth table for $p \rightarrow q$, we conclude that a statement $p \rightarrow q$ and its inverse $\sim p \rightarrow \sim q$ are *not* logically equivalent. The inverse of $p \rightarrow q$, however, is logically equivalent to the converse of $p \rightarrow q$ (see Examples 2 and 3). That is, $\sim p \rightarrow \sim q \equiv q \rightarrow p$.

EXAMPLE 4

Determine whether $\sim(p \rightarrow q)$ is logically equivalent to $\sim p \rightarrow \sim q$.

Solution   The truth table for $\sim p \rightarrow \sim q$ was shown in Example 3. We construct the truth table for $\sim(p \rightarrow q)$.

p	q	$\sim(p \rightarrow q)$	
T	T	**F**	T
T	F	**T**	F
F	T	**F**	T
F	F	**F**	T
1	2	**4**	3

We conclude that $\sim(p \rightarrow q)$ and $\sim p \rightarrow \sim q$ are not logically equivalent. ■

The conditional might appear as part of a longer proposition.

EXAMPLE 5

Construct a truth table for $\sim(p \wedge q) \rightarrow r$.

Solution

p	q	r	$\sim(p \wedge q) \rightarrow r$			
T	T	T	F	T	**T**	T
T	T	F	F	T	**T**	F
T	F	T	T	F	**T**	T
T	F	F	T	F	**F**	F
F	T	T	T	F	**T**	T
F	T	F	T	F	**F**	F
F	F	T	T	F	**T**	T
F	F	F	T	F	**F**	F
1	2	3	5	4	**7**	6	■

Truth tables may be used to study the logic of an argument. A prosecutor addressing a jury might say "if the defendant is guilty, he should be convicted of first degree murder. The defendant is guilty. Therefore you should convict him of first degree murder." The lawyer has presented an argument. That is, from the first two statements made, you are asked to draw a conclusion.

> **Definition**
> An *argument* is a sequence of propositions, one of which (called the *conclusion*) is claimed to follow from the others (called *premises*).

The conclusion is often indicated by a key word, such as therefore, hence, thus, so, or consequently.

EXAMPLE 6

Identify the premises and conclusion of the following argument: If the sum of the digits of a number is divisible by 3, then the number is divisible by 3. The sum of the digits in 693 (6 + 9 + 3 = 18) is divisible by 3. Therefore, 693 is divisible by 3.

Solution   There are two premises.
(1) If the sum of the digits of a number is divisible by 3, then the number is divisible by 3.
(2) The sum of the digits in 693 is divisible by 3.
The conclusion follows the word "therefore." The conclusion is that 693 is divisible by 3. ■

Arguments are usually given in symbolic form by writing the premises under one another, separated from the conclusion by a line. The argument of Example 6 follows:

s: The sum of the digits of a number is divisible by 3.

n: The number is divisible by 3.

$$s \rightarrow n$$

$$\frac{s}{n}$$

Symbolic logic allows us to determine whether arguments are based on sound reasoning. An argument is said to be *valid* if the conclusion does indeed follow from the premises. The validity of an argument concerns the reasoning used and not the truth value of the conclusion. The examples that follow are chosen to illustrate that truth value and validity are two different concepts.

> **Definition**
> An argument consisting of premises $p_1, p_2, \ldots, p_n$ and conclusion c is *valid* if $(p_1 \wedge p_2 \wedge \ldots \wedge p_n) \rightarrow c$ is a tautology.

If one or more of the premises are false, then the conjunction $(p_1 \wedge p_2 \wedge \ldots \wedge p_n)$ is false. If the antecedent of the conditional $(p_1 \wedge p_2 \wedge \ldots \wedge p_n)$

$\rightarrow c$ is false, the truth value of the conditional is true regardless of the truth value of the consequent. Rather than check every possible combination of truth values for the conditional $(p_1 \wedge p_2 \wedge . . \wedge p_n) \rightarrow c$ to see if it is a tautology, we need only check those combinations for which all of the premises are true. Thus, we can say that an argument is valid if the conclusion is true whenever all the premise are true. The procedure can be written as an algorithm.

To determine whether an argument is valid:

1. Construct a truth table with each component, each premise, and the conclusion as column headings.
2. Determine the truth value of each premise.
3. Strike out any row that has one or more false premises.
4. If the conclusion is true in every row left, the argument is valid.

EXAMPLE 7

Determine whether the following argument is valid. If the last digit of a number is 5, then the number is divisible by 5. The last digit of the number 22 is a 5. (Remember: truth $\neq$ validity.) Therefore the number 22 is divisible by 5.

Solution l: the last digit of the number is 5
    f: the number is divisible by 5

We write: $l \rightarrow f$

$$\frac{l}{f}$$

We construct a truth table.

		Premises	Conclusion	
l	f	$(l \rightarrow f)$	l	f
T	T	T	T	T
~~T~~	~~F~~	~~F~~	~~T~~	~~F~~	first premise is false: strike this row
~~F~~	~~T~~	~~T~~	~~F~~	~~T~~	second premise is false: strike this row
~~F~~	~~F~~	~~T~~	~~F~~	~~F~~	second premise is false: strike this row

The only remaining row is the first row. Since the conclusion is true in this row, we say that the argument is valid. ∎

EXAMPLE 8

Determine whether the following argument is valid. If the sum of the digits of a number is divisible by 3, then the number is divisible by 3.
7 (which is the sum of $1 + 6$) is not divisible by 3.
Therefore, the number 16 is not divisible by 3.

Solution s: the sum of the digits of a number is divisible by 3
n: the number is divisible by 3

We write: $s \rightarrow n$

$$\frac{\sim s}{\sim n}$$

We construct a truth table.

		Premises		Conclusion
s	n	$(s \rightarrow n)$	$\sim s$	$\sim n$
T	T	T	F	F	second premise is false: strike this row
T	F	F	F	T	both premises are false: strike this row
F	T	T	T	Ⓕ	
F	F	T	T	T	

The conclusion in the third row is false, so the argument is *not* valid. ■

EXAMPLE 9   Determine whether the following argument is valid. If the last digit of a number is 5, then the number is divisible by 5. The number 10 is divisible by 5. Therefore, the last digit in the number 10 is 5.

Solution l: the last digit of a number is a 5
n: the number is divisible by 5

The argument is written as: $l \rightarrow n$

$$\frac{n}{l}$$

We construct a truth table.

l	n	$(l \rightarrow n)$	n	l
T	T	T	T	T	
T	F	F	F	T	both premises are false: strike this row
F	T	T	T	F	
F	F	T	F	F	second premise is false: strike this row

The conclusion is false in the third row, so the argument is *not* valid. ■

Example 6 shows a valid argument with a true conclusion.
Example 7 shows a valid argument with a false conclusion.
Example 8 shows an invalid argument with a true conclusion.
Example 9 shows an invalid argument with a false conclusion.
Any argument that takes the form $p \rightarrow q$

$$\frac{p}{q}$$ is valid. Examples 6 and 7 follow this

pattern. This form of argument may be called by its Latin name, *modus ponens*. You should not confuse it with the invalid argument $p \rightarrow q$

$$\frac{q}{p}.$$

Example 9 illustrates this form.

Another form of valid argument is called *modus tollens:* $p \rightarrow q$

$$\frac{\sim q}{\sim p}.$$

You should not confuse it with the invalid argument $p \rightarrow q$

$$\frac{\sim p}{\sim q}.$$

Example 8 illustrates this form.

3.5 EXERCISES

Construct truth tables for each expression. See Examples 1–3.

1. $\sim p \rightarrow q$ **2.** $p \rightarrow \sim q$ **3.** $q \rightarrow \sim p$ **4.** $\sim q \rightarrow p$

Show that each pair of propositions is logically equivalent. See Example 4.

5. $p \rightarrow q \equiv \sim p \vee q$ **6.** $\sim p \rightarrow q \equiv p \vee q$

7. $p \rightarrow \sim q \equiv \sim p \vee \sim q$ **8.** $\sim(p \rightarrow q) \equiv p \wedge \sim q$

Construct a truth table for each expression. See Example 5.

9. $(p \wedge q) \rightarrow r$ **10.** $p \wedge (q \rightarrow r)$

11. $p \vee (q \rightarrow r)$ **12.** $(p \vee q) \rightarrow r$

13. $(p \wedge q) \rightarrow \sim r$ **14.** $(p \vee \sim q) \rightarrow r$

15. $(p \wedge \sim q) \rightarrow \sim r$ **16.** $r \rightarrow (\sim p \vee q)$

17. $(r \rightarrow \sim p) \vee q$ **18.** $\sim r \rightarrow (q \wedge p)$

Determine whether each argument is valid. See Examples 7–9.

19. If your desk is cluttered, then you are a genius. Your desk is cluttered. Therefore you are a genius.

20. If your desk is cluttered, then you are a genius. Your desk is not cluttered. Therefore you are not a genius.

21. If your home is immaculate, then you are a dull person. Your home is not immaculate. Therefore you are not a dull person.

22. If your home is immaculate, then you are a dull person. You are not a dull person. Therefore your home is not immaculate.

23. If the denominator is zero, then the division is not defined. The denominator is zero. Therefore the division is not defined.

24. If the denominator is zero, then the division is not defined. The denominator is not zero. Therefore the division is defined.

✳**25.** A statement of the form "*p* if and only if *q*" is called a *biconditional* and is written $p \leftrightarrow q$. The abbreviation "iff" is often used for "if and only if." The statement "*p* if and only if *q*" is equivalent to the statement "*p* if *q* and *p* only if *q*." Construct a truth table for $p \leftrightarrow q$.

✳**26.** The word "or" may be used in the exclusive sense. The symbol for exclusive or is $\veebar$. Thus, $p \veebar q$ means "*p* or *q*, but not both." Construct a truth table for $p \veebar q$.

✳**27.** Express $p \veebar q$ (See problem 26 above) in two different ways without using the $\veebar$ symbol.

➤**28.** Use pseudocode or a flowchart to show an algorithm for reading the records of a company's employees and counting the number who have accumulated at least twenty sick days or twenty vacation days, but not both.

3.6 Decisions

The IF . . .THEN statement can be used in computer programming when a decision is required. The single alternative decision structure takes the form IF (condition) THEN (action). The action is a statement or group of statements written in a particular programming language. Some actions available in one language may not be available in others. For that reason, we will not examine the action in detail. The condition, however, is often a mathematical comparison of two quantities. The symbols used in such a comparison are called *relational operators*.

Relational Operators

Algebraic symbols	BASIC and Pascal symbols
=	=
<	<
≤	<=
>	>
≥	>=
≠	<>

Thus, we might write IF (X < 3) THEN (action) or IF (N = 2) THEN (action). If the condition is true, the action will be performed. If the condition is false, it will not.

In computer programming, the connectives AND, OR, and NOT are known as *Boolean operators*. Boolean operators and relational operators can be combined in the same statement. Boolean operators take priority over relational operators. We might write (X > 4) OR (X < 1) in BASIC or Pascal. The parentheses are necessary. Without them, X > 4 OR X < 1 would be interpreted as X > (4 OR X) < 1, which is meaningless. The algebraic expression 2 < X

≤ 3 is written as (2 < X) AND (X < = 3) in BASIC and Pascal. The word AND is necessary. The two inequalities cannot be combined as they are in algebra.

EXAMPLE 1

Write each expression in BASIC or Pascal.

(a)   X > 3 or X < 0
(b)   2 ≤ X < 4
(c)   X = 1 or X ≠ 3

Solution
(a)   (X > 3) OR (X < 0)
(b)   (2 < = X) AND (X < 4)
(c)   (X = 1) OR (X <> 3) ■

Boolean and relational operators are often used in computer programming to form the condition in an IF . . . THEN statement. For example, we might have IF (X < 3) OR (Y > 2) THEN (action).

EXAMPLE 2

If X is a positive integer, determine the values of X for which the action will be performed.

(a)   IF (X > = 2) THEN . . . .
(b)   IF (X <> 100) THEN . . . .

Solution
(a)   X > = 2 means X ≥ 2, so X ∈ {2, 3, 4, 5, . . .}.
(b)   X <> 100 means X ≠ 100, so X ∈ {1, 2, 3, 4, . . . 99, 101, 102, 103, 104, . . .}.
Notice that 100 is not included in this set. ■

EXAMPLE 3

If X is a nonnegative integer, determine the values of X for which the action will be performed.

(a)   IF (X < 3) OR (X > 5) THEN . . . .
(b)   IF (X > = 2) AND (X < = 7) THEN . . . .

Solution
(a)   (X < 3) OR (X > 5) means X ∈ {0, 1, 2} ∪ {6, 7, 8, . . .}, so X ∈ {0, 1, 2, 6, 7, 8, . . .}.
(b)   (X > = 2) AND (X < = 7) means X ∈ {2, 3, 4, 5, . . .} ∩ {0, 1, 2, 3, 4, 5, 6, 7}, so X ∈ {2, 3, 4, 5, 6, 7}. ■

The double alternative decision structure, available in some languages, takes the form IF (condition) THEN (action 1) ELSE (action 2). If the condition is true, action 1 will be performed. If the condition is false, action 2 will be performed.

EXAMPLE 4

If X = 3, determine which action will be performed.

(a)   IF (X > 5) THEN (action 1) ELSE (action 2).

(b)   IF (X < 3) THEN (action 1) ELSE (action 2).

(c)   IF (X <= 5) THEN (action 1) ELSE (action 2).

Solution

(a)   If X = 3, the condition X > 5 is false; action 2 will be performed.

(b)   If X = 3, the condition X < 3 is false; action 2 will be performed.

(c)   If X = 3, the condition X <= 5 is true; action 1 will be performed. ■

EXAMPLE 5

If X = 1, determine which action will be performed.

(a)   IF (X > 3) OR (X < 0) THEN (action 1) ELSE (action 2).

(b)   IF (X < 5) AND (X >= 0) THEN (action 1) ELSE (action 2).

(c)   IF (X = 1) OR (X = 2) THEN (action 1) ELSE (action 2).

Solution

(a)   If X = 1, (X > 3) OR (X < 0) is false; action 2 will be performed.

(b)   If X = 1, (X < 5) AND (X >= 0) is true; action 1 will be performed.

(c)   If X = 1, (X = 1) OR (X = 2) is true; action 1 will be performed. ■

3.6 EXERCISES

Write each algebraic expression in BASIC or Pascal. See Example 1.

1. A < 3 or B > 5

2. A ≥ C and C ≤ B

3. A > 0 or A < 5, or C > −3

4. 0 < A < 5, or C > −3

5. A > 0, or −3 < C < 5

6. A > 0 and C > 5, and B < −3

7. X ≠ 0 or Y ≠ 0

8. C = 3, and X ≠ 0 or Y ≠ 0

9. C = 3 and X ≠ 0, or Y ≠ 0

10. P ≥ Q, or Q < 5 and R = 7

If X is a nonnegative integer, determine the values of X for which the action will be performed. See Examples 2 and 3.

11. IF (X < 3) AND (X > 0) THEN . . . .

12. IF (X < 5) AND (X >= 0) THEN . . . .

13. IF (X < 2) AND (X > 7) THEN . . . .

14. IF (X < 6) OR (X > 10) THEN . . . .

15. IF (X < 1) OR (X >= 7) THEN . . . .

16. IF ((X >= 0) AND (X < 3)) OR (X > 5) THEN . . . .

17. IF (X > 0) AND ((X < 3) OR (X > 5)) THEN . . . .

18. IF (X > 0) AND ((X < 3) OR (X > 7)) THEN . . . .

19. IF (X < 3) AND (X > 0) THEN . . . .

20. IF (X < 3) OR (X > 0) THEN . . . .

If X = 2, determine which action will be performed. See Examples 4 and 5.

21. IF (X <> 2) THEN (action 1) ELSE (action 2).

22. IF (X >= 3) THEN (action 1) ELSE (action 2).

23. IF (X < 7) THEN (action 1) ELSE (action 2).

24. IF (X > 0) THEN (action 1) ELSE (action 2).

If X = 5, determine whether action 1 or action 2 will be performed. See Examples 4 and 5.

25. IF (X < 1) OR (X > 3) THEN (action 1) ELSE (action 2).

26. IF (X > 1) AND (X <= 3) THEN (action 1) ELSE (action 2).

27. IF (X = 3) OR (X = 5) THEN (action 1) ELSE (action 2).

28. IF (X > 7) AND (X < 3) THEN (action 1) ELSE (action 2).

29. IF (X > 3) AND (X < 7) THEN (action 1) ELSE (action 2).

30. IF (X = 5) OR (X = −5) THEN (action 1) ELSE (action 2).

✳**31.** Consider the following algorithm to compute a sales commission. It uses "nested" decision structures to avoid repeating a check of sales amount. Rewrite it without using a nested decision structure.
Get SALES.
If SALES ≤ 200
   Then assign COMMISSION 15% of SALES
   Else If SALES ≤ 400
      Then assign COMMISSION 30 + 10% of SALES over 200
      Else assign COMMISSION 50 + 8% of SALES over 400.
   ENDIF.
ENDIF.
Show COMMISSION.

➥**32.** Use pseudocode or a flowchart to show an algorithm with nested decision structures to compute taxes according to the following scale:
3 percent of the first $2000, 4 percent of the next $2000, 5 percent of the next $2000, and 6 percent of any amount over $6000.

3.7   Boolean Algebra

When Boole studied the relationship between properties of sets and logic, he noticed a similarity to the properties of real numbers.

Closure properties:

For sets: If A and B are sets, then $A \cup B$ is a set.
       If A and B are sets, then $A \cap B$ is a set.
For logic: If p and q are propositions, then $p \vee q$ is a proposition.
       If p and q are propositions, then $p \wedge q$ is a proposition.
For real numbers: If a and b are real numbers, then $a + b$ is a real number.
       If a and b are real numbers, then $a \cdot b$ is a real number.

Commutative properties:

For sets: If A and B are sets, then $A \cup B = B \cup A$.
       If A and B are sets, then $A \cap B = B \cap A$.
For logic: If p and q are propositions, then $p \vee q \equiv q \vee p$.
       If p and q are propositions, then $p \wedge q \equiv q \wedge p$.
For real numbers: If a and b are real numbers, then $a + b = b + a$.
       If a and b are real numbers, then $ab = ba$.

Associative properties:

For sets: $(A \cup B) \cup C = A \cup (B \cup C)$
       $(A \cap B) \cap C = A \cap (B \cap C)$
For logic: $(p \vee q) \vee r \equiv p \vee (q \vee r)$
       $(p \wedge q) \wedge r \equiv p \wedge (q \wedge r)$
For real numbers: $(a + b) + c = a + (b + c)$
       $(a \cdot b) \cdot c = a \cdot (b \cdot c)$

Boole realized that the analogy does not hold for all of the field properties of real numbers. There is only one distributive property for real numbers. Multiplication distributes over addition, but not vice versa. For sets and logic, there are two distributive properties.

Distributive properties:

For sets: $A \cup (B \cap C) = (A \cup B) \cap (A \cup C)$ and
       $A \cap (B \cup C) = (A \cap B) \cup (A \cap C)$
For logic: $p \vee (q \wedge r) \equiv (p \vee q) \wedge (p \vee r)$ and
       $p \wedge (q \vee r) \equiv (p \wedge q) \vee (p \wedge r)$
For real numbers: $a \cdot (b + c) = (a \cdot b) + (a \cdot c)$

Thus, Boole discovered that the algebra of real numbers is not the only algebra that can be developed. You will see (Chapter 8) that the algebra of matrices is an example of yet another algebra—one in which the commutative property fails to hold.

The other properties for Boolean algebra follow.

Idempotent properties:

For sets: $A \cup A = A$ For logic: $p \vee p \equiv p$

$A \cap A = A$ $p \wedge p \equiv p$

Complementation properties:

For sets: $A \cup A' = U$ For logic: $p \vee {\sim}p \equiv T$

$A \cap A' = \varnothing$ $p \wedge {\sim}p \equiv F$

De Morgan's laws:

For sets: $(A \cup B)' = A' \cap B'$ For logic: ${\sim}(p \vee q) \equiv {\sim}p \wedge {\sim}q$

$(A \cap B)' = A' \cup B'$ ${\sim}(p \wedge q) \equiv {\sim}p \vee {\sim}q$

Dual complementation property:

For sets: $(A')' = A$ For logic: ${\sim}({\sim}p) \equiv p$

Laws of absorption:

For sets: $A \cup (A \cap B) = A$ For logic: $p \vee (p \wedge q) \equiv p$

$A \cap (A \cup B) = A$ $p \wedge (p \vee q) \equiv p$

Special properties:

For sets: $A \cup U = U$ For logic: $p \vee T \equiv T$

$A \cap U = A$ $p \wedge T \equiv p$

$A \cup \varnothing = A$ $p \vee F \equiv p$

$A \cap \varnothing = \varnothing$ $p \wedge F \equiv F$

$\varnothing' = U$ ${\sim}F \equiv T$

$U' = \varnothing$ ${\sim}T \equiv F$

In *Boolean algebra,* which is the algebra of both sets and logic, a $+$ sign is used instead of $\cup$ or $\vee$, and $\cdot$ is used instead of $\cap$ or $\wedge$. Also, 1 is used instead of U or T, and 0 is used instead of $\varnothing$ or F. The "1" and "0" are merely symbols and are not the 1 and 0 of the real number system. Thus, one statement can express a property for both sets *and* logic.

Notation for Boolean Algebra:

Boolean algebra	Sets	Logic
$+$	$\cup$	$\vee$
$\cdot$	$\cap$	$\wedge$
$'$	$'$	$\sim$
0	$\varnothing$	F
1	U	T

Using this notation, we write the properties of Boolean algebra:

Properties of Boolean Algebra:

Commutative Properties:

$A + B = B + A$ $\qquad\qquad A \cdot B = B \cdot A$

Associative Properties:

$(A + B) + C = A + (B + C)$ $\qquad (A \cdot B) \cdot C = A \cdot (B \cdot C)$

Distributive Properties:

$A \cdot (B + C) = A \cdot B + A \cdot C$ $\qquad A + (B \cdot C) = (A + B) \cdot (A + C)$

Idempotent Properties:

$A + A = A$ $\qquad\qquad A \cdot A = A$

Complementation Properties:

$A + A' = 1$ $\qquad\qquad A \cdot A' = 0$

De Morgan's Laws:

$(A + B)' = A' \cdot B'$ $\qquad\qquad (A \cdot B)' = A' + B'$

Dual Complementation Property:

$(A')' = A$

Laws of Absorption:

$A \cdot (A + B) = A$ $\qquad\qquad A + (A \cdot B) = A$

Properties involving 0 and 1:

$A + 1 = 1$ $\qquad\qquad A \cdot 0 = 0$
$0' = 1$ $\qquad\qquad 1' = 0$
$A + 0 = A$ $\qquad\qquad A \cdot 1 = A$

These properties can be verified using Venn diagrams and truth tables.

EXAMPLE 1

Verify the commutative property $a + b = b + a$ for sets and logic.

Solution   For sets, we verify that $A \cup B = B \cup A$ using Venn diagrams. See Figures 3.1 and 3.2. Since the two diagrams are identical, the sets are the same.

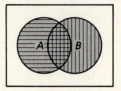

Figure 3.1

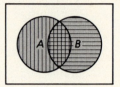

Figure 3.2

For logic, we verify that $p \vee q \equiv q \vee p$ using truth tables. Since the final columns are identical, the propositions are logically equivalent.

p	q	$p \vee q$
T	T	T **T** T
T	F	T **T** F
F	T	F **T** T
F	F	F **F** F
1	2	3 **5** 4

p	q	$q \vee p$
T	T	T **T** T
T	F	F **T** T
F	T	T **T** F
F	F	F **F** F
1	2	3 **5** 4

3.7 EXERCISES

For each statement of a Boolean algebra property, name the property and verify the property for both sets and logic. See Example 1.

1. $A \cdot B = B \cdot A$.

2. $(A + B) + C = A + (B + C)$

3. $(A \cdot B) \cdot C = A \cdot (B \cdot C)$

4. $A \cdot (B + C) = (A \cdot B) + (A \cdot C)$

5. $A + (B \cdot C) = (A + B) \cdot (A + C)$

6. $A + A = A$

7. $A \cdot A = A$

8. $A + A' = 1$

9. $A \cdot A' = 0$

10. $(A + B)' = A' \cdot B'$

11. $(A \cdot B)' = A' + B'$

12. $(A')' = A$

13. $A \cdot (A + B) = A$

14. $A + (A \cdot B) = A$

For each statement of a Boolean algebra property, verify the property for both sets and logic. See Example 1.

15. $A + 1 = 1$

16. $A \cdot 0 = 0$

17. $1' = 0$

18. $0' = 1$

19. $A \cdot 1 = A$

20. $A + 0 = A$

✳**21.** In Boolean algebra $X \cdot X = X$. In the algebra of real numbers, for what values of X does $X \cdot X$ equal X?

✳**22.** In the algebra of real numbers, it is true that if $A \neq 0$ and $A \cdot X = A \cdot Y$, then $X = Y$. Is the statement true in Boolean algebra?

✳**23.** In the algebra of real numbers, it is true that if $X \cdot Y = 0$, then $X = 0$ or $Y = 0$. Is the statement true in Boolean algebra?

✳**24.** What symbol in set notation corresponds to $\rightarrow$ in symbolic logic?

3.8   Circuits

In 1938, Claude Shannon discovered that Boolean algebra has an important application in the design of electrical circuits. Figure 3.3 represents part of an electrical circuit. (The battery or other power source is not shown since only the switches are related to Boolean algebra.) The lines represent *leads* or wires that conduct electric current. Current cannot flow from one end of the circuit to the other if the switch in the middle is open. We say the switch is "off."

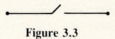

Figure 3.3

If we turn the switch "on" (i.e., close the switch), then current will flow. See Figure 3.4.

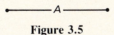

Figure 3.4

Sometimes the switch will be open and sometimes it will be closed. The circuit is drawn with a gap where the switch would be and is labeled with a letter of the alphabet to indicate a variable switch (i.e., one that can be on or off). See Figure 3.5.

$$\bullet\!\!-\!\!-\!\!-\!\!A\!\!-\!\!-\!\!-\!\!\bullet$$

Figure 3.5

Two or more switches may occur in the same circuit. Switches may be combined in *series* (end to end) or in *parallel* (side by side) as indicated in Figures 3.6 and 3.7. In a series circuit, to go from one end of the circuit to the other, current must flow through both A **and** B. In a parallel circuit, to go from one end of the circuit to the other, current must flow through A **or** B (inclusive or).

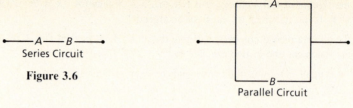

Series Circuit

Figure 3.6

Parallel Circuit

Figure 3.7

Switches that operate so that one is closed when the other is open are said to be in *opposite states*. The notation of Boolean algebra is used with switching circuits. Thus $A \cdot B$ or AB denotes two switches in series (current must flow through A **and** B), $A + B$ denotes two switches in parallel (current must flow through A **or** B), and A' denotes a switch in the opposite state from A (if current flows through A, it does **not** flow through A').

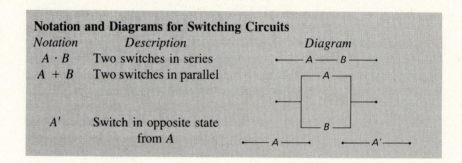

Notation and Diagrams for Switching Circuits

Notation	*Description*	*Diagram*
$A \cdot B$	Two switches in series	
$A + B$	Two switches in parallel	
A'	Switch in opposite state from A	

EXAMPLE 1   Write a Boolean expression for the switching circuit shown.

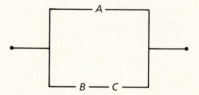

Solution   Current flowing along the upper path must flow through A.

Current flowing along the lower path must flow through B **and** C. Since current may flow along the upper path **or** the lower path, we have $A + (BC)$. ∎

EXAMPLE 2   Draw the switching circuit represented by $(A + B)C$.

Solution   Current may flow through either A **or** B, **and** then, in either case, it must flow through C. We have

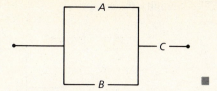

If 0 represents a switch through which current does not flow, while 1 represents a switch through which current flows, all of the properties of Boolean algebra apply to switching circuits. These properties may be used to simplify a circuit. That is, it may be possible to use fewer switches to achieve the same result. If fewer switches are used, the manufacturer's production costs will be less.

To simplify a switching circuit:
1. Write a Boolean expression for the circuit.
2. Apply the properties of Boolean algebra to the expression.
3. Draw the simplified circuit.

EXAMPLE 3

Simplify the circuit:

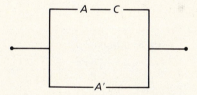

Solution The circuit is represented by $(AC) + A'$.

$(AC) + A' = (A + A')(C + A')$ by a distributive property

$\qquad = 1 (C + A')$ by the complementation property: $A + A' = 1$

$\qquad = C + A'$ since $1X = X$

Thus, the circuit could be drawn as:

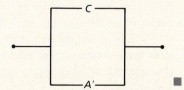

EXAMPLE 4   Simplify the circuit:

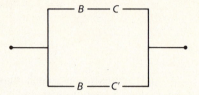

Solution   The circuit can be represented as $(BC) + (BC')$.

$(BC) + (BC') = B(C + C')$ by a distributive property

$\qquad\qquad\quad = B \cdot 1 \qquad$ by the complementation property: $C + C' = 1$

$\qquad\qquad\quad = B \qquad\quad$ since $X \cdot 1 = X$

Thus, the circuit could be drawn as ●——— B ———● . ■

Circuitry for digital computers is designed to respond to two voltage levels. The two levels may differ from machine to machine, but Boolean algebra is used to describe the circuits. If "1" is associated with the higher voltage and "0" is associated with the lower voltage, the assignment is called the *positive logic* convention. The basic elements in the circuits are called *gates,* because they direct the flow of logic signals just as switches direct electric current. The symbols used for such circuits are shown in the following table.

Gate Symbols

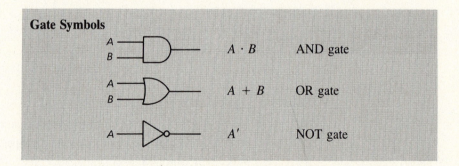

A drawing that illustrates the interconnection of logic gates is called a *logic diagram*. The value of the output voltage (often called simply the *output*) of an AND gate is "1" if all of the input voltages (or *inputs*) have value "1." The output of an OR gate has the value "1" if at least one of the inputs has value "1." Thus, to list the output for all possible inputs is much like constructing a truth table, where 1's are used instead of T's and 0's are used instead of F's. (In computer and electronics applications, the 0's are usually listed before the 1's. We have listed the 1's first to emphasize the analogy with truth tables.)

EXAMPLE 5

Determine the output for all possible inputs in the logic diagram shown.

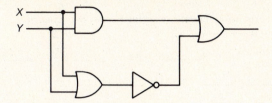

Solution The output of the circuit can be represented as $(XY) + (X + Y)'$. There are four possible inputs for X and Y. They are shown in the table that follows.

X	Y	$(XY) + (X + Y)'$			
1	1	1	**1**	1	0
1	0	0	**0**	1	0
0	1	0	**0**	1	0
0	0	0	**1**	0	1
1	2	3	**6**	4	5 ▧

EXAMPLE 6

Determine the output for all possible inputs in the logic diagram shown.

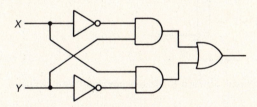

The output for the circuit can be represented as $XY' + X'Y$. The four possible inputs are shown in the table that follows.

X	Y	$X \cdot Y' + X' \cdot Y$							
1	1	1	0	0	**0**	0	0	1
1	0	1	1	1	**1**	0	0	0
0	1	0	0	0	**1**	1	1	1
0	0	0	0	1	**0**	1	0	0
1	2	3	5	4	**9**	6	8	7 ▧

3.8 EXERCISES

Write a Boolean expression for each switching circuit. See Example 1.

1.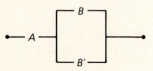

2.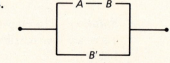

3.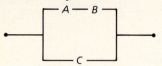

4.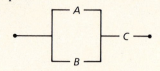

5.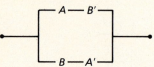

6.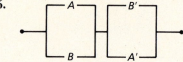

7.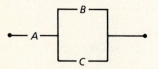

8.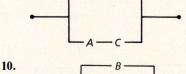

9.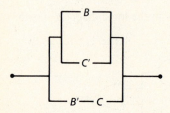

10.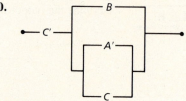

Draw a switching circuit for each Boolean expression. See Example 2.

11. $A + B'C$

12. $(A + B')C$

13. $AC + B'C$

14. $(A + B)(A + C)$

15. $A + (BC)$

16. $A(A + B)$

17. $A + AB$

18. $(A + B')(A' + B)$

19. $(AB')(AB)$

20. $(A + BC) + C$

Simplify each circuit. See Examples 3 and 4.

21. $B(AB + B)$

22. $AB(A + B')$

23. $A + B(A + B')$

24. $A + B + A'B'$

25. $B(A + C) + C$

26. $A' + AB + C$

27. $(A + BC)C'$

28. $AB' + BC + B$

29. $(A + B)(B + AC)$

30. $ABC + ABC'$

Determine the output for all possible inputs for each logic diagram. See examples 5 and 6.

31.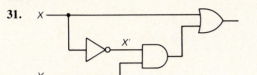

32.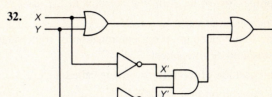

33.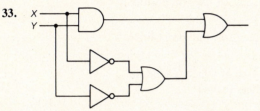

34.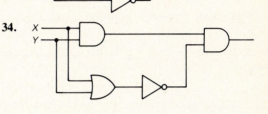

35.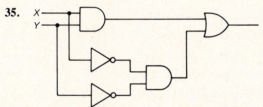

36.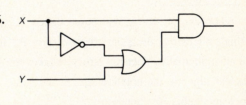

37.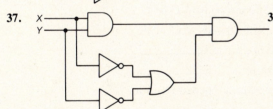

38.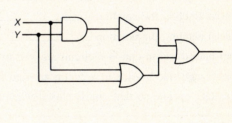

✳39. (a) Draw a switching circuit for the Boolean expression $(A + B)C + [A' (B' + C) + B']$.

(b) Simplify the circuit, and draw the simplified circuit.

CHAPTER REVIEW

A _____ is a declarative sentence that is either true or false, but not both.

"And," "or," "not," and "if . . . then" are called _____
_____ .

A proposition of the form $p \wedge q$ is called a _____ .

A proposition of the form $p \lor q$ is called a _____ .

A proposition of the form $\sim p$ is called a _____ .

Two propositions are _____ _____ if they have the same truth values for every combination of truth values for the simple propositions used as components.

A proposition of the form $p \to q$ is called a _____ .

The proposition $q \to p$ is the _____ of the proposition $p \to q$.

The proposition $\sim p \to \sim q$ is the _____ of the proposition $p \to q$.

The proposition $\sim q \to \sim p$ is the _____ of the proposition $p \to q$.

An _____ is a sequence of propositions, one of which (called the _____) is claimed to follow from the others (called _____ _____).

The symbols $=$, $<$, $\leq$, $>$, and $\geq$ are called _____ operators.

The single alternative decision structure takes the form _____ .

The double alternative decision structure takes the form _____ .

In Boolean algebra, the notation _____ corresponds to $\cup$ or $\lor$.

In Boolean algebra, the notation _____ corresponds to $\cap$ or $\land$.

In Boolean algebra, the notation _____ corresponds to U or T.

In Boolean algebra, the notation _____ corresponds to $\emptyset$ or F.

Switches placed end-to-end in a circuit are said to be in _____ .

Switches placed side-by-side in a circuit are said to be in _____ .

Switches that operate so that one is closed when the other is open are said to be in _____ _____ .

CHAPTER TEST

For problems 1–5, write each compound proposition in symbolic form using the given letters.

   s: It has 64K of RAM.
   t: It has 128K of RAM.
   d: It has two disk drives.
   c: It has a color monitor.

1. It has 64K of RAM and a color monitor.

2. It has 128K of RAM and two disk drives.

3. It has two disk drives, but not a color monitor.

4. It has 64K or 128K of RAM, and two disk drives.

5. It has 128K of RAM, and two disk drives or a color monitor.

6. Construct a truth table for $\sim(p \vee \sim q)$.

7. Construct a truth table for $\sim p \wedge q$.

8. Use De Morgan's laws to rewrite $\sim(p \vee \sim q)$.

Determine whether the sentences are logically equivalent.

9. **(a)** You cannot be too rich or too thin.

   (b) You cannot be too rich, and you cannot be too thin.

10. **(a)** You cannot be too rich or too thin.

    (b) You cannot be too rich or you cannot be too thin.

Insert parentheses to show the correct order of connectives.

11. $p \vee \sim q \wedge r$

12. $\sim p \wedge \sim q \vee r$

13. Construct a truth table for $(p \wedge \sim q) \vee \sim r$.

14. Is $p \vee (\sim q \wedge r) \equiv (p \vee \sim q) \wedge (p \vee r)$?

15. Is $(\sim p \vee q) \vee (p \wedge \sim q)$ a tautology?

Write the symbolic form of each proposition using the letters given.

16. If you peel onions under water, then you will not cry.

    p: you peel onions under water

    c: you cry

17. If parsley is washed with hot water, then it will retain its flavor and is easier to chop.

    h: parsley is washed with hot water

    r: parsley retains its flavor

    c: parsley is easier to chop

18. Insert parentheses to show the correct order for the connectives: $p \wedge \sim q \rightarrow r \wedge s$.

19. Write the inverse and negation of: If you heat lemons before using them, then there will be twice as much juice.

20. Write the converse and contrapositive of: If you beat egg whites in an aluminum pan, they will darken.

21. Construct a truth table for $\sim p \rightarrow \sim q$.

22. Show that $\sim p \rightarrow \sim q \equiv p \vee \sim q$

Construct a truth table for each statement.

23. $(\sim p \wedge q) \rightarrow r$ **24.** $p \rightarrow (\sim q \wedge r)$

25. Determine whether the argument is valid:

    If you add a little salt to the starch, the iron will not stick.

    You did not add salt to the starch.

    Therefore the iron will stick.

Write each algebraic expression in BASIC or Pascal.

26. X > 3 and X ≤ 5

27. A > 0, or A ≤ 7 and B > 2

28. If X is a nonnegative integer, for what values of X will the action be performed?
IF (X < 7) AND (X > 5) THEN (action).

If X = 3, determine whether action 1 or action 2 will be performed.

29. IF (X <> 4) THEN (action 1) ELSE (action 2).

30. IF (X > 5) OR (X < 1) THEN (action 1) ELSE (action 2).

For each statement of a Boolean algebra property, name the property and verify it for both sets and logic.

31. $(A + B) + C = A + (B + C)$

32. $A(B + C) = AB + AC$

33. $(A + B)' = A' \cdot B'$

34. $A + A' = 1$

35. $A(A + B) = A$

36. Write a Boolean expression for

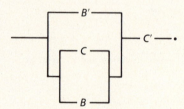

Draw a switching circuit to illustrate each Boolean expression.

37. $BC' + B'C$

38. $(B + C')(B' + C)$

39. Simplify the circuit $(A + B'C)B$

40. Determine the output for all possible inputs for the logic diagram in the following figure.

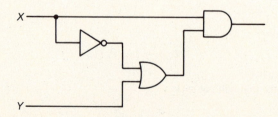

⫸**41.** The flowchart in the following figure shows an algorithm that accepts a list of numbers A as input and counts those that are not (less than or equal to 5) or not (greater than or equal to 2). Use De Morgan's law to simplify it.

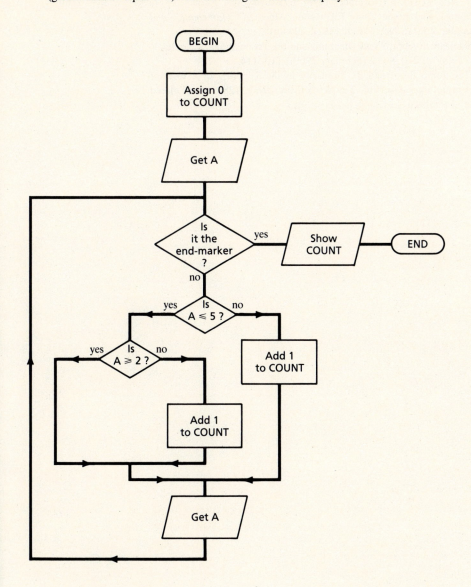

FOR FURTHER READING

Bookman, Jack. "Why 'False → False' is True—A Discovery Explanation," *The Mathematics Teacher,* Vol. 71 No. 8 (November 1978), 675–676.

Fitting, Melvin. "Propositional Logic Using Elementary Algebra," *Mathematics and Computer Education,* Vol. 16 No. 3 (Fall 1982), 204–207.

Gardner, Martin. "Boolean Algebra, Venn Diagrams and the Propositional Calculus," *Scientific American,* Vol. 220 No. 2 (February 1969), 110–114.

Gardner, Martin. "The Jump Proof and its Similarity to the Toppling of a Row of Dominoes," *Scientific American,* Vol. 236 No. 5 (May 1977), 128–135. (Colored hat problems are generalized.)

Biography   George Boole

George Boole was born November 2, 1815. His father was a cobbler who earned only a meager living. Boole believed that a classical education would enable him to live a more prosperous life. He taught himself Latin, French, and German by the time he was 14. When he was 15, his father's business was failing, and he helped support his parents. He became a teacher at an elementary school for boys. Though he had little time to pursue his own education, he spent his spare time studying math.

In 1835, he opened his own school and began to publish some of his work. In 1847, he wrote a short piece titled "Mathematical Analysis of Logic, Being an Essay Towards a Calculus of Deductive Reasoning." It is possible that this pamphlet was written in defense of his friend Augustus De Morgan, who was involved in a controversy concerning his own work. Boole gave a much simpler treatment of logic in this essay than any of his predecessors had given.

Even though he had no college degree, Boole was appointed in 1849 to the position of professor of mathematics at Queen's College in Cork, Ireland. His colleagues considered it odd that when he met people from a social background similar to his own, he would invite them to his home to discuss science.

An Investigation of the Laws of Thought on which are Founded the Mathematical Theories of Logic and Probabilities (often called simply *The Laws of Thought*), which explained and extended his system of logic, was published in 1854. In November of 1864, as he walked two miles through a cold rain to go to his classes, he caught pneumonia. He died on December 8.

After Boole's death, Mary Everest Boole, his wife since 1855 and mother of his five daughters (also niece of Sir George Everest, for whom Mt. Everest is named), wrote a pamphlet called "Boole's Psychology," which applied some of his ideas to the education of young children.

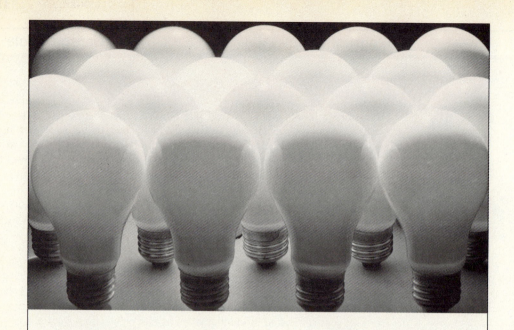

Chapter Four
Numeration Systems

Because a light bulb is either on or off, it operates like a digit in the binary number system used by computers. The system has only two digits, 0 and 1. The circuitry for a computer is generally designed to respond to a high voltage level as a 1 and a low voltage level as a 0.

4.1 Base Two Integers

When the "new math" was popular, many math teachers made a distinction between number and numeral. A *number* is an abstract idea; a *numeral* is a symbol for a number. There is only one number "four," but there are many ways to write a symbol for that number. The ancient Egyptians, for example, wrote $99\cap\cap\cap\cap\cap$ꟾꟾꟾ for 256. Each one of these symbols has a meaning, and the number represented when they are combined is the sum of their values. Such a system is said to be *additive*. If our present system were additive, 256 would equal $2 + 5 + 6$, or 13. Instead, the "2" in this numeral means 2 hundreds because of its position. The "2" in 625, however, means 2 tens. The numerals 256 and 625 would have the same value in an additive system, but they have different values in our system. We say that each digit has a *place value* based on its position.

   With place values of ones, tens, hundreds, thousands, and so on, we are using powers of ten. The first few powers of ten are listed in the following table.

Powers of 10

$10^0 = 1$ $10^1 = 10$ $10^2 = 100$ $10^3 = 1000$ $10^4 = 10,000$

   Our numeration system is called a *decimal* or *base 10* system. This system gets its name from the number of symbols that can be used as digits and their place values.

Base Ten or Decimal System of Numeration

1. There are ten symbols to use as digits: 0, 1, 2, 3, 4, 5, 6, 7, 8, and 9.

2. The place values are powers of the base 10.

Consider the numeral 36,824. If we label each digit with its place value, we have: $\dfrac{3}{10^4} \dfrac{6}{10^3}, \dfrac{8}{10^2} \dfrac{2}{10^1} \dfrac{4}{10^0}$.

   Computers use a *binary* or *base 2* system. In it, the symbols 0 and 1 have place values dependent upon their positions. The first few powers of two are listed in the following table.

Powers of 2

$2^0 = 1$ $2^1 = 2$ $2^2 = 4$ $2^3 = 8$ $2^4 = 16$ $2^5 = 32$ $2^6 = 64$

This system also gets its name from the number of symbols that can be used as digits (called *bits* as a contraction for "binary digits") and their place values.

Base Two or Binary System of Numeration

1. There are two symbols to use as bits: 0 and 1.
2. The place values are powers of the base 2.

Consider the base two numeral 101101. If we label each bit with its place value, we have: $\dfrac{1}{32}\ \dfrac{0}{16}\ \dfrac{1}{8}\ \dfrac{1}{4}\ \dfrac{0}{2}\ \dfrac{1}{1}$. It should be read as "one zero one one zero one." Since there are no special names (such as "hundreds" or "thousands") for base two place values, we will refer to "groups of 32" or "groups of 16," and so on. The base two numeral 101101, then, represents

$$1 \text{ group of } \mathbf{32}\ =\ 32$$
$$0 \text{ groups of } \mathbf{16}\ =\ 0$$
$$1 \text{ group of } \mathbf{8}\ \ =\ 8$$
$$1 \text{ group of } \mathbf{4}\ \ =\ 4$$
$$0 \text{ groups of } \mathbf{2}\ =\ 0 \text{ and}$$
$$1 \text{ group of } \mathbf{1}\ \ =\ 1 \text{ for a total of 45.}$$

When we are considering numerals in two or more bases simultaneously, the base will be written as a subscript to avoid confusion. 101101_2 means forty-five, as shown in the example above, but 101101_{10} means one-hundred-one thousand, one-hundred-one. If the subscript is omitted, the base is understood to be ten.

EXAMPLE 1

Convert 110101_2 to a base ten numeral.

Solution   We label the bits of this numeral with their place values.

Both the digits and place values are known. $\dfrac{1}{32}\ \ \dfrac{1}{16}\ \ \dfrac{0}{8}\ \ \dfrac{1}{4}\ \ \dfrac{0}{2}\ \ \dfrac{1}{1}$

We have 1 group of **32** = 32
1 group of **16** = 16
0 groups of **8** = 0
1 group of **4** = 4
0 groups of **2** = 0
1 group of **1** = 1
for a total of 53_{10}.

Therefore we say $110101_2 = 53_{10}$. ∎

EXAMPLE 2

Convert 1000001_2 to a base ten numeral.

Solution   Label the bits of this numeral with their place values.

Both the digits and place values are known.
$$\frac{1}{64}\ \frac{0}{32}\ \frac{0}{16}\ \frac{0}{8}\ \frac{0}{4}\ \frac{0}{2}\ \frac{1}{1}$$

We have 1 group of **64** = 64
      0 groups of **32** = 0
      0 groups of **16** = 0
      0 groups of **8** = 0
      0 groups of **4** = 0
      0 groups of **2** = 0
      1 group of **1** = 1
         for a total of 65_{10}.

Therefore we say $1000001_2 = 65_{10}$. ■

A computer uses the binary system to store numbers. It is a convenient system since numbers can be represented using only two voltage levels. One level represents 0 while another level represents 1. It is important to understand how the binary system is related to our base ten system. We use base ten for input, but a computer converts the numerals to base two for computation, and then converts them back to base ten for output.

You have seen several examples of conversion from base two to base ten. To convert a base ten numeral to a base two numeral, it is again necessary to think of powers of two. For example, 98_{10} must be converted to a numeral having place values of (from right to left) 1, 2, 4, 8, 16, 32, and 64. Even though we don't know what the bits are, we do know that the numeral will have seven bits. The eighth bit would represent a place value that is larger than necessary to write 98_{10}. Thus we have a numeral of the form $\frac{-}{64}\ \frac{-}{32}\ \frac{-}{16}\ \frac{-}{8}\ \frac{-}{4}\ \frac{-}{2}\ \frac{-}{1}$. To fill in these blanks, we ask "how many groups of 64 are there in 98?" Since $98 \div 64 = 1$ R 34, we put a "1" in the 64's position. Then the 34 left must be distributed over the remaining positions. Since there is one group of 32 in 34 ($34 \div 32 = 1$ R 2), we put a "1" in the 32's position. There are no groups of 16 in 2, no groups of 8 in 2, and no groups of 4 in 2. Therefore we put a "0" in each of these three positions. There is one group of 2 in 2 ($2 \div 2 = 1$ R 0), so we put a "1" in the 2's position. Since we have "used up" the 98 (that is, the remainder is 0), we put a "0" in the 1's position. Thus, we have $\frac{1}{64}\ \frac{1}{32}\ \frac{0}{16}\ \frac{0}{8}\ \frac{0}{4}\ \frac{1}{2}\ \frac{0}{1}$. That is, $98_{10} = 1100010_2$.

EXAMPLE 3

Convert 27_{10} to a base two numeral.

Solution   Label the digits of the base two numeral with their place values. Since we don't know yet what the digits are, we must leave them blank. Place values larger than 16 are unnecessary. We have:

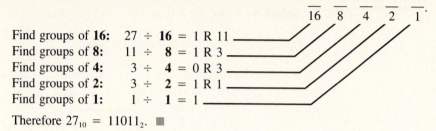

Find groups of **16:** $27 \div \mathbf{16} = 1 \text{ R } 11$
Find groups of **8:** $11 \div \mathbf{8} = 1 \text{ R } 3$
Find groups of **4:** $3 \div \mathbf{4} = 0 \text{ R } 3$
Find groups of **2:** $3 \div \mathbf{2} = 1 \text{ R } 1$
Find groups of **1:** $1 \div \mathbf{1} = 1$

Therefore $27_{10} = 11011_2$. ■

EXAMPLE 4

Convert 38_{10} to a base two numeral.

Solution   Label the digits of the base two numeral. The digits are blank, and place values larger than 32 are unnecessary.

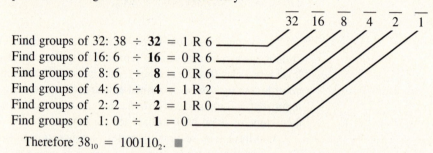

Find groups of 32: $38 \div \mathbf{32} = 1 \text{ R } 6$
Find groups of 16: $6 \div \mathbf{16} = 0 \text{ R } 6$
Find groups of 8: $6 \div \mathbf{8} = 0 \text{ R } 6$
Find groups of 4: $6 \div \mathbf{4} = 1 \text{ R } 2$
Find groups of 2: $2 \div \mathbf{2} = 1 \text{ R } 0$
Find groups of 1: $0 \div \mathbf{1} = 0$

Therefore $38_{10} = 100110_2$. ■

The conversions from binary to decimal and from decimal to binary have much in common. Later you will study octal (base 8) and hexadecimal (base 16) numeration systems. The following conversion rule will be helpful.

> **To convert a numeral to base ten or from base ten:**
> 1. Label the digits of the *nondecimal* numeral with their place values, even if the digits of that numeral are blank.
> 2. If the *digits* of the nondecimal numeral are known (that is, not blank), multiply each digit by its place value. The sum of these products is the base ten numeral.
> 3. If the *digits* of the nondecimal numeral are blank, divide the base ten numeral by the largest place value used as a label. The quotient is the leftmost digit of the numeral sought. Divide the remainder by the next place value to get the next digit, and so on.

EXAMPLE 5

Convert 101110_2 to a base ten numeral.

Solution Label the digits of the base two numeral.

$$\underline{1}\ \underline{0}\ \underline{1}\ \underline{1}\ \underline{1}\ \underline{0}$$
$$32\ 16\ 8\ 4\ 2\ 1$$

The digits are known, so we have: $32 + 0 + 8 + 4 + 2 + 0$ or 46_{10}. ■

EXAMPLE 6

Convert 54_{10} to a base two numeral.

Solution Label the digits of the base two numeral:

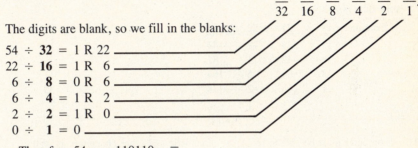

The digits are blank, so we fill in the blanks:

$$54 \div \mathbf{32} = 1 \text{ R } 22$$
$$22 \div \mathbf{16} = 1 \text{ R } 6$$
$$6 \div \mathbf{8} = 0 \text{ R } 6$$
$$6 \div \mathbf{4} = 1 \text{ R } 2$$
$$2 \div \mathbf{2} = 1 \text{ R } 0$$
$$0 \div \mathbf{1} = 0$$

Therefore $54_{10} = 110110_2$. ■

EXAMPLE 7

Convert 10101110_2 to a base ten numeral.

Solution Label the digits of the base two numeral.

$$\underline{1}\ \underline{0}\ \underline{1}\ \underline{0}\ \underline{1}\ \underline{1}\ \underline{1}\ \underline{0}$$
$$128\ 64\ 32\ 16\ 8\ 4\ 2\ 1$$

The digits are known, so we have: $128 + 32 + 8 + 4 + 2$ or 174_{10}. ■

4.1 EXERCISES

Convert each binary numeral to a decimal numeral. See Examples 1, 2, 5, and 7.

1. 110	**2.** 101	**3.** 111	**4.** 1000
5. 1011	**6.** 1010	**7.** 1100	**8.** 1101
9. 110111	**10.** 100101	**11.** 111010	**12.** 101100
13. 100011	**14.** 11011011	**15.** 10010110	**16.** 10110011
17. 10111101	**18.** 10101001	**19.** 10111011	**20.** 11001100

Convert each decimal numeral to a binary numeral. See Examples 3, 4, and 6.

21. 9	**22.** 16	**23.** 18	**24.** 22
25. 35	**26.** 41	**27.** 47	**28.** 52
29. 53	**30.** 59	**31.** 63	**32.** 65

33. 66   **34.** 79   **35.** 81   **36.** 96
37. 110   **38.** 128   **39.** 198   **40.** 240

41. Use pseudocode or a flowchart to show an algorithm for converting any binary numeral to a decimal numeral. (Hint: Start with the rightmost bit, and use an iteration structure to double each succeeding place value.)

42. Two examples of a multiplication procedure called "Russian Peasant Multiplication" are shown below. The first number is repeatedly halved (remainders are ignored) in the first column, and the second number is repeatedly doubled in the second column. When the number in the first column is even, the number in the second column is deleted. The sum of the remaining numbers in the second column is the product of the original two numbers. Explain why this procedure works.

Example:
13	×	21		22	×	41
13		21		~~22~~		~~41~~
~~6~~		~~42~~		11		82
3		84		5		164
1		168		~~2~~		~~328~~
		273		1		656
						902

43. A computer must be able to recognize alphabetic and other characters as well as numbers. Characters must be translated into a code using 0's and 1's. It is convenient if the code for numeric characters corresponds to their binary values, but you should understand that these codes are not used for doing arithmetic.

   a. Write the digits 0–9, using four bits for each.

   b. How many other characters could be represented using four bits?

   c. If six bits were used, how many other characters could be represented?

44. A commonly used eight-bit (or seven-bit) code is ASCII (American Standard Code for Information Interchange), in which

   a. The four-bit representation of a numeric character is prefixed by 0011 (or 011).

   b. The characters A–O are numbered in order with A = 0001, and are prefixed by 0100 (or 100).

   c. The characters P–Z are numbered in order with P = 0000, and are prefixed by 0101 (or 101).

   Write the eight-bit ASCII code for the characters 0–9 and A–Z.

45. Another eight-bit code is EBCDIC (Extended Binary Coded Decimal Interchange Code), in which

   a. The four-bit representation of a numeric character is prefixed by 1111.

   b. The characters A–I are numbered in order with A = 0001, and are prefixed by 1100.

   c. The characters J–R are numbered in order with J = 0001, and are prefixed by 1101.

   d. The characters S–Z are numbered in order with S = 0010, and are prefixed with 1110.

   Write the EBCDIC code for the characters 0–9 and A–Z.

4.2 Base Two Fractions

In our decimal system we can write fractions smaller than one by using a decimal point. To the right of the point, each place value is a fraction with one as the numerator and a power of ten as the denominator. At this point, we drop the distinction between number and numeral. The word "number" will refer to both numbers and numerals. Consider 3614.6234. If we label each digit with its place value, we have:

$$\frac{3}{1000} \ \frac{6}{100} \ \frac{1}{10} \ \frac{4}{1} \ . \ \frac{6}{1/10} \ \frac{2}{1/100} \ \frac{3}{1/1000} \ \frac{4}{1/10,000}$$

We can also use the binary system to write fractions smaller than one. To the right of the point (called a *binary point*), each place value is a fraction with one as the numerator and a power of two as the denominator. Table 4.1 lists powers of two.

Table 4.1

	Powers of Two	
N	$(2)^N$	$(1/2)^N$
0	1	
1	2	0.5
2	4	0.25
3	8	0.125
4	16	0.0625
5	32	0.03125
6	64	0.015625
7	128	0.0078125
8	256	0.00390625
9	512	0.001953125
10	1024	0.0009765625
11	2048	0.00048828125
12	4096	0.000244140625
13	8192	0.0001220703125
14	16,384	0.00006103515625
15	32,768	0.000030517578125

Consider 11.01_2. If we label its digits with their place values, we have

$$\frac{1}{2} \ \frac{1}{1} \ . \ \frac{0}{1/2} \ \frac{1}{1/4}$$

This number represents 1 group of **2** $= 2$
1 group of **1** $= 1$
0 groups of **1/2** $= 0.0$
1 group of **1/4** $= 0.25$
for a total of 3.25_{10}.

EXAMPLE 1

Convert 101.101_2 to base ten.

Solution   Label the digits of the base two number with their place values:

This number represents

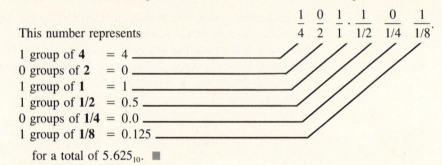

| $\frac{1}{4}$ | $\frac{0}{2}$ | $\frac{1}{1}$. | $\frac{1}{1/2}$ | $\frac{0}{1/4}$ | $\frac{1}{1/8}$ |

1 group of **4** = 4
0 groups of **2** = 0
1 group of **1** = 1
1 group of **1/2** = 0.5
0 groups of **1/4** = 0.0
1 group of **1/8** = 0.125

for a total of 5.625_{10}. ∎

EXAMPLE 2

Convert 11.00101_2 to base ten.

Solution   Label the digits of the base two number with their place values:

$$\frac{1}{2} \quad \frac{1}{1} . \quad \frac{0}{1/2} \quad \frac{0}{1/4} \quad \frac{1}{1/8} \quad \frac{0}{1/16} \quad \frac{1}{1/32}.$$

This number represents $2 + 1 + 1/8 + 1/32 = 2 + 1 + 0.125 + 0.03125 = 3.15625_{10}$. ∎

Place values can be used to convert base ten fractions to base two fractions, but the arithmetic is often tedious. The following algorithm is used instead. In the next chapter, you will learn why it works.

To convert a decimal fraction to a binary fraction:

1. Multiply the fraction by two and remove the integer portion of the product. The integer is the first digit to the right of the binary point.
2. Multiply the remaining decimal fraction by two and remove the integer. The integer is the next digit to the right in the binary fraction.
3. Repeat step two until
   a. the remaining fraction is zero, or
   b. a repeating pattern emerges.

EXAMPLE 3

Convert 0.1_{10} to a binary fraction.

Solution   We multiply by two:

    $0.1 \times 2 = 0.2$ The integer is 0.

    $0.2 \times 2 = 0.4$ The integer is 0.

    $0.4 \times 2 = 0.8$ The integer is 0.

    $0.8 \times 2 = 1.6$ The integer is 1.

    $0.6 \times 2 = 1.2$ The integer is 1.

    0.2×2 will start a repeating pattern.

The four steps shown in boldface type will repeat. Notice that the first line is not included in the repeating pattern. Thus, we have $0.1_{10} = 0.0\overline{0011}_2$. The bar above the digits 0011 indicates that this pattern repeats infinitely. ■

EXAMPLE 4

Convert 0.75_{10} to a binary fraction.

Solution   We multiply by two:

    $0.75 \times 2 = 1.5$ The integer is 1.

    $0.5 \ \ \times 2 = 1.0$ The integer is 1.

The remaining fraction is 0, so we are through. That is, $0.75_{10} = 0.11_2$. ■

EXAMPLE 5

Convert 0.35_{10} to a binary fraction.

Solution   We multiply by two:

    $0.35 \times 2 = 0.7$ The integer is 0.

    $0.7 \times 2 = 1.4$ The integer is 1.

    $0.4 \times 2 = 0.8$ The integer is 0.

    $0.8 \times 2 = 1.6$ The integer is 1.

    $0.6 \times 2 = 1.2$ The integer is 1.

    $0.2 \times 2 = 0.4$ The integer is 0.

0.4×2 starts a repeating pattern, so $0.35_{10} = 0.01\overline{0110}_2$. ■

Programmers often use similar algorithms to convert integers from one base to another.

> **To convert a binary integer to a decimal integer:**
> 1. Multiply the leftmost digit by two and add the next digit to the right.
> 2. Multiply the sum by two and add the next digit.
> 3. Repeat step two until the last digit has been added.
>   The sum is the decimal integer.

EXAMPLE 6

Convert 101101_2 to a decimal integer.

Solution   For the explanation, we spread out the bits of the number.

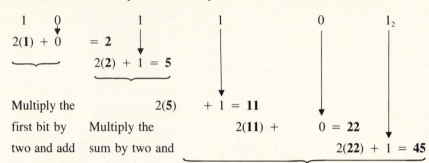

the next bit.   add the next bit.   Repeat step two until the last bit is added. ■

EXAMPLE 7

Convert 110101_2 to a decimal integer.

Solution

$2(1)\ + 1 = 3$

$2(3)\ + 0 = 6$

$2(6)\ + 1 = 13$

$2(13) + 0 = 26$

$2(26) + 1 = 53$ Therefore, $110101_2 = 53_{10}$. ■

To convert a decimal integer to a binary integer:

1. Divide the decimal integer by two and remove the remainder. The remainder is the ones digit of the binary integer.
2. Divide the quotient by two and remove the remainder. The remainder is the next digit to the left in the binary integer.
3. Repeat step two until the quotient is zero.

EXAMPLE 8

Convert 45_{10} to a binary integer.

Solution

Begin here

$$
\begin{array}{r|l}
0 & \text{R } 1 \\
2\overline{)\ 1} & \text{R } 0 \\
2\overline{)\ 2} & \text{R } 1 \\
2\overline{)\ 5} & \text{R } 1 \\
2\overline{)11} & \text{R } 0 \\
2\overline{)22} & \text{R } 1 \\
2\overline{)45} &
\end{array}
$$

Read the remainders from top to bottom to get the binary integer 101101. ■

4.2 EXERCISES

Convert each binary number to a decimal number. See Examples 1 and 2.

1. 11.011	**2.** 10.0101	**3.** 11.0011	**4.** 110.0101
5. 110.0111	**6.** 110.01001	**7.** 110.10001	**8.** 101.00011

Convert each decimal number to a binary number. See Examples 3–5.

9. 0.3	**10.** 0.25	**11.** 0.35	**12.** 0.7
13. 0.225	**14.** 7.625	**15.** 8.375	**16.** 17.875

Convert each binary number to decimal. See Examples 6 and 7.

17. 110111	**18.** 100101	**19.** 101101
20. 111100	**21.** 100100	**22.** 101111

Convert each decimal number to binary. See Example 8.

23. 34	**24.** 65	**25.** 79
26. 122	**27.** 185	**28.** 544

✳**29.** After data is received by a computer and translated into code such as ASCII or EBCDIC (See 4.1 Exercises, problems 44 and 45), numeric data must undergo another translation into the binary number system before calculations are done.

a. Write 17 using the ASCII code, and write it as a binary number.

b. Write 32 using the EBCDIC code, and write it as a binary number.

➠**30.** Use pseudocode or a flowchart to show an algorithm for converting a binary fraction to a decimal fraction.

4.3   Binary Arithmetic

Now that you know how the computer converts a numeral to binary, the next step is to see how computations are done in binary. In Section 2.4, you saw the addition and multiplication tables for base ten. The tables for base two are much shorter (see Tables 4.2 and 4.3).

<table>
<tr><td colspan="3">Table 4.2</td><td colspan="3">Table 4.3</td></tr>
<tr><td colspan="3">Binary Addition Table</td><td colspan="3">Binary Multiplication Table</td></tr>
<tr><td>+</td><td>0</td><td>1</td><td>×</td><td>0</td><td>1</td></tr>
<tr><td>0</td><td>0</td><td>1</td><td>0</td><td>0</td><td>0</td></tr>
<tr><td>1</td><td>1</td><td>10</td><td>1</td><td>0</td><td>1</td></tr>
</table>

The only entry that looks different from base 10 computation is the result of $1 + 1 = 10$. It looks strange at first, but remember that 10_2 means 2_{10}, which is exactly what you would expect when adding $1 + 1$. Binary addition of larger numbers is done column by column.

> **To add binary numbers:**
> 1. Line up the binary points.
> 2. Add column by column; carry when necessary.

In the examples that follow, the subscripts are omitted. All computations are in binary.

EXAMPLE 1

Perform the following binary addition: $1010 + 111$

Solution   Line up the binary points. In each of these two numbers, the point is understood to be at the end.

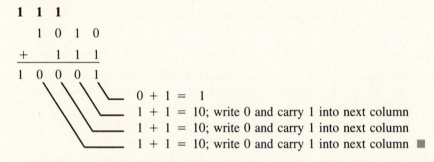

$$
\begin{array}{cccc}
\mathbf{1} & \mathbf{1} & \mathbf{1} & \\
1 & 0 & 1 & 0 \\
+ & 1 & 1 & 1 \\
\hline
1 & 0 & 0 & 0 & 1
\end{array}
$$

$0 + 1 = 1$
$1 + 1 = 10$; write 0 and carry 1 into next column
$1 + 1 = 10$; write 0 and carry 1 into next column
$1 + 1 = 10$; write 0 and carry 1 into next column ■

EXAMPLE 2

Perform the following binary addition: $1010.11 + 101.011$.

Solution   Line up the binary points, and add column by column.

$$
\begin{array}{ccccccc}
\mathbf{1} & \mathbf{1} & \mathbf{1} & \mathbf{1} & \mathbf{1} & & \mathbf{1} & \\
 & 1 & 0 & 1 & 0 & . & 1 & 1 \\
+ & & 1 & 0 & 1 & . & 0 & 1 & 1 \\
\hline
1 & 0 & 0 & 0 & 0 & . & 0 & 0 & 1
\end{array}
$$ ■

In a very long problem, you may encounter several 1's in a column. The associative property is useful:

$$1 + 1 + 1 = (1 + 1) + 1 =$$
$$10 + 1 = 11.$$

For four 1's,

$$1 + 1 + 1 + 1 = (1 + 1) + (1 + 1) =$$
$$10 + 10 = 100.$$

EXAMPLE 3

Perform the following binary addition: 1011 + 111.

Solution   Line up the binary points, and add column by column.

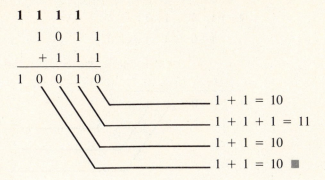

1 1 1 1
```
    1  0  1  1
  + 1  1  1
 ─────────────
1  0  0  1  0
```

$1 + 1 = 10$

$1 + 1 + 1 = 11$

$1 + 1 = 10$

$1 + 1 = 10$ ■

EXAMPLE 4

Perform the following binary addition: 1110.01 + 110.10.

Solution   Line up the binary points, and add column by column.

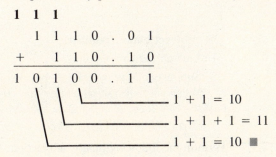

1 1 1
```
   1  1  1  0  .  0  1
 +    1  1  0  .  1  0
 ──────────────────────
1  0  1  0  0  .  1  1
```

$1 + 1 = 10$

$1 + 1 + 1 = 11$

$1 + 1 = 10$ ■

When you have four or more 1's in a single column, you may find it helpful to carry into the next *two* columns. (We could do this in base ten also, but the numbers are rarely large enough to require it.)

EXAMPLE 5

Perform the following binary addition: 1011 + 111 + 10.

Solution

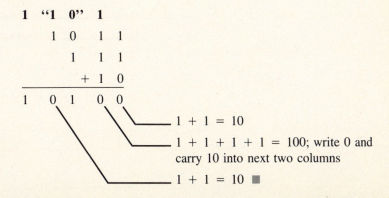

1 "1 0" 1
```
     1  0  1  1
     1     1  1
   +    1  0
 ───────────────
1  0  1  0  0
```

$1 + 1 = 10$

$1 + 1 + 1 + 1 = 100$; write 0 and carry 10 into next two columns

$1 + 1 = 10$ ■

EXAMPLE 6

Perform the following binary addition: 1011.01 + 11.11 + 10.01.

Solution

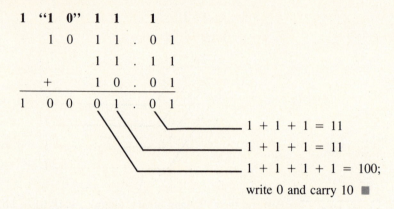

$$1 + 1 + 1 = 11$$
$$1 + 1 + 1 = 11$$
$$1 + 1 + 1 + 1 = 100;$$
write 0 and carry 10 ∎

Subtraction is done similarly to subtraction in base 10.

To subtract binary numbers:
1. Line up the binary points.
2. Subtract column by column; borrow as necessary.
   When you borrow, remember that $10 - 1 = 1$ in base two.

EXAMPLE 7

Perform the following binary subtraction: 1011 − 10.

Solution

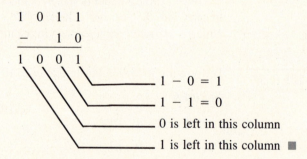

$$1 - 0 = 1$$
$$1 - 1 = 0$$
0 is left in this column
1 is left in this column ∎

EXAMPLE 8

Perform the following binary subtraction: 1010.01 − 101.01.

Solution

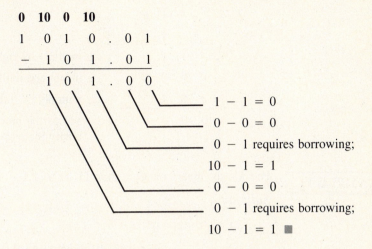

0   10   0   10
1   0   1   0 . 0 1
−   1   0   1 . 0 1
───────────────────────
         1   0   1 . 0 0

1 − 1 = 0

0 − 0 = 0

0 − 1 requires borrowing;

10 − 1 = 1

0 − 0 = 0

0 − 1 requires borrowing;

10 − 1 = 1 ▨

EXAMPLE 9

Perform the following binary subtraction: 1101.01 − 110.1.

Solution

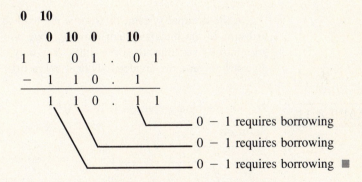

0   10
         0   10   0   10
1   1   0   1 . 0   1
−   1   1   0 . 1
───────────────────────
         1   1   0 . 1   1

0 − 1 requires borrowing

0 − 1 requires borrowing

0 − 1 requires borrowing ▨

To multiply in the binary system, set up the problem just as you would in base 10. Be sure to add the partial products in base two.

EXAMPLE 10

Perform the binary multiplication 110.1 × 101.

Solution

```
          1  1  0  .  1
             1  0     1
```
```
          1  1  0     1 __ multiply by rightmost 1
       0  0  0  0        __ indent and multiply by 0
    1  1  0  1           __ multiply by leftmost 1
 1  0  0  0  0  0  .  1 __ add partial products and count off one
                           binary place ■
```

EXAMPLE 11

Perform the binary multiplication 11.10 × 100.

Solution

```
  1  1  .  1  0
  ×           1     0     0 __ let terminal 0's "overhang"
  1  1     1  0  .  0     0 __ bring down 0's and multiply by 1;
                             count off two binary places ■
```

In the decimal system, division is the hardest of the four basic operations to learn. In the binary system, however, we don't have to guess whether 16 goes into 97 six times or seven times. The only choices are 1 or 0. Either the divisor goes into the dividend (1) or it doesn't (0).

EXAMPLE 12

Perform the binary division 1010 ÷ 10.

Solution

```
                      ┌──────────  1 ÷ 10 = 0, so we leave the first position blank
        /       ┌──────────  10 ÷ 10 = 1, so we put a 1 over the 10
            1  0  1
    10) 1  0  1  0
       -1  0           __ 1 × 10 = 10; subtract and bring down next digit
            1          __ 1 ÷ 10 = 0, so put 0 next in the quotient
          -0           __ 0 × 10 = 0; subtract and bring down next digit
            1  0 __ 10 ÷ 10 = 1, so put 1 next in quotient
          -1  0 __ 1 × 10 = 10; subtract
            0 __ there is no remainder ■
```

EXAMPLE 13   Perform the binary division $10100 \div 11$.

Solution

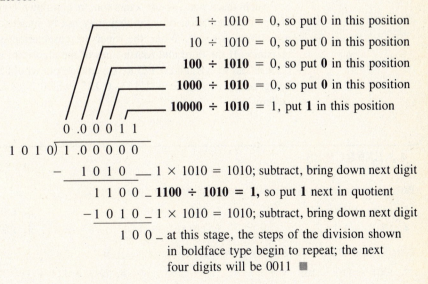

$1 \div 11 = 0$, so leave this position blank

$10 \div 11 = 0$, so leave this position blank

$101 \div 11 = 1$, so put 1 over the 101

1 1 0 R 10

11) 1 0 1 0 0

-1 1   __ $1 \times 11 = 11$; subtract and bring down next digit

1 0 0   __ $100 \div 11 = 1$, so put 1 next in quotient

-1 1   __ $1 \times 11 = 11$; subtract and bring down next digit

1 0 __ $10 \div 11 = 0$, so put 0 next in quotient

-0 __ $0 \times 11 = 0$; subtract

10 __ the remainder is 10 ▪

EXAMPLE 14   Perform the binary division $1 \div 1010$.

Solution   Since 1010 won't go into 1, we use the binary point and annex zeroes.

$1 \div 1010 = 0$, so put 0 in this position

$10 \div 1010 = 0$, so put 0 in this position

$\mathbf{100 \div 1010} = 0$, so put **0** in this position

$\mathbf{1000 \div 1010} = 0$, so put **0** in this position

$\mathbf{10000 \div 1010} = 1$, put **1** in this position

0 . 0 0 0 1 1

1 0 1 0) 1 . 0 0 0 0 0

$-$ 1 0 1 0 __ $1 \times 1010 = 1010$; subtract, bring down next digit

1 1 0 0 _ **$1100 \div 1010 = 1$,** so put **1** next in quotient

-1 0 1 0 _ $1 \times 1010 = 1010$; subtract, bring down next digit

1 0 0 _ at this stage, the steps of the division shown in boldface type begin to repeat; the next four digits will be 0011 ▪

If we convert Example 14 to a decimal problem, we have $1 \div 10$. Thus, we see that 0.1_{10} is a repeating binary fraction. That is, $0.1_{10} = 0.0\overline{0011}_2$, where the pattern 0011 repeats infinitely. Repeating binary fractions may cause unexpected things to happen in computer programs. For example, consider what would happen if the following algorithm were implemented on a computer.

Assign X a value of 0.
Assign CTR a value of 0.
While CTR is less than 10,
    Add 0.1 to X.
    Add 1 to CTR.
    ENDWHILE.
If X = 1, then show message "ONE."
    Otherwise show message "NOT ONE."
    ENDIF.

The computer would calculate

$$X = 0 + 0.1 + 0.1 + 0.1 + 0.1 + 0.1 + 0.1 + 0.1 + 0.1 + 0.1 + 0.1$$

We would expect X to have a value of 1 after these additions, but it has a value of 1.000000119209. The decimal number 0.1 is converted to a binary number rounded off to a finite number of binary places. The error introduced by rounding off this binary number is compounded in the execution of the program when 0.1 is added ten times. When the value of X is compared to 1, the numbers are not equal, so the computer would print the message "NOT ONE."

Such discrepancies are common in computer computations, because the binary representation of a number may not be exact, even if the decimal representation is exact. When the binary representation is exact, there is no discrepancy. If the third and fourth lines of the algorithm were changed to "While CTR is less than 8, add 0.125 to X," the output would be "ONE." The number being added each time is 1/8 = 0.125, which is exactly 0.001 in the binary system.

4.3 EXERCISES

Perform each addition using binary arithmetic. See Examples 1 and 2.

1. 1101 + 101

2. 1011 + 110

3. 110.01 + 11.011

4. 100.1 + 1.11

5. 110.101 + 11.01

6. 101.011 + 10.11

Perform each addition using binary arithmetic. See Examples 3–6.

7. 1001 + 110 + 101

8. 1011 + 111 + 11

9. 10111.011 + 1100.1 + 110.01 + 11

10. 11011.01 + 1011.1 + 1001 + 101.001

Perform each subtraction using binary arithmetic. See Examples 7–9.

11. 1110 − 101

12. 1001 − 111

13. 10.1011 − 1.0110

14. 1.1011 − 0.0101

15. 101.01 − 11.11

16. 10.001 − 0.11

17. $101.101 - 0.011$

18. $10.00 - 1.01$

19. $100 - 1.01$

20. $1000 - 1.11$

Perform each multiplication using binary arithmetic. See Examples 10 and 11.

21. 111×11

22. 110×11

23. 1.01×1.1

24. 101×101

25. 10.1×0.01

26. 11.1×1.01

27. 10.11×1.01

28. 10.01×0.011

29. 110.1×1.011

30. 101.1×10.01

Perform each division using binary arithmetic. See Examples 12 and 13.

31. $1100 \div 11$

32. $1110 \div 10$

33. $10100 \div 100$

34. $10101 \div 111$

35. $1111 \div 10$

36. $1001 \div 10$

37. $1010 \div 11$

38. $100010 \div 101$

39. $101001 \div 110$

40. $110000 \div 111$

✳**41.** A number is palindromic when its digit are the same read from right to left as when read from left to right, such as 15651. Give an example that is palindromic:

(a) when written in decimal, but not in binary.

(b) when written in binary, but not in decimal.

(c) when written in both decimal and binary.

✳**42.** Choose a decimal number that is not palindromic. Reverse the digits and add to the original number. Keep repeating the procedure with the resulting number until a palindromic number is obtained.

Example: 57   (number)

        + 75   (digits reversed)

         132   (sum)

        +231   (digits of sum reversed)

         363   (number is palindromic). Give an example that is palindromic:

(a) in fewer steps when written in decimal than in binary.

(b) in fewer steps when written in binary than in decimal.

(c) in the same number of steps whether written in binary or decimal.

✳**43.** Consider the logic diagram called a *half-adder*. For all possible inputs, determine the output for both the sum and carry paths. Do the results agree with binary addition as shown in Table 4.2?

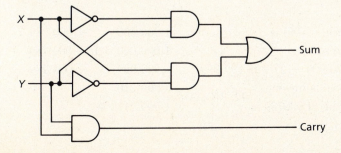

4.4   Complementary Arithmetic

One of the first computing machines was a mechanical device built by Blaise Pascal in the seventeenth century. He used the concept of subtraction by complements. Let us consider this technique in base ten using six-digit numbers just as Pascal did. The same method is used by the computer to do subtraction in the binary system.

Since we are using six-digit numbers, if a number does not have six digits, we simply prefix enough zeroes to make it six digits.

> **Definition**
>
> The complement of a six-digit decimal number x is $1,000,000 - x$.

The complement of 000107 is $1,000,000 - 000107$. If we simultaneously subtract 1 and add 1, we have added zero, the identity, so we have: $1,000,000 - 1 - 000,107 + 1$. But $1,000,000 - 1 = 999,999$, so we have $999,999 - 000,107 + 1$. Now the subtraction can be done without borrowing. The result is $999,892 + 1 = 999,893$. Although you would not do "pencil and paper" subtraction this way, it is useful in machine subtraction, because it requires no borrowing. We need to know that $1,000,000 - 1 = 999,999$, but this part of the procedure is the same for every problem, so the machine does not have to be programmed to borrow.

EXAMPLE 1

Find the complement of 113.

Solution

The complement of 113 is	1,000,000	$-$ 000,113.
Subtract 1 and add 1:	1,000,000 $-$ **1**	$-$ 000,113 $+$ **1.**
Replace (1,000,000 $-$ 1) with 999,999:	999,999	$-$ 000,113 $+$ 1.
Subtract without borrowing:	999,886	$+$ 1.
The complement of 000,113 is	999,887. ■	

EXAMPLE 2

Find the complement of 789.

Solution

The complement of 789 is	1,000,000	$-$ 000,789.
Subtract 1 and add 1:	1,000,000 $-$ **1**	$-$ 000,789 $+$ **1.**
Replace (1,000,000 $-$ 1) with 999,999:	999,999	$-$ 000,789 $+$ 1.
Subtract without borrowing:	999,210	$+$ 1.
The complement of 000789 is	999,211. ■	

Rather than subtract a number, we can add its complement.

Thus, $\begin{array}{r} 000140 \\ -000113 \end{array}$ becomes $\begin{array}{r} 000140 \\ +999887 \end{array}$, because 999887 is the complement of 000113 (see Example 1).

Since the complement of x is 1,000,000 $-$ x, when we add the complement of x to a number, we are adding an extra 1,000,000. If we then subtract 1,000,000 (by simply removing the leading 1), we have the correct six-digit answer to the original subtraction.

EXAMPLE 3

Subtract by adding the complement: 000140 $-$ 000113.

Solution   Since the complement of 000113 = 999,887 (see Example 1),

$$\begin{array}{r} \mathbf{1111} \\ 000140 \\ +999887 \\ \hline 000027 \end{array}$$

$\begin{array}{r} 000140 \\ -000113 \end{array}$ becomes   where the 1 carried into the leftmost column is dropped. ■

EXAMPLE 4

Subtract by adding the complement: 001259 $-$ 000789.

Solution   Since the complement of 000789 = 999211 (see Example 2),

$$\begin{array}{r} \mathbf{11}\mathbf{1} \\ 001259 \\ +999211 \\ \hline 000470 \end{array}$$

$\begin{array}{r} 001259 \\ -000789 \end{array}$ becomes   where the 1 carried into the leftmost column is dropped. ■

We used six-digit numbers in the examples above. The same principles work, however, for any number of digits n, if we use $10^n - x$ as the complement of x. For binary numbers, we will use eight bits.

Definition
The complement of an eight-bit binary number x is 100000000 $-$ x.

The complement of 00010011 is	100000000	$-$ 00010011.
Subtract 1 and add 1:	100000000 -1	$-$ 00010011 $+$ **1**.
Replace (100000000 $-$ 1) by 11111111:	11111111	$-$ 00010011 $+$ 1.
(This happens in every problem.)		
Subtract without borrowing:		**11101100** $+$ 1
The complement of 00010011 is		**11101101.**

Examine the third and fourth lines of the example. In base two, taking $11111111 - x$ simply changes 1's to 0's and 0's to 1's. We will take this shortcut in the future and bypass the steps done in the first three lines. The last step is to add the 1. The resulting number is known as the two's complement. (The number obtained before adding 1 is called the one's complement, and it is possible to develop a subtraction rule using it also. The two's complement, however, is more commonly used by computers.)

To find the two's complement of a binary number:
1. Change the 1's to 0's and the 0's to 1's.
2. Add 1.

EXAMPLE 5

Find the two's complement of 00010011.

Solution   Change the digits: 11101100.
                Add 1:   11101101. ∎

EXAMPLE 6

Find the two's complement of 01011010.

Solution   Change the digits: 10100101.
                Add 1:   10100110. ∎

In the binary system, as in the decimal system, a subtraction problem can be replaced with an equivalent addition problem by using complementary arithmetic.

To subtract binary numbers using complementary arithmetic:
1. Replace the subtrahend (the number to be subtracted) with its complement.
2. Add this complement to the minuend (the number from which the subtrahend is subtracted).

EXAMPLE 7

Perform the binary subtraction by using the two's complement: $01111100 - 00010011$.

Solution   Find the complement of 00010011: $11101100 + 1 = 11101101$.

$$\begin{array}{r} \mathbf{11111} \\ 01111100 \end{array}$$

01111100
−00010011   becomes   +11101101   (the complement of 00010011)
                                01101001   where the 1 in the leftmost column is

dropped. ∎

EXAMPLE 8   Perform the binary subtraction by using the two's complement: 01101101 − 01011011.

Solution   Find the complement of 01011011: 10100100 + 1 = 10100101.

$$\begin{array}{ll}
& \textbf{11 11 1} \\
01101101 & 01101101 \\
\underline{-01011011} \quad \text{becomes} & \underline{+\ 10100101} \quad \text{(the complement of 01011011)} \\
& 00010010 \quad \text{where the 1 in the leftmost column is}
\end{array}$$

dropped. ■

4.4 EXERCISES

Find the two's complement of each binary number. See Examples 5 and 6.

1. 01001010   **2.** 00111011

3. 01100110   **4.** 00010001

5. 00000001   **6.** 11111111

7. 10110101   **8.** 11000100

9. 11100111   **10.** 10100110

Perform each subtraction by using complementary arithmetic. See Examples 7 and 8.

11. 01110110 − 00001001   **12.** 00111100 − 00000111

13. 01011011 − 00111011   **14.** 01101101 − 00101101

15. 01001001 − 00010110   **16.** 01000000 − 00000001

17. 01010101 − 00101010   **18.** 01010111 − 00111000

19. 01110111 − 00011011   **20.** 01001010 − 00000101

Work each problem below by borrowing. See Section 4.3, Examples 7–9.

21. 01110110 − 00001001. Compare with problem 11.

22. 00111100 − 00000111. Compare with problem 12.

23. 01011011 − 00111011. Compare with problem 13.

24. 01101101 − 00101101. Compare with problem 14.

25. 01001001 − 00010110. Compare with problem 15.

Work each problem below by converting to base 10 before subtracting.

26. 01000000 − 00000001. Compare with problem 16.

27. 01010101 − 00101010. Compare with problem 17.

28. 01010111 − 00111000. Compare with problem 18.

29. 01110111 − 00011011. Compare with problem 19.

30. 01001010 − 00000101. Compare with problem 20.

✴**31.** Subtraction can be done by other methods as well as by borrowing and by using complements. The four following examples illustrate one such method.

$$4 \ ^6\not{7} \ 3 \qquad \ ^3\not{4} \ 0 \ 5 \qquad \ ^5\not{6} \ ^0\not{1} \ 4 \qquad \ ^2\not{3} \ 0 \ 2$$
$$\underline{-3 \ 2 \ ^3\not{1}} \qquad \underline{- \ 2 \ ^5\not{8} \ 3} \qquad \underline{- \ 2 \ ^1\not{9} \ ^2\not{8}} \qquad \underline{- \ 1 \ 2 \ ^3\not{1} \ ^4\not{8}}$$
$$1 \ 4 \ 6 \qquad \ \ 1 \ 5 \ 2 \qquad \ \ \ 3 \ 1 \ 6 \qquad \ \ \ 1 \ \ \ 2 \ 6$$

(a) Explain how to do subtraction this way.

(b) Explain why it works.

(c) Try it with binary numbers.

4.5   Negative Numbers

We have used 0's and 1's to represent positive numbers. It is also possible to write negative numbers using only 0's and 1's. In the last section, you learned that it is possible to subtract a number by adding its complement. This procedure is similar to the one for subtracting signed numbers: to subtract a number, add its opposite. This relationship between subtraction and negative numbers makes it plausible that we can represent negative numbers by using the two's complement.

To write a negative number in base two:

1. Write its absolute value in base two.

2. Find the two's complement of the absolute value.

EXAMPLE 1

Write -59 as an eight-bit binary number.

Solution   Write 59 in binary: $59_{10} = 0 \ 0 \ 1 \ 1 \ 1 \ 0 \ 1 \ 1_2$.
Find the complement of 59: $1 \ 1 \ 0 \ 0 \ 0 \ 1 \ 0 \ 0 + 1 = 1 \ 1 \ 0 \ 0 \ 0 \ 1 \ 0 \ 1$.
Therefore $-59_{10} = 1 \ 1 \ 0 \ 0 \ 0 \ 1 \ 0 \ 1_2$. ∎

EXAMPLE 2

Write -83 as an eight-bit binary number.

Solution   Write 83 in binary: $83_{10} = 0 \ 1 \ 0 \ 1 \ 0 \ 0 \ 1 \ 1_2$.
Find the complement of 83: $1 \ 0 \ 1 \ 0 \ 1 \ 1 \ 0 \ 0 + 1 = 1 \ 0 \ 1 \ 0 \ 1 \ 1 \ 0 \ 1$.
Therefore $-83_{10} = 1 \ 0 \ 1 \ 0 \ 1 \ 1 \ 0 \ 1_2$. ∎

From Example 1, we know that $-59_{10} = 11000101_2$. But we must also be able to determine that $11000101_2 = -59_{10}$ and not $128 + 64 + 4 + 1 = 197_{10}$. The first (leftmost) digit is used to indicate sign. A 0 indicates a positive;

a 1 indicates a negative. To convert a number from base 2 to base 10, we first evaluate the sign bit. If it is a 0, we evaluate the number based on its place values. If it is a 1, the number is negative and is therefore the complement of some positive number. Before we can evaluate the place values, we need to know the positive number. Since the negative of a negative is positive, we take the complement of the complement to get the positive.

> **To evaluate signed numbers:**
> 1. If the first (leftmost) bit is a 0, evaluate the place values. The number is positive.
> 2. If the first bit is a 1, find the complement of the number and evaluate its place values. The number is negative.

EXAMPLE 3

Convert 11000101_2 to base 10.

Solution

Since the first bit is a 1, the number is negative.
We find its complement: $00111010 + 1 = 00111011$.
This binary number represents $32 + 16 + 8 + 2 + 1 = 59_{10}$.
Therefore $11000101_2 = -59_{10}$. ∎

EXAMPLE 4

Convert 10100100_2 to base 10.

Solution

Since the first bit is a 1, the number is negative.
We find its complement: $01011011 + 1 = 01011100$.
This binary number represents $64 + 16 + 8 + 4 = 92_{10}$.
Therefore $10100100_2 = -92_{10}$. ∎

EXAMPLE 5

Convert 01000101_2 to base 10.

Solution   Since the first bit is a 0, we know the number is positive. Evaluate its place values. The number represents: $64 + 4 + 1 = 69_{10}$. Therefore $01000101_2 = 69_{10}$. ∎

Computers can handle numbers much larger than those that can be written in eight bits. We have used only eight bits to make the examples simpler. If we use sixteen bits, as is common for microcomputers, we can write 32,768 positive numbers: 0000000000000000_2 through 0111111111111111_2 (i.e., 0_{10} through 32767_{10}). We can also write 32,768 negative numbers: 1111111111111111_2 through 1000000000000000_2 (-1_{10} through $-32,768_{10}$).

In an integer arithmetic problem, each individual operand may belong to the set of integers within the range -32768 to 32767, but this set is not closed under addition, subtraction, multiplication, or division. When the sum, difference, or product of two integers falls outside the set, the computer prints an error message. The word *overflow* is often used to describe the occurrence of a result outside the allowed set of values. Integer division will not produce an error message, for it is performed as described in Section 2.8

In performing arithmetic operations with signed numbers, it is important to use the same number of bits throughout the problem. The answer will have the same number of bits as well. We will continue to use eight bits. If there happens to be a "carry" into the ninth column, it is simply omitted from the final answer.

To add decimal integers using binary arithmetic:

1. Write each number as an eight-bit binary number.
2. Add, omitting any carry into the ninth position.
3. The resulting eight-bit number is the sum.

EXAMPLE 6

Perform the decimal addition $-59 + 83$ in the binary system. Check the answer against the expected result.

Solution

$-59_{10} = 11000101_2$ (see Example 1).

$83_{10} = \underline{01010011_2}$ (see Example 2).

The sum is 00011000_2. The carry into the ninth column is omitted.

Check: This number is positive. It represents $16 + 8 = 24_{10}$. Since $-59_{10} + 83_{10} = 24_{10}$, the answer is correct. ∎

EXAMPLE 7

Perform the decimal addition $35 + (-66)$ in the binary system. Check the answer against the expected result.

Solution

$$35_{10} = 00100011_2. \quad 66_{10} = 01000010_2.$$

Therefore, $-66_{10} = \underline{10111110_2}$, the complement of 66_{10}.

The sum is $\qquad 11100001_2$.

Check: The number is the negative of $00011111_2 = 16 + 8 + 4 + 2 + 1 = 31_{10}$. That is, $11100001_2 = -31_{10}$. Since $35_{10} + (-66_{10}) = -31_{10}$, the answer is correct. ∎

EXAMPLE 8

Perform the decimal addition $-25 + (-13)$ in the binary system. Check the answer against the expected result.

Solution

$25_{10} = 00011001_2$, so $-25_{10} = 11100111_2$.

Since $13_{10} = 00001101_2$, $-13_{10} = 11110011_2$.

Thus, we have $-25_{10} = 11100111_2$.

$$-13_{10} = \underline{11110011_2}.$$

The sum is 11011010_2.

Check: The number is the negative of $00100110_2 = 32 + 4 + 2 = 38_{10}$. That is, $11011010_2 = -38_{10}$. Since $-25_{10} + (-13_{10}) = -38_{10}$, the answer is correct. ■

4.5 EXERCISES

Write each number as an eight-bit binary number. See Examples 1 and 2.

1. 37	**2.** -37	**3.** 45
4. -45	**5.** 52	**6.** -52
7. 78	**8.** -78	**9.** 91
10. -91	**11.** 123	**12.** -123

Write each number as a decimal number. See Examples 3–5.

13. 00101101	**14.** 11010011	**15.** 00110110
16. 11001010	**17.** 01010101	**18.** 00110011
19. 10011001	**20.** 11100011	**21.** 10101010
22. 10000000	**23.** 11111111	**24.** 00010001

Perform each decimal addition in the binary system. Use eight bits. Check your answer by converting to decimal. See Examples 6–8.

25. $-26 + 34$	**26.** $26 + (-34)$	**27.** $-26 + (-34)$	**28.** $26 + 34$
29. $45 + 21$	**30.** $-45 + 21$	**31.** $45 + (-21)$	**32.** $-45 + (-21)$
33. $-17 + 32$	**34.** $17 + 32$	**35.** $-17 + (-32)$	**36.** $17 + (-32)$

➤**37.** Use pseudocode or a flowchart to show an algorithm for converting a signed decimal number x ($-128 \le x \le 127$) to an eight-bit binary number.

➤**38.** Use pseudocode or a flowchart to show an algorithm for converting an eight-bit signed binary number to a decimal number.

4.6 Octal, Hexadecimal, and Decimal Conversions

You have seen how we can represent both integers and fractions using the binary system. The binary numerals, however, are often inconvenient for humans to use, because a large number of bits are required to write a relatively small number. Base 10 allows us to write numbers more compactly, but it is an inconvenient system for a computer to use. A system based on 8's, called the *octal system,* produces compact numerals like our base ten system, yet it is more compatible with the computer's base two system, since $2^3 = 8$. Like the decimal and binary systems, it gets its name from the number of symbols and their place values.

Base Eight or Octal System of Numeration
1. There are eight symbols to use as digits: 0, 1, 2, 3, 4, 5, 6, and 7.
2. The place values are powers of the base 8 (i.e., 1, 8, 64, 512, 4096, and so on).

The octal system is not as common as the hexadecimal system (base 16). Using a base 16 system, we can write large numbers very compactly, and because $2^4 = 16$, this system is also compatible with the computer's base two system.

Base Sixteen or Hexadecimal System of Numeration
1. There are 16 symbols to use as digits:
   0, 1, 2, 3, 4, 5, 6, 7, 8, 9, A, B, C, D, E, and F.
2. The place values are powers of the base 16 (i.e., 1, 16, 256, 4096, and so on).

We cannot use 10 for ten, because in base 16, 10 means sixteen. Likewise, 11 means seventeen, 12 means eighteen, 13 means nineteen, 14 means twenty, and 15 means twenty-one. Thus, it is necessary to introduce six new symbols. The letters A, B, C, D, E, and F are used to represent ten, eleven, twelve, thirteen, fourteen, and fifteen respectively. The conversion rules that you learned in Sections 4.1 and 4.2 are valid in these systems also.

Conversion Rules

1. To convert an integer or fraction *to* base ten *from* another base or to convert an integer *from* base 10 *to* another base:
   a. Label the digits of the **nondecimal** number with their place values, even if the digits of that number are blank.
   b. If the **digits** of the nondecimal number are known, multiply each digit by its place value. The sum of these products is the base 10 number.
   c. If the digits of the nondecimal number are blank, divide the base 10 number by the largest place value used as a label. The quotient is the leftmost digit of the number sought. Divide the remainder by the next place value to get the next digit, and so on.

2. To convert a fraction *from* base 10 *to* another base *n:*
   a. Multiply the decimal fraction by *n* and "remove" the integer portion of the product. The integer portion is the first digit to the right of the point in the base *n* fraction.
   b. Multiply the remaining decimal fraction by *n* and "remove" the integer, which is the next digit to the right in the base *n* fraction.
   c. Repeat step b until:
      1. The remaining fraction is 0 or
      2. A repeating pattern emerges.

EXAMPLE 1

Convert 42.3_8 to base 10.

Solution   Label the digits of the base 8 numeral with their place values.

$$\frac{4}{8}\frac{2}{1}.\frac{3}{1/8}$$ The number represents $(4 \times \mathbf{8}) + (2 \times \mathbf{1}) + (3 \times \mathbf{1/8})$
$$\qquad\qquad\qquad\qquad 32 \quad + \quad 2 \quad + \quad 3/8$$
$$\qquad\qquad\qquad\qquad\qquad 34.375.$$

Thus, $42.3_8 = 34.375_{10}$. ■

EXAMPLE 2

Convert 73_{10} to base 8.

Solution   Label the digits of the base 8 numeral with their place values. We don't know yet what the digits are, so we leave them blank.

$$\overline{}\ \overline{}\ \overline{}$$
$$64\ \ 8\ \ 1$$

Find groups of 64: $73 \div \mathbf{64} = 1$ R 9.
Find groups of 8: $9 \div \mathbf{8} = 1$ R 1.
Find groups of 1: $1 \div \mathbf{1} = 1$.

Thus, $73_{10} = 111_8$. ■

EXAMPLE 3

Convert 0.75_{10} to base 8.

Solution   We multiply by 8: $0.75 \times 8 = 6.00$. The integer portion is 6. The fraction portion is 0, so $0.75_{10} = 0.6_8$. ∎

EXAMPLE 4

Convert $C3.8_{16}$ to base 10.

Solution   Label the digits of the base 16 numeral.

$$\underset{16}{C}\ \underset{1}{3}\ .\ \underset{1/16}{8}$$ The number represents $(12 \times \mathbf{16}) + (3 \times \mathbf{1}) + (8 \times \mathbf{1/16})$
or $\quad 192 \quad + \quad 3 \quad + \quad 0.5$
or $\quad\quad\quad\quad\quad 195.95$

Thus $C3.8_{16} = 195.5_{10}$. ∎

EXAMPLE 5

Convert 97.4_{10} to base 16.

Solution   Label the digits of the base 16 integer. We don't yet know what the digits are, so we leave them blank.

$\underset{16}{\underline{}}\ \underset{1}{\underline{}}\ .$ Find groups of 16: $97 \div \mathbf{16} = 6$ R 1.
Find groups of 1: $1 \div \mathbf{1} = 1$.

Thus $97_{10} = 61_{16}$.

To convert the fraction 0.4, we multiply by 16: $0.4 \times 16 = 6.4$ The integer portion is 6, and 0.4×16 establishes a repeating pattern. That is, $0.4_{10} = 0.\overline{6}_{16}$. Therefore $97.4_{10} = 61.\overline{6}_{16}$. ∎

The algorithms from Section 4.2 may also be used for conversion of integers.

> **To convert a base *n* integer to a base 10 integer:**
> 1. Multiply the leftmost digit by *n* and add the next digit to the right.
> 2. Multiply the sum by *n* and add the next digit.
> 3. Repeat step 2 until the last digit has been added.
>   The sum is the decimal integer.

EXAMPLE 6

Convert 217_8 to base 10.

Solution

$8(2) + 1 =\ 17$.
$8(17) + 7 = 143$. Therefore, $217_8 = 143_{10}$. ∎

EXAMPLE 7

Convert 98_{10} to base 8.

Solution

Begin here:

$$
\begin{array}{r}
0 \ \text{R} \ 1 \\
8\overline{)\ 1} \ \text{R} \ 4 \\
8\overline{)12} \ \text{R} \ 2 \\
8\overline{)98}
\end{array}
$$

Therefore, $98_{10} = 142_8$. ∎

EXAMPLE 8

Convert $2CC_{16}$ to base 10.

Solution

$16 \times 2 + 12 = 44.$

$16 \times 44 + 12 = 716.$ Therefore, $2CC_{16} = 716_{10}$. ∎

EXAMPLE 9

Convert 427_{10} to base 16.

Solution

Begin here:

$$
\begin{array}{r}
0 \ \text{R} \ 1 \\
6\overline{)\ 1} \ \text{R} \ 10 = A \\
16\overline{)\ 26} \ \text{R} \ 11 = B \\
16\overline{)427}
\end{array}
$$

Therefore, $427_{10} = 1AB_{16}$. ∎

4.6 EXERCISES

Convert each octal number to decimal. See Examples 1 and 6.

1. 72	**2.** 1234	**3.** 204	**4.** 324	**5.** 11.11
6. 12.23	**7.** 42.61	**8.** 7.05	**9.** 1010.101	**10.** 1007.64

Convert each hexadecimal number to decimal. See Examples 4 and 8.

11. 144	**12.** 234	**13.** 10B	**14.** 2BD	**15.** 1A.C
16. B2.2	**17.** 5.A	**18.** FD.1	**19.** 1E3.41	**20.** 2BF.88

Convert each decimal number to octal. See Examples 2, 3, and 7.

21. 34	**22.** 40.5	**23.** 59.25	**24.** 80	**25.** 612
26. 60	**27.** 80.375	**28.** 36.0625	**29.** 559	**30.** 445

Convert each decimal number to hexadecimal. See Examples 5 and 9.

31. 34	**32.** 40.5	**33.** 59.25	**34.** 80	**35.** 612
36. 60	**37.** 80.375	**38.** 36.0625	**39.** 559	**40.** 445

✳**41.** Based on what you have learned about the ten's complement and the two's complement, write:

(a) the eight's complement of 743_8 using four digits.

(b) the sixteen's complement of $7A4F_{16}$ using six digits.

✳42. If "0" is used to signify a positive number, what digit would signify a negative number:

(a) in base 10?

(b) in base 8?

(c) in base 16?

(d) in base n?

✳43. Using four digits and complementary arithmetic, work each problem.

$$23_{10}$$
$$-45_{10}$$

$$27_8$$
$$-55_8$$

$$17_{16}$$
$$-2D_{16}$$

➡44. Use pseudocode or a flowchart to show an algorithm for converting a base B integer to a decimal integer.

➡45. Use pseudocode or a flowchart to show an algorithm for converting a decimal integer to a base B integer.

➡46. Use pseudocode or a flowchart to show an algorithm for converting a decimal fraction to a base B fraction.

✳47. The ASCII and EBCDIC codes (See 4.1 Exercises, problems 44 and 45) are often given in hexadecimal digits.

a. Write the ASCII representation of the characters 0–9 and A–Z in hexadecimal.

b. Write the EBCDIC representation of the characters 0–9 and A–Z in hexadecimal.

✳48. a. Find a two-digit octal number such that if the digits are reversed, the new number is half of the original.

b. Is it possible to find such a number in the decimal system?

4.7 Octal, Hexadecimal, and Binary Conversions

Because $2^3 = 8$, there is a relationship between base 2 and base 8 that makes conversions between these two bases convenient. Any octal digit is equivalent to a three-bit binary number, as shown in Table 4.4.

Table 4.4

Octal	Binary
0	000
1	001
2	010
3	011
4	100
5	101
6	110
7	111

We use the fact that three binary digits may represent an octal digit in the following conversion rule:

> **To convert back and forth between binary and octal:**
> 1. If the digits in the given number are binary digits, group them in threes, starting at the binary point. Replace each group by a single octal digit equivalent in value.
> 2. If the digits in the given number are octal digits, replace each digit by a group of three binary digits equivalent in value.

Consider 11101101_2. To convert it to octal, we notice that the binary point is at the end of this number. Starting there, we consider groups of three digits at a time: 011 101 101. Since the leftmost group is not a complete group of three, we fill in the missing digit with a 0. Each of these groups can be replaced by a single octal digit. As shown in Table 4.2, the leftmost group represents 3, the middle group represents 5, and the rightmost group represents 5. Therefore, $11101101_2 = 355_8$.

EXAMPLE 1

Convert 111.01101_2 to an octal numeral.

Solution Starting at the binary point, group the digits into three's: $111.$ 011 010. (Since the rightmost group is not a complete group of three, we fill in the missing digit with a 0.) Each of these groups can be replaced by a single octal digit. The left group represents 7, the middle group represents 3, and the right group represents 2. Thus, $111.01101_2 = 7.32_8$. ∎

EXAMPLE 2

Convert 401_8 to a binary numeral.

Solution Since it takes three binary digits to represent one octal digit, the number will have three groups of binary digits. Each octal digit will be replaced by one of these groups. The leftmost octal digit (4) is written 100, the middle digit (0) is written 000, and the rightmost digit (1) is written 001. Thus, $401_8 = 100000001_2$. ∎

EXAMPLE 3

Convert 67.1_8 to a binary numeral.

Solution The binary numeral will have two groups of three digits on the left side of the binary point, and one group of three on the right. The 6 is written as 110, 7 is written as 111, and 1 is written as 001. Thus $67.1_8 = 110111.001_2$. ∎

Since $2^4 = 16$, we could group binary digits in fours to convert to or from hexadecimal. See Table 4.5.

Table 4.5

Hexadecimal	Binary
0	0000
1	0001
2	0010
3	0011
4	0100
5	0101
6	0110
7	0111
8	1000
9	1001
A	1010
B	1011
C	1100
D	1101
E	1110
F	1111

We use the fact that four binary digits may represent one hexadecimal digit in the following conversion rule.

To convert back and forth between binary and hexadecimal:

1. If the digits in the given number are binary digits, group them in fours, starting at the binary point. Replace each group by a single hexadecimal digit equivalent in value.

2. If the digits in the given number are hexadecimal digits, replace each digit by a group of four binary digits equivalent in value.

EXAMPLE 4

Convert 11101101_2 to a hexadecimal numeral.

Solution   The binary point is at the end. Therefore, the groups are 1110 1101. The left group represents 14. But each group of four binary digits must be replaced by a single hexadecimal digit. Remember that 14 is written as E in base 16. The righthand group represents 13, which is written as D. Therefore, $11101101_2 = ED_{16}$. ∎

EXAMPLE 5

Convert 111.01101_2 to a hexadecimal numeral.

Solution Starting   at   the   binary   point,   the   groups   are: $\underline{0111}$. $\underline{0110}$ $\underline{1000}$ (where the rightmost group has been made a complete group of four by filling in with 0's). The leftmost group represents 7, the middle group represents 6, and the rightmost group represents 8. Therefore, 111.01101_2 $= 7.68_{16}$. ∎

EXAMPLE 6

Convert $1A.C_{16}$ to a binary numeral.

Solution   There will be two groups of four binary digits on the left side of the binary point and one group of four binary digits on the right side of the binary point. The 1 is written as 0001, A is written as 1010, and C is written as 1100. Therefore, $1A.C_{16} = 00011010.1100_2$. ∎

To convert from octal to hexadecimal or from hexadecimal to octal, it is convenient to use binary.

> **To convert back and forth between octal and hexadecimal:**
> 1. Convert the number to binary by using groups of binary digits.
>   (a) Each octal digit is equivalent to a three-bit binary number.
>   (b) Each hexadecimal digit is equivalent to a four-bit binary number.
> 2. Regroup the binary digits and convert to the other base.

EXAMPLE 7

Convert 37.7_8 to a hexadecimal numeral.

Solution

$37.7_8 = 011$ 111 . 111_2.
$011111.111_2 = 0001$ 1111 . $1110_2 = 1F.E_{16}$. ∎

EXAMPLE 8

Convert $3.B_{16}$ to an octal numeral.

Solution

$3.B_{16} = 0011$. $1011_2 = 011$. 101 $100_2 = 3.54_8$. ∎

4.7 EXERCISES

Convert each binary number to octal. See Example 1.

1. 101110   **2.** 100111   **3.** 1011010.111   **4.** 1000101.101

5. 1101011.10   **6.** 1011001.01   **7.** 1010.1011   **8.** 11.11011

Convert each octal number to binary. See Examples 2 and 3.

9. 62   **10.** 76   **11.** 22.3   **12.** 162.5

13. 3.04   **14.** 2.56   **15.** 12.475   **16.** 17.031

Convert each binary number to hexadecimal. See Examples 4 and 5.

17. 101110   **18.** 100111   **19.** 1011010.111   **20.** 1000101.101

21. 1101011.10   **22.** 1011001.01   **23.** 1010.1011   **24.** 11.11011

Convert each hexadecimal number to binary. See Example 6.

25. 93   **26.** C8   **27.** A3.8   **28.** B9.6

29. 2.DE   **30.** F.74   **31.** 1B.0F   **32.** C.402

Convert each octal number to hexadecimal. See Example 7.

33. 42   **34.** 62.5   **35.** 12.47   **36.** 17.03

Convert each hexadecimal number to octal. See Example 8.

37. 3.D   **38.** F.2   **39.** A.2B   **40.** 1B.0F

✳**41.** (a) Convert 110010_2 to a base 4 number.

     (b) Convert 231_4 to a base 2 number.

✳**42.** (a) Convert 786_9 to a base 3 number.

     (b) Convert 102112_3 to a base 9 number.

✳**43.** Consider the number $111111111_2 = (256 + 128 + 64) + (32 + 16 + 8) + (4 + 2 + 1)$. Use the distributive property on each group of three addends, and explain how the procedure relates the binary number to its octal equivalent.

✳**44.** When each digit of a decimal number is translated into a four-bit binary number, the result is a code known as BCD (binary coded decimal). Write the following numbers in BCD, and write them as binary numbers.

     a. 7

     b. 17

     c. 77

     d. 13

✳**45.** What is the next number in the sequence?

     12, 13, 14, 15, 20, 22, 30, 40

4.8 Octal and Hexadecimal Arithmetic

In order to do arithmetic in bases 8 and 16, we construct addition and multi-plication tables (see Tables 4.6 and 4.7). These tables are used in the same way as the base 10 tables in Section 2.4

Table 4.6

Octal Addition Table

+	0	1	2	3	4	5	6	7
0	0	1	2	3	4	5	6	7
1	1	2	3	4	5	6	7	10
2	2	3	4	5	6	7	10	11
3	3	4	5	6	7	10	11	12
4	4	5	6	7	10	11	12	13
5	5	6	7	10	11	12	13	14
6	6	7	10	11	12	13	14	15
7	7	10	11	12	13	14	15	16

Table 4.7

Octal Multiplication Table

×	0	1	2	3	4	5	6	7
0	0	0	0	0	0	0	0	0
1	0	1	2	3	4	5	6	7
2	0	2	4	6	10	12	14	16
3	0	3	6	11	14	17	22	25
4	0	4	10	14	20	24	30	34
5	0	5	12	17	24	31	36	43
6	0	6	14	22	30	36	44	52
7	0	7	16	25	34	43	52	61

To add octal numbers:
1. Line up the octal points.
2. Add column by column; carry as necessary.

EXAMPLE 1

Perform the following octal addition: 752 + 46.

Solution

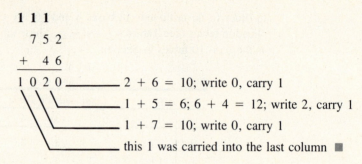

```
  1 1 1
    7 5 2
  +   4 6
  1 0 2 0
```
____ 2 + 6 = 10; write 0, carry 1

____ 1 + 5 = 6; 6 + 4 = 12; write 2, carry 1

____ 1 + 7 = 10; write 0, carry 1

____ this 1 was carried into the last column ∎

EXAMPLE 2

Perform the following octal addition: 607 + 25.

Solution

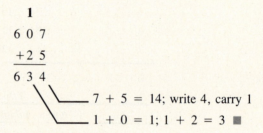

```
    1
  6 0 7
 +2 5
  6 3 4
```
____ 7 + 5 = 14; write 4, carry 1

____ 1 + 0 = 1; 1 + 2 = 3 ∎

EXAMPLE 3

Perform the following octal addition: 72.34 + 1.6

Solution   Line up the octal points:

```
    1
  7 2 . 3 4
 +1 . 6
  7 4 . 1 4
```
____ 3 + 6 = 11; write 1, carry 1

____ 1 + 2 = 3; 3 + 1 = 4 ∎

Octal multiplication is performed just as decimal or binary multiplication is performed. The only difference is that we use the "number facts" from the octal tables.

> **To multiply two octal numbers:**
> 1. Line up the rightmost digits of the two numbers.
> 2. Multiply column by column, indenting each time.
> 3. Add the partial products, and count off the total number of octal places that were involved in the problem to locate the octal point.

EXAMPLE 4

Perform the following octal multiplication: 34×6.

Solution

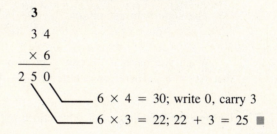

```
          3
          3 4
        × 6
      ‾‾‾‾‾‾
      2 5 0
```

$6 \times 4 = 30$; write 0, carry 3

$6 \times 3 = 22$; $22 + 3 = 25$ ∎

EXAMPLE 5

Perform the following octal multiplication: 127×13.

Solution

```
        1 2 7
      × 1 3          multiply by 3: 3 × 7 = 25; write 5, carry 2
      ‾‾‾‾‾          3 × 2 = 6; 6 + 2 = 10; write 0, carry 1
      4 0 5          3 × 1 = 3; 3 + 1 = 4
    + 1 2 7          indent and multiply by 1:
    ‾‾‾‾‾‾‾                  1 × 7 = 7; 1 × 2 = 2; 1 × 1 = 1
    1 6 7 5          add the partial products ∎
```

EXAMPLE 6

Perform the following octal multiplication: 34.5×0.12.

Solution

```
      3 4 . 5
      0. 1  2         multiply by 2: 2 × 5 = 12; write 2, carry 1
      ‾‾‾‾‾‾          2 × 4 = 10; 10 + 1 = 11; write 1, carry 1
        7 1  2        2 × 3 = 6; 6 + 1 = 7
      3 4 5           indent and multiply by 1

      ‾‾‾‾‾‾
      4 .3 6  2       add the partial products and count off three
                      octal places ∎
```

We can also use the tables to do subtraction and division by reading them differently. To subtract 3 from 11 using the addition table, locate 3 on the left, and read across that row until you see 11. The 6 at the top of that column is the answer.

> **To use an addition table for subtraction:**
> 1. Find the subtrahend (the number being subtracted) in the lefthand column.
> 2. Read across that row until you find the minuend (the number from which you are subtracting).
> 3. The number in the top row over the minuend is the difference.

When you have to "borrow" in an octal subtraction problem, remember that $10_8 = 8_{10}$. Therefore in base 8, $10 - 1 = 7$.

EXAMPLE 7

Perform the following octal subtraction: $752 - 46$.

Solution

$$
\begin{array}{r}
4\ 12 \\
7\ 5\ 2 \\
-4\ 6 \\
\hline
7\ 0\ 4
\end{array}
$$

— $2 - 6$ requires borrowing; $12 - 6 = 4$
— $4 - 4 = 0$ ■

EXAMPLE 8

Perform the following octal subtraction: $607 - 25$.

Solution

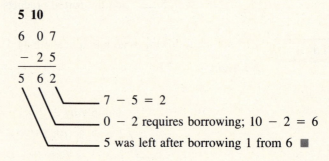

$$
\begin{array}{r}
5\ 10 \\
6\ \ 0\ 7 \\
-\ 2\ 5 \\
\hline
5\ \ 6\ 2
\end{array}
$$

— $7 - 5 = 2$
— $0 - 2$ requires borrowing; $10 - 2 = 6$
— 5 was left after borrowing 1 from 6 ■

EXAMPLE 9

Perform the following octal subtraction: $10 - 3.2$

Solution

$$
\begin{array}{r}
\text{0} \quad \text{7} \\
\text{10} \quad \text{10} \\
\text{1} \quad \text{0} \cdot \text{0} \\
- \quad \text{3} \cdot \text{2} \\
\hline
\text{4} \cdot \text{6}
\end{array}
$$

      $0 - 2$ requires borrowing; $10 - 2 = 6$

      $7 - 3 = 4$ ∎

To do hexadecimal arithmetic, we first construct hexadecimal addition and multiplication tables (see Tables 4.8 and 4.9). Then we perform the operations just as we would in base 10, but we use the hexadecimal tables to get our "number facts."

Table 4.8

Hexadecimal Addition Table																
+	0	1	2	3	4	5	6	7	8	9	A	B	C	D	E	F
0	0	1	2	3	4	5	6	7	8	9	A	B	C	D	E	F
1	1	2	3	4	5	6	7	8	9	A	B	C	D	E	F	10
2	2	3	4	5	6	7	8	9	A	B	C	D	E	F	10	11
3	3	4	5	6	7	8	9	A	B	C	D	E	F	10	11	12
4	4	5	6	7	8	9	A	B	C	D	E	F	10	11	12	13
5	5	6	7	8	9	A	B	C	D	E	F	10	11	12	13	14
6	6	7	8	9	A	B	C	D	E	F	10	11	12	13	14	15
7	7	8	9	A	B	C	D	E	F	10	11	12	13	14	15	16
8	8	9	A	B	C	D	E	F	10	11	12	13	14	15	16	17
9	9	A	B	C	D	E	F	10	11	12	13	14	15	16	17	18
A	A	B	C	D	E	F	10	11	12	13	14	15	16	17	18	19
B	B	C	D	E	F	10	11	12	13	14	15	16	17	18	19	1A
C	C	D	E	F	10	11	12	13	14	15	16	17	18	19	1A	1B
D	D	E	F	10	11	12	13	14	15	16	17	18	19	1A	1B	1C
E	E	F	10	11	12	13	14	15	16	17	18	19	1A	1B	1C	1D
F	F	10	11	12	13	14	15	16	17	18	19	1A	1B	1C	1D	1E

Table 4.9

Hexadecimal Multiplication Table

×	0	1	2	3	4	5	6	7	8	9	A	B	C	D	E	F
0	0	0	0	0	0	0	0	0	0	0	0	0	0	0	0	0
1	0	1	2	3	4	5	6	7	8	9	A	B	C	D	E	F
2	0	2	4	6	8	A	C	E	10	12	14	16	18	1A	1C	1E
3	0	3	6	9	C	F	12	15	18	1B	1E	21	24	27	2A	2D
4	0	4	8	C	10	14	18	1C	20	24	28	2C	30	34	38	3C
5	0	5	A	F	14	19	1E	23	28	2D	32	37	3C	41	46	4B
6	0	6	C	12	18	1E	24	2A	30	36	3C	42	48	4E	54	5A
7	0	7	E	15	1C	23	2A	31	38	3F	46	4D	54	5B	62	69
8	0	8	10	18	20	28	30	38	40	48	50	58	60	68	70	78
9	0	9	12	1B	24	2D	36	3F	48	51	5A	63	6C	75	7E	87
A	0	A	14	1E	28	32	3C	46	50	5A	64	6E	78	82	8C	96
B	0	B	16	21	2C	37	42	4D	58	63	6E	79	84	8F	9A	A5
C	0	C	18	24	30	3C	48	54	60	6C	78	84	90	9C	A8	B4
D	0	D	1A	27	34	41	4E	5B	68	75	82	8F	9C	A9	B6	C3
E	0	E	1C	2A	38	46	54	62	70	7E	8C	9A	A8	B6	C4	D2
F	0	F	1E	2D	3C	4B	5A	69	78	87	96	A5	B4	C3	D2	E1

EXAMPLE 10

Perform the following hexadecimal addition: 1A3 + 27.

Solution

$$
\begin{array}{r}
1\ A\ 3 \\
+\ 2\ 7 \\
\hline
1\ C\ A
\end{array}
$$

3 + 7 = A

A + 2 = C ▪

EXAMPLE 11

Perform the following hexadecimal addition: 1A.B + C.D.

Solution

$$
\begin{array}{r}
1\ 1 \\
1\ A\ .\ B \\
+\ C\ .\ D \\
\hline
2\ 7\ .\ 8
\end{array}
$$

B + D = 18; write 8, carry 1

1 + A = B; B + C = 17; write 7, carry 1

1 + 1 = 2 ▪

Hexadecimal multiplication is performed just like decimal, binary, or octal multiplication if we use the hexadecimal tables. Line up the rightmost digits, multiply column by column, indenting each time. Add the partial products, and count off the total number of hexadecimal places involved in the problem to locate the hexadecimal point.

EXAMPLE 12

Perform the hexadecimal multiplication: 1AB × 3.

Solution

$$
\begin{array}{r}
2\ 2 \\
1\ A\ B \\
\times\quad 3 \\
\hline
5\ 0\ 1
\end{array}
$$

3 × B = 21; write 1, carry 2

3 × A = 1E; 1E + 2 = 20; write 0, carry 2

3 × 1 = 3; 3 + 2 = 5 ■

EXAMPLE 13

Perform the following hexadecimal multiplication: 2.C3 × 1A.

Solution

$$
\begin{array}{r}
2\ .\ C\ 3 \\
\times\ 1\ A \\
\hline
1\ B\quad 9\ E \\
2\ C\quad 3 \\
\hline
4\ 7\ .\ C\ E
\end{array}
$$

multiply by A: A × 3 = 1E; write E, carry 1

A × C = 78; 78 + 1 = 79; write 9, carry 7

A × 2 = 14; 14 + 7 = 1B

indent and multiply by 1

add the partial products and count off two places ■

In hexadecimal subtraction, you must not forget that $10_{16} = 16_{10}$. When you borrow, you have $10 - 1 = F$. You may use the hexadecimal addition table to do subtraction.

EXAMPLE 14

Perform the following hexadecimal subtraction: 1A3 − 27.

Solution

$$
\begin{array}{r}
9\ 13 \\
1\ A\ 3 \\
-\ 2\ 7 \\
\hline
1\ 7\ C
\end{array}
$$

3 − 7 requires borrowing; 13 − 7 = C

9 − 2 = 7 ■

EXAMPLE 15   Perform the following hexadecimal subtraction: 1A.B − C.D.

Solution

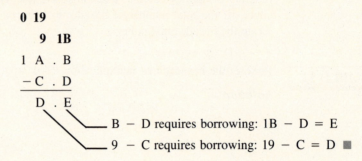

           0 19

              9 1B

         1 A . B

       − C . D

          D . E

                 B − D requires borrowing: 1B − D = E

                 9 − C requires borrowing: 19 − C = D ∎

4.8 EXERCISES

Perform each octal addition. See Examples 1–3.

1. 643 + 57   **2.** 516 + 36   **3.** 63.45 + 2.7

4. 3.76 + 1.542   **5.** 0.01 + 12.746

Perform each octal multiplication. See Examples 4–6.

6. 45 × 5   **7.** 236 × 21   **8.** 25.4 × 0.23

9. 23.7 × 4.6   **10.** 2.11 × 3.07

Perform each octal subtraction. See Examples 7–9.

11. 643 − 57   **12.** 516 − 36   **13.** 63.45 − 2.7

14. 23.26 − 3.5   **15.** 10.00 − 0.647

Perform each hexadecimal addition. See Examples 10 and 11.

16. 2BC + 38   **17.** 3CD + 25   **18.** 2B.C + E.D

19. 7.83 + 2.C   **20.** 19.A + F.02

Perform each hexadecimal multiplication. See Examples 12 and 13.

21. 2BC × 2   **22.** 3D4 × 2B   **23.** 34C × 2.1

24. 4F.2 × 3.E   **25.** E.15 × 10.C

Perform each hexadecimal subtraction. See Examples 14 and 15.

26. 2B2 − 38   **27.** 3CD − 25   **28.** 2B.C − E.D

29. 7.83 − 2.C   **30.** 19.A − F.02

❋**31.** Divide 7435_8 by 21_8.

❋**32.** Divide $7D30E_{16}$ by $2A1_{16}$.

❋**33.** Write the following fractions as repeating decimals.

    a. 1/9 (base 10)

    b. 1/7 (base 8)

    c. 1/F (base 16)

CHAPTER REVIEW

A symbol for a number is called a _____.

A _____ or base ten system of numeration has 10 symbols to use as digits, and the place values are powers of _____.

A _____ or base two system of numeration has 2 symbols to use as digits, and the place values are powers of _____.

The word _____ is a contraction for "binary digit."

The _____ of an eight-bit binary number x is $100000000 - x$.

A negative binary number has the digit _____ as the leftmost bit.

When the sum, difference, or product of two integers falls outside the range of integers that can be represented by that computer, the word _____ is used to describe the condition.

An _____ or base eight system of numeration has 8 symbols to use as digits, and the place values are powers of _____.

A _____ or base sixteen system of numeration has 16 symbols to use as digits, and the place values are powers of _____.

In base sixteen, the number ten is represented by the digit _____.

Any octal digit can be represented by _____ binary digits.

Any hexadecimal digit can be represented by _____ binary digits.

CHAPTER TEST

Convert each number to the indicated base.

1. 11100101_2 = _____ 10

2. 10011101_2 = _____ 10

3. 55_{10} = _____ 2

4. 153_{10} = _____ 2

5. 201_{10} = _____ 2

6. 101.1001_2 = _____ 10

7. 11.01011_2 = _____ 10

8. 111.1101_2 = _____ 10

9. 0.1875_{10} = _____ 2

10. 7.15_{10} = _____ 2

Perform each operation using binary arithmetic.

11. $110.101 + 10.11$

12. $1101 + 101 + 111$

13. $110.110 - 10.101$

14. 101.11×1.01

15. $100011 \div 100$

16. Find the two's complement of each number.

(a) 00101101   **(b)** 11011011

17. Perform the subtraction using complementary arithmetic.
$01101001 - 00101011$

18. Perform the subtraction using complementary arithmetic.
$00110011 - 00001111$

19. Perform the subtraction by borrowing.
$01101001 - 00101011$

20. Convert to base ten and subtract.
$00110011 - 00001111$

Write each number as an eight-bit binary number.

21. (a) 63   **(b)** -63

22. (a) 103   **(b)** -103

23. Write each number as a signed decimal number.

(a) 00010111   **(b)** 11100110

Perform each addition in the binary system. Use eight bits. Check your answer by converting to decimal.

24. $34 + (-19)$

25. $-34 + (-19)$

Convert each number to the indicated base.

26. $463_8 = $ _____ 10

27. $2BC_{16} = $ _____ 10

28. $2B.C_{16} = $ _____ 10

29. $211.75_{10} = $ _____ 8

30. $211.75_{10} = $ _____ 16

31. $11000111.001_2 = $ _____ 8

32. $57.34_8 = $ _____ 2

33. $11000111.001_2 = $ _____ 16

34. $1A7.E6_{16} = $ _____ 2

35. $D2.7C_{16} = $ _____ 8

Perform the indicated operation.

36. $23.54_8 + 134.7_8$

37. $35.2_8 \times 1.6_8$

38. $32.41_8 - 1.7_8$

39. $1A.7_{16} \times 0.2B_{16}$

40. $2F8_{16} - A9_{16}$

41. The following flowchart shows an algorithm that detects a palindrome when a three-digit decimal integer is entered. Modify it to detect a palindrome when a four-digit binary integer is entered. Use binary arithmetic.

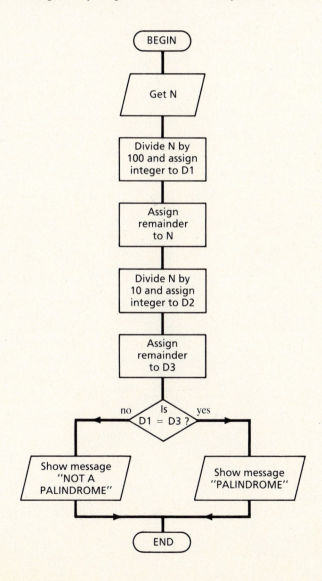

FOR FURTHER READING

Bronson, Gary and Karl Lyon. ''Two's Complement Numbers Revisited,'' *BYTE*, Vol. 10 No. 6 (June 1985), 230–231. (A method for evaluating negative numbers in two's complement form is given.)

Donahue, Richard J. ''Computing Palindromic Sums by Computer,'' *The Mathematics Teacher*, Vol. 77 No. 4 (April 1984), 269–271. (A BASIC program to determine the palindromic sum of any two-digit number is developed.)

Golden, Frederic. ''Big Dimwits and Little Geniuses,'' *Time* (January 3, 1983), 30–32. (The history of computers is discussed.)

Zirkel, Gene. ''An Extension to Changing Bases,'' *The MATYC Journal*, Vol 15 No. 3 (Fall 1981), 224.

Biography   ## Augusta Ada Byron

Augusta Ada Byron, for whom the computer language ADA is named, was born December 10, 1815. Her parents separated about a month later, and her father, Lord Byron (the poet), left England. As a child, she was taught by tutors. Later, she was helped by Augustus De Morgan.

In 1833, she saw Charles Babbage's Difference Engine, which he had worked on for 10 years with government support. It was designed to do calculations for navigation and astronomy, but it was never completed. Babbage constantly changed the plans, and they got more and more elaborate until in 1834, he began to work on a machine he called the Analytical Engine. Unlike the Difference Engine, the Analytical Engine was to be programmable.

Ada Byron was fascinated by the project. She believed that the design would work, and she encouraged Babbage. She is said to have suggested that the binary system be used, but the Analytical Engine used the decimal system. Consequently, 100 years passed before binary was actually used in a machine.

In 1835, Ada married William, eighth Lord King, and when he became an Earl, she became the Countess of Lovelace. She continued to work with Babbage and translated a paper about his project from French into English. She added notes of her own and nearly tripled the length of the paper now called "Observations on Mr. Babbage's Analytical Engine." This paper contains a nearly complete program, which Lady Lovelace wrote with Babbage's help. Although the program lacks some important concepts of modern programming, she is often cited as the "first computer programmer."

When government funds for Babbage's project were cut off in 1842, Lady Lovelace devised a betting system to try to win money for the project. She lost more than she won.

She was often sick, and on November 27, 1852, Lady Lovelace died. Although Babbage lived until 1871, he never completed the Analytical Engine.

Chapter Five
Polynomials

The variables used in computer programming to write polynomials and other algebraic expressions may be thought of as the names of storage locations. As a program is executed, the contents of each location might change, just as the contents of mailboxes might change.

5.1   Exponents

Algebra is sometimes described as "generalized arithmetic." That is, addition, subtraction, multiplication, and division are used with letters as well as numbers. The letters (or *variables*) represent numbers whose values we may not know. We begin our review of algebra by examining the rules of exponentiation.

An exponent is a symbol used to indicate a repeated multiplication. The expression x^p means that p x's are to be multiplied together. The x in this expression is called the *base*, and the p is called the *exponent*. (Sometimes we say "x is raised to the power p.") For example, $7^2 = 7(7) = 49$; $3^4 = 3(3)(3)(3) = 81$; $x^5 = x(x)(x)(x)(x)$.

> **Definition**
>
> x^p means that x is to be used as a factor p times if x is a real number (or a variable which represents a real number) and p is a positive integer.

In the following examples, we will develop some rules that are shortcuts to enable us to simplify algebraic expressions efficiently.

EXAMPLE 1

Simplify $7^2 (7^3)$.

Solution   Because the first expression represents two factors of 7, and the second represents three factors of 7, there are a total of five 7's in the product: $7^2 (7^3) = 7(7)(7)(7)(7) = 7^5$. To take a shortcut, we add the exponents. That is $7^2 (7^3) = 7^{2+3} = 7^5$. ∎

EXAMPLE 2

Simplify x^5/x^3.

Solution   There are five factors of x in the numerator. There are three factors of x in the denominator; these three factors in the denominator "cancel" three of the x's in the numerator, leaving two factors of x or x^2. To take a shortcut, we subtract exponents. That is, $x^5/x^3 = x^{5-3} = x^2$. ∎

If we generalize the concepts involved in the previous examples, we have two of the rules for working with exponents.

> **Exponent Rules**
> If x is a real number and m and n are positive integers:
>
> $$x^m (x^n) = x^{m+n} \qquad (1)$$
>
> $$\frac{x^n}{x^n} = x^{m-n} \text{ if } x \neq 0 \qquad (2)$$

Rule 1 requires that both factors have the same base. The product also has this base. If the two factors have different bases, rule 1 does not apply. For example, $2^3 (3^2)$ is written as $8(9) = 72$.

The definition we have used for exponents makes sense only if the exponents are positive integers. The rules, however, can be extended to nonpositive integers.

EXAMPLE 3

Simplify x^4/x^4.

Solution   Using rule 2, we have $x^4/x^4 = x^0$. We also know that if $x \neq 0$, $x^4/x^4 = 1$. So that there will not be two different answers to the same problem, mathematicians agree that $x^0 = 1$. (The argument holds for any value of $x \neq 0$ and any exponent.) ■

EXAMPLE 4

Simplify $7^2/7^5$.

Solution   By rule 2, $7^2/7^5 = 7^{2-5} = 7^{-3}$.

But we also know that $7^2/7^5 = \dfrac{7(7)}{7(7)(7)(7)(7)} = \dfrac{1}{7^3}$.

So that there will not be two different answers to the same problem, mathematicians agree that $7^{-3} = 1/7^3$. In general, $x^{-n} = 1/x^n$. ■

Thus, we have developed two more rules.

Exponent Rules
If x is a nonzero real number:

$$x^0 = 1 \tag{3}$$

$$x^{-n} = 1/x^n \text{ if } n \text{ is a positive integer} \tag{4}$$

Since a negative exponent can occur in either the numerator or denominator of a fraction, you should know how to simplify either case. No zero or negative exponents should appear in the simplified form.

EXAMPLE 5

Simplify x^{-2}/y^{-3}.

Solution

$x^{-2} = 1/x^2$; $y^{-3} = 1/y^3$.

Therefore, $x^{-2}/y^{-3} = 1/x^2 \div 1/y^3 = 1/x^2 \, (y^3/1) = y^3/x^2$. ■

We could say that any *factor* with a negative exponent crosses the fraction bar, and the exponent becomes positive. Generalizing this result, we have rule 5.

Exponent Rule
If x and y are nonzero real numbers and m and n are positive integers:

$$\frac{x^{-m}}{y^{-n}} = \frac{y^n}{x^m} \qquad (5)$$

Exponent rules 1–5 were stated for exponents that are positive integers. All five rules, however, can be used when the exponents are zero or negative integers as well as when they are positive integers.

EXAMPLE 6

Simplify each expression:

(a) $3x^3 (4x^{-5})$

(b) $\dfrac{6y^4}{3y}$

(c) 2^0

Solution

(a)   We use the commutative and associative properties to reorder and regroup the factors as $3(4)(x^3)(x^{-5}) = 12x^{3-5} = 12x^{-2}$ (by rule 1). By rule 4, we have $\dfrac{12}{x^2}$ as the final answer.

(b) $\dfrac{6}{3} = 2$ and $\dfrac{y^4}{y} = y^3$ by rule 2. Therefore, $\dfrac{6y^4}{3y} = 2y^3$.

(c) $2^0 = 1$ by rule 3. ∎

EXAMPLE 7

Simplify:

(a) $\dfrac{8a^{-3}b^4}{4x^2y^{-2}}$

(b) $\dfrac{x^4}{x^{-3}}$

Solution

(a) We know that $\dfrac{8}{4} = 2$. By rule 5, a^{-3} crosses the fraction bar to become a^3, and y^{-2} crosses the fraction bar to become y^2. Thus, we have $\dfrac{2b^4y^2}{a^3x^2}$.

(b) We can use rule 2 to get $x^{4 - (-3)} = x^7$.
Or we can use rule 5 to get $x^4(x^3) = x^7$. ■

Example 7(b) shows that there may be more than one correct way to work a problem. We arrive at the same answer, however, regardless of the method we choose. In each case, we could justify the reasoning by citing one of the exponent rules.

5.1 EXERCISES

Use rule 1 to simplify each expression. See Examples 1 and 6(a).

1. $2x^3(3x^2)$ **2.** $(3^4)(3^7)$ **3.** $2y^2(y^4)(7y)$

4. $(3a^2b)(4ab^2)$ **5.** $6x^3(8x^6)$

Use rule 2 to simplify each expression. See Examples 2 and 6(b).

6. $\dfrac{x^7}{x^4}$ **7.** $\dfrac{4x^3y^6}{2x^2y^4}$ **8.** $\dfrac{8a^5b^2}{4a^3b}$

9. $\dfrac{12y^4}{24y^2}$ **10.** $\dfrac{7z^3}{14z^2}$

Use rule 3 or rule 4 to simplify each expression. See Examples 3 and 4.

11. 2^0 **12.** 10^0 **13.** 2^{-4} **14.** 8^{-2} **15.** 10^{-3}

Simplify each expression. There should be no zero or negative exponents in the final answer. See Examples 5–7.

16. $5x^4(3x^{-6})$ **17.** $7x^{-3}(8x^2)$ **18.** $4a^5b^{-4}(5a^{-3}b^2)$

19. $9x^3y^{-2}z(3^{-2}y^5)$ **20.** $3a^{-2}b^4(6a^{-3}b^2c)$ **21.** $\dfrac{4x^2y}{2xy^2}$

22. $\dfrac{10a^2bc}{5ab^2c}$ **23.** $\dfrac{12xy^2}{3xz}$ **24.** $\dfrac{2^{-3}a}{3b^{-2}}$

25. $\dfrac{5a^{-2}b}{10a^2b}$ **26.** $\dfrac{5ab^{-1}}{10a^2b}$ **27.** $\dfrac{36x^{-3}y}{9xy^{-2}}$

28. $\dfrac{48x^{-3}y}{12xy^{-2}}$ **29.** $\dfrac{2x^0y}{3xy^0}$ **30.** $\dfrac{3^2ab^{-1}}{2^{-3}x^{-2}y}$

▶**31.** The English mathematician G. H. Hardy once visited the Indian mathematician Ramanujan in the hospital. Hardy remarked, "I thought the number of my taxicab was 1729. It seemed to me rather a dull number." Ramanujan replied, "It is a very interesting number. It is the smallest number expressible as the sum of two cubes in two different ways."†

(a) What is the largest number that could be one of the cubes used to express 1729?

(b) Use pseudocode or a flowchart to show an algorithm for producing all numbers less than or equal to 1729 that can be expressed as the sum of two cubes. Hint: Use nested loops (i.e., an iteration within an iteration).

5.2   Additional Exponent Rules

Parentheses are used to clarify algebraic expressions. We now develop exponent rules for use with expressions that have parentheses.

EXAMPLE 1

Simplify $(x^2)^3$.

Solution $(x^2)^3 = x^2(x^2)(x^2)$. By rule 1, we have x^6. To use a shortcut, we multiply exponents. That is, $(x^2)^3 = x^{2(3)} = x^6$. ∎

EXAMPLE 2

Simplify $(xy)^2$.

Solution $(xy)^2 = (xy)(xy)$. Using the commutative and associative properties to reorder and regroup the factors, we have $x(x)(y)(y) = x^2y^2$. To use a shortcut, we apply the exponent to each *factor* individually. That is, $(xy)^2 = x^2y^2$. ∎

EXAMPLE 3

Simplify $\left(\dfrac{x}{y}\right)^2$.

Solution $\left(\dfrac{x}{y}\right)^2 = \left(\dfrac{x}{y}\right)\left(\dfrac{x}{y}\right) = \dfrac{x^2}{y^2}$. To use a shortcut, we apply the exponent to both numerator and denominator. ∎

The concepts involved in the previous examples can be generalized to give the following exponent rules.

†Snow, C. P. foreword to *A Mathematician's Apology* by G. H. Hardy, Cambridge University Press (1967), 37.

Exponent Rules

If x and y are real numbers and m and n are integers:

$$(x^m)^n = x^{mn} \qquad (6)$$

$$(xy)^n = x^n y^n \qquad (7)$$

$$(x/y)^n = x^n/y^n \text{ where } y \neq 0 \qquad (8)$$

Rule 1 and rule 6 are often confused, but they are used in different circumstances. Rule 1 applies when two separate quantities, each having its own exponent, are multiplied together; rule 6 applies when there is one quantity with two exponents.

You have seen how to simplify expressions with integral exponents. If we allow fractional exponents, we don't want to abandon the old rules; they should still be valid under the extended range of exponents. Consider the problem $x^{1/2}(x^{1/2})$. By rule 1, $x^{1/2}(x^{1/2}) = x^1 = x$. To ask, "What does $x^{1/2}$ mean?", we could ask, "What number times itself is x?" The answer is $\sqrt{x}$. We then say that $x^{1/2}$ must mean $\sqrt{x}$. By generalizing this example, we have for any positive integer n, $x^{1/n} = \sqrt[n]{x}$ when $\sqrt[n]{x}$ is defined.

Exponent Rule

If x is a real number and n is a positive integer:

$$x^{1/n} = \sqrt[n]{x} \text{ when } \sqrt[n]{x} \text{ is defined.} \qquad (9)$$

EXAMPLE 4

Simplify each expression:

(a) $25^{1/2}$

(b) $-(25^{1/2})$

(c) $(-25)^{1/2}$

Solution

(a) $(25)^{1/2} = \sqrt{25}$. Although there are two square roots of 25 (5 and -5), the $\sqrt{}$ symbol means the positive square root only. Hence, $\sqrt{25} = 5$.

(b) $-(25^{1/2})$ means to find $25^{1/2}$ and prefix the negative sign. Hence, we have $-\sqrt{25} = -5$.

(c) $(-25)^{1/2}$ means $\sqrt{-25}$, which is not a real number, because there is no real number that can be multiplied by itself to obtain -25. ■

EXAMPLE 5

Simplify each expression:

(a) $8^{1/3}$

(b) $-(8^{1/3})$

(c) $(-8)^{1/3}$

Solution

(a) $8^{1/3}$ means $\sqrt[3]{8} = 2$.

(b) $-(8^{1/3})$ means to find $8^{1/3}$ and prefix the negative sign. Hence, we have $-\sqrt[3]{8} = -2$.

(c) $(-8)^{1/3}$ means $\sqrt[3]{-8}$, which is -2, because $(-2)(-2)(-2) = -8$. ∎

The tenth and last exponent rule allows us to simplify an expression having a fractional exponent with a numerator other than 1.

EXAMPLE 6

Simplify $8^{2/3}$.

Solution By rule 6, we have $8^{2/3} = (8^{1/3})^2 = (\sqrt[3]{8})^2 = 2^2 = 4$. We could also say, $8^{2/3} = (8^2)^{1/3} = \sqrt[3]{(8^2)} = \sqrt[3]{64} = 4$. ∎

> **Exponent Rule**
> If x is a real number and m and n are integers such that $n > 0$,
>
> $$x^{m/n} = (\sqrt[n]{x})^m = \sqrt[n]{(x^m)} \text{ where } \sqrt[n]{x} \text{ is defined.} \qquad (10)$$

The fractional exponent consists of two parts: the denominator, which tells us which root to take, and the numerator, which tells us which power to use. You should take the root first whenever possible. For example, consider $64^{4/3}$. If we take the root first, we have $64^{4/3} = (\sqrt[3]{64})^4 = 4^4 = 256$. But if we take the power first, we have $64^{4/3} = \sqrt[3]{(64^4)} = \sqrt[3]{16,777,216}$, which is also 256, but you probably wouldn't know that.

EXAMPLE 7

Simplify $7^{3/5}$.

Solution This time we can't take the root first, because we don't know what $\sqrt[5]{7}$ is, but we do know that it is defined in the real number system. Therefore we take the power first, and say $7^{3/5} = \sqrt[5]{(7^3)} = \sqrt[5]{343}$. ∎

The following examples review the exponent rules of this section.

EXAMPLE 8

Simplify:

(a) $(3^2)^4$

(b) $(2xy)^3$

(c) $(2/3)^2$

Solution

(a) $(3^2)^4 = 3^8$ by rule 6.

(b) $(2xy)^3 = 2^3x^3y^3 = 8x^3y^3$ by rule 7.

(c) $(2/3)^2 = 2^2/3^2 = 4/9$ by rule 8. ∎

EXAMPLE 9

Simplify:

(a) $(3a^2b)^3$

(b) $4(x^{-2}y)^{-1}$

(c) $\left(\dfrac{2a^{-1}b^2}{3^2xy^{-2}}\right)^3$

Solution

(a) By rule 7, $(3a^2\,b)^3 = 3^3(a^2)^3b^3$, which is $27a^6\,b^3$.

(b) By rule 7, $4(x^{-2}y)^{-1} = 4(x^{-2})^{-1}(y)^{-1}$, which is $4x^2\,y^{-1}$ by rule 6. Finally, we have $\dfrac{4x^2}{y}$.

(c) $\left(\dfrac{2a^{-1}b^2}{3^2xy^{-2}}\right)^3 = \dfrac{2^3a^{-3}b^6}{3^6x^3y^{-6}}$, which is $\dfrac{2^3b^6y^6}{3^6a^3x^3}$. ∎

EXAMPLE 10

Simplify. You may assume that all variables represent positive real numbers.

(a) $\left(\dfrac{4}{9x^4}\right)^{1/2}$

(b) $\left(\dfrac{8a^3}{27}\right)^{-2/3}$

(c) $\left(\dfrac{5}{7}\right)^{2/3}$

Solution

(a) $\left(\dfrac{4}{9x^4}\right)^{1/2} = \dfrac{4^{1/2}}{(9x^4)^{1/2}} = \dfrac{\sqrt{4}}{\sqrt{9x^4}} = \dfrac{2}{3x^2}$

(b) $\left(\dfrac{8a^3}{27}\right)^{-2/3} = \dfrac{(8a^3)^{-2/3}}{27^{-2/3}} = \dfrac{27^{2/3}}{(8a^3)^{2/3}} = \dfrac{(\sqrt[3]{27})^2}{(\sqrt[3]{8a^3})^2} = \dfrac{3^2}{(2a)^2} = \dfrac{9}{4a^2}$

(c) $\left(\dfrac{5}{7}\right)^{2/3} = \dfrac{5^{2/3}}{7^{2/3}} = \dfrac{\sqrt[3]{(5^2)}}{\sqrt[3]{(7^2)}} = \dfrac{\sqrt[3]{25}}{\sqrt[3]{49}}$. ∎

5.2 EXERCISES

Simplify each expression using rule 6. See Example 1.

1. $(a^2)^3$ **2.** $(3^4)^5$ **3.** $(2^3)^4$ **4.** $(x^2)^4$ **5.** $(5^3)^3$

Simplify each expression using rule 7. See Example 2.

6. $(xy)^3$ **7.** $(3a)^2$ **8.** $(5xy)^3$ **9.** $(4b)^2$ **10.** $(ab)^5$

Simplify each expression using rule 8. See Example 3.

11. $\left(\dfrac{2}{x}\right)^3$ **12.** $\left(\dfrac{3}{4}\right)^2$ **13.** $\left(\dfrac{y}{6}\right)^5$ **14.** $\left(\dfrac{2}{3}\right)^3$ **15.** $\left(\dfrac{a}{b}\right)^4$

Simplify each expression using rule 9. See Examples 4 and 5.

16. $16^{1/2}$ **17.** $(-16)^{1/2}$ **18.** $-(16^{1/2})$ **19.** $27^{1/3}$ **20.** $(-27)^{1/3}$

21. $-(27^{1/3})$ **22.** $(64^{1/2})$ **23.** $(64^{1/3})$ **24.** $(-64)^{1/3}$ **25.** $(-64)^{1/2}$

Simplify each expression using rule 10. See Examples 6 and 7.

26. $64^{2/3}$ **27.** $27^{2/3}$ **28.** $4^{2/3}$ **29.** $5^{3/4}$ **30.** $81^{3/4}$

Simplify each expression. There should be no negative exponents in the final answer. See Examples 8 and 9.

31. $(x^2y)^4$ **32.** $(2a^5b)^3$ **33.** $3(x^2y^{-1})^3$ **34.** $2(a^5b)^3$

35. $(5xy)^{-1}$ **36.** $5(xy)^{-1}$ **37.** $\left(\dfrac{3b}{4}\right)^2$ **38.** $\left(\dfrac{2x}{y}\right)^3$

39. $\left(\dfrac{4b^2}{5}\right)^3$ **40.** $\left(\dfrac{2x^2}{7}\right)^2$ **41.** $\left(\dfrac{x}{y^3}\right)^{-1}$ **42.** $\dfrac{3(ab)^{-2}}{5}$

43. $\dfrac{3ab^{-2}}{5}$ **44.** $\left(\dfrac{5ab^{-1}}{c^2}\right)^3$ **45.** $\left(\dfrac{2ab^{-1}}{c^2}\right)^{-3}$ **46.** $\left(\dfrac{7xy^{-1}}{3^{-1}ab}\right)^{-2}$

47. $\left(\dfrac{4a^{-1}b}{3a^{-2}b}\right)^2$ **48.** $\left(\dfrac{7xy^{-1}}{3^{-1}x^2y^0}\right)^0$ **49.** $\dfrac{2^{-1}ab^2}{3a^{-2}b^{-1}}$ **50.** $\dfrac{2a^{-1}b^2}{3^{-1}xy^{-3}}$

Simplify each expression. You may assume that all variables represent positive real numbers. See Example 10.

51. $\left(\dfrac{16}{25}\right)^{1/2}$ **52.** $\left(\dfrac{27}{64}\right)^{-1/3}$ **53.** $\left(\dfrac{x^4}{16}\right)^{3/4}$

54. $\left(\dfrac{25}{49b^2}\right)^{-1/2}$ **55.** $\left(\dfrac{a^{-2}}{9}\right)^{3/2}$ **56.** $\left(\dfrac{x^{-3}}{8y^6}\right)^{1/3}$

Simplify each expression.

✳**57.** $\sqrt{5}\sqrt[3]{5}$

✳**58.** $\sqrt{3}\sqrt[3]{2}$

✳**59.** Which is larger $\sqrt[10]{10}$ or $\sqrt[3]{2}$?

5.3   Addition and Multiplication of Algebraic Expressions

In this chapter, the words *term* and *factor* are used often. In an expression that contains addition (or subtraction, since to subtract x, we add $-x$), the operands are called *terms*. In an expression that contains multiplication, the operands are called *factors*. Thus, we say that $x + y$ has two terms, but xy has two factors.

The distinction between terms and factors is important in an expression such as $2x + 3y$, which contains both addition and multiplication. If we knew the values of x and y, we could evaluate this expression following the order of operations. The last operation performed would be addition, so we say that this expression has two terms. Each term by itself has two factors. In the expression $(x + 2)(x - 1)$, if we knew the value of x, the last operation to be performed would be multiplication, so we say this expression has two factors. Each factor has two terms.

> **Definition**
> *Terms* are algebraic expressions that are to be added. *Factors* are algebraic expressions that are to be multiplied.

EXAMPLE 1

Determine the number of terms or factors in each expression.

(a) $2xy - 3$

(b) $x^2 - 4x + 1$

(c) $(x + 2)(3 - 4y)$

(d) $x(x - 1)(x - 2)$

Solution

(a) $2xy - 3$ has two terms. They are $2xy$ and -3. (Subtracting 3 is equivalent to adding -3.)

(b) $x^2 - 4x + 1$ has three terms. They are x^2, $-4x$, and 1.

(c) $(x + 2)(3 - 4y)$ has two factors. They are $(x + 2)$ and $(3 - 4y)$.

(d) $x(x - 1)(x - 2)$ has three factors. They are x, $(x - 1)$, and $(x - 2)$. ■

We often refer to the *like terms* of an algebraic expression. Like terms must contain the same variables raised to identical powers. For example, we say that in the expression $3x + 4y - x + 5y$, $3x$ and $-x$ are like terms, and $4y$ and $5y$ are like terms.

Definition
Like terms of an algebraic expression are terms that contain identical variables raised to identical powers.

EXAMPLE 2

In each expression, name the like terms.

(a) $3x^2 + x - 1 + 4x^2$

(b) $3x + 4y - 3y - 1$

(c) $4ab + 2bc + 7ab$

(d) $3a^2b + 3ab^2 - a^2b$

Solution

(a) $3x^2$ and $4x^2$ are like terms because both contain x to the second power.

(b) $4y$ and $-3y$ are like terms because both contain y to the first power.

(c) $4ab$ and $7ab$ are like terms because both contain ab.

(d) $3a^2b$ and $-a^2b$ are like terms because both contain a^2 and b. ■

You may use the distributive property to simplify an addition or subtraction problem. For example, $3x + 4x = (3 + 4)x = 7x$. As a shortcut, we say that since $3x$ and $4x$ are like terms, they may be combined to yield $7x$.

To add or subtract algebraic expressions:
1. Locate like terms.
2. Add or subtract as indicated.
3. The sum will have the same variables and exponents as the terms.

EXAMPLE 3

Add the following algebraic expressions.

(a) $3x + 4y - x + 5y$

(b) $3x^2 + x - 1 + 4x$

(c) $4ab + 2bc + 7ab$

(d) $3a^2b + 3ab - a^2b$

Solution

(a) $3x + 4y - x + 5y = 3x - x + 4y + 5y = 2x + 9y$

(b) $3x^2 + x - 1 + 4x = 3x^2 + x + 4x - 1 = 3x^2 + 5x - 1$

(c) $4ab + 2bc + 7ab = 4ab + 7ab + 2bc = 11ab + 2bc$

(d) $3a^2b + 3ab - a^2b = 3a^2b - a^2b + 3ab = 2a^2b + 3ab$ ■

Addition and subtraction problems often involve grouping symbols. We follow the standard order of operations, performing the operations within grouping symbols first *if possible*. In algebraic expressions, however, unless the terms enclosed by grouping symbols are like terms, we cannot combine them. We may have to reorder and regroup, as we did in Example 3, to put like terms together.

EXAMPLE 4

Add the following algebraic expressions:

(a) $(x^2 + x - 1) + (2x^2 + 3x - 4)$

(b) $(2x + y - 1) + (3y - 4z + 2)$

Solution

(a)   Since the operations within parentheses cannot be performed, we reorder and regroup to put like terms together: $x^2 + 2x^2 + x + 3x - 1 - 4 = 3x^2 + 4x - 5$.

(b)   Since the operations within parentheses cannot be performed, we reorder and regroup to put like terms together: $2x + y + 3y - 4z - 1 + 2 = 2x + 4y - 4z + 1$. ■

In a subtraction problem, we may need to use the distributive property before we combine like terms. That is, an expression like $-(x + 3)$ can be written as $-1(x + 3)$ and is therefore equal to $-x - 3$.

EXAMPLE 5

Subtract the following algebraic expressions:

(a) $(x^2 + x - 1) - (2x^2 + 3x + 4)$

(b) $(2x + y - 1) - (3y - 4z + 2)$

Solution

(a)   Since the operations within parentheses cannot be performed, we use the distributive property: $x^2 + x - 1 - 2x^2 - 3x - 4$. Then we can combine like terms to get $- x^2 - 2x - 5$.

(b)   Use the distributive property: $2x + y - 1 - 3y + 4z - 2$. Combine like terms to get $2x - 2y + 4z - 3$. ■

Multiplication of algebraic expressions is based on the distributive property. For example, $2(x + 3) = 2x + 6$. When both of the factors have two or more terms, each term of the first factor distributes over the second factor. For example,

$$(x + 2)(x - 3) = x(x - 3) + 2(x - 3)$$
$$= x^2 - 3x + 2x - 6$$
$$= x^2 - x - 6.$$

In long problems especially, it is helpful to collect like terms as we do the multiplication. We might write $(x + 2)(x - 3)$ as

$$x^2 - 3x$$
$$\underline{+ 2x - 6}$$
$$x^2 - x - 6.$$

> **To multiply two algebraic expressions:**
> 1. Multiply each term of the first factor by the second factor using the distributive property. Collect like terms in columns as the multiplication is done.
> 2. Combine like terms.

EXAMPLE 6

Multiply $(x - 5)(x + 4)$.

Solution

$$(x - 5)(x + 4) = x^2 + 4x \qquad \text{from } x(x + 4)$$
$$\underline{- 5x - 20} \quad \text{from } -5(x + 4)$$
$$x^2 - x - 20 \quad \text{by combining like terms} \ \blacksquare$$

EXAMPLE 7

Multiply $(2x - 1)(3x + 1)$.

Solution

$$(2x - 1)(3x + 1) = 6x^2 + 2x \qquad \text{from } 2x(3x + 1)$$
$$\underline{- 3x - 1} \quad \text{from } -1(3x + 1)$$
$$6x^2 - \ x - 1 \quad \text{by combining like terms} \ \blacksquare$$

EXAMPLE 8

Multiply $(x + 2)(x^2 - x + 1)$

Solution

$$(x + 2)(x^2 - x + 1) = x^3 - x^2 + x \qquad \text{from } x(x^2 - x + 1)$$
$$\underline{+ 2x^2 - 2x + 2} \quad \text{from } 2(x^2 - x + 1)$$
$$x^3 + x^2 - x + 2 \quad \text{by combining like terms}$$

x's are in one column

x^2's are in one column $\ \blacksquare$

EXAMPLE 9   Multiply $(x^2 + x - 1)(x^2 - x + 1)$.

Solution

$$(x^2 + x - 1)(x^2 - x + 1) = x^4 - x^3 + x^2 \qquad \begin{array}{l} \text{from} \\ x^2(x^2 - x + 1) \end{array}$$

$$+ \; x^3 - x^2 + \quad x \qquad \begin{array}{l} \text{from} \\ x(x^2 - x + 1) \end{array}$$

$$\underline{\qquad\qquad - x^2 + \quad x - 1 \quad} \begin{array}{l} \text{from} \\ -1(x^2 - x + 1) \end{array}$$

$$x^4 \qquad\qquad - x^2 + 2x - 1 \quad \begin{array}{l} \text{by combining} \\ \text{like terms} \end{array} \blacksquare$$

EXAMPLE 10   Multiply $(x + 2)^3$.

Solution

$$(x + 2)^3 = (x + 2)(x + 2)(x + 2)$$
$$(x + 2)[(x + 2)(x + 2)] \qquad \text{by grouping the last factors}$$
$$(x + 2)[x^2 + 2x + 2x + 4] \quad \text{from the multiplication}$$
$$(x + 2)[x^2 + 4x + 4]$$
$$x^3 + 4x^2 + \quad 4x \qquad\qquad \text{from } x(x^2 + 4x + 4)$$
$$\underline{\quad + 2x^2 + \quad 8x + 8 \quad} \qquad \text{from } 2(x^2 + 4x + 4)$$
$$x^3 + 6x^2 + 12x + 8 \qquad \text{by combining like terms} \quad \blacksquare$$

For longer problems, we follow the standard order of operations.

EXAMPLE 11   Simplify each expression:

(a) $2(3x + 3) + (3x - 1)$

(b) $x(2x + 3) - (3x - 1)$

(c) $(2x + 3)(3x - 1) + 1$

Solution

(a) $2(3x + 3) + (3x - 1)$ since operations within parentheses cannot be done

$\qquad\quad 6x + 6 + 3x - 1$ from the distributive property

$\qquad\quad 6x + 3x + 6 - 1$ by reordering and regrouping like terms

$\qquad\qquad\quad 9x + 5$ by combining like terms

(b) $x(2x + 3) - (3x - 1)$ since operations within parentheses cannot be done

$\qquad\quad 2x^2 + 3x - 3x + 1$ from the distributive property

$\qquad\quad 2x^2 + 3x - 3x + 1$ by grouping like terms

$\qquad\quad 2x^2 + 1$ by combining like terms

(c) $(2x + 3)(3x - 1) + 1$ since operations within parentheses cannot be done

$6x^2 - 2x + 9x - 3 + 1$ from $(2x + 3)(3x - 1)$

$6x^2 - 2x + 9x - 3 + 1$ by grouping like terms

$6x^2 + 7x - 2$ by combining like terms ■

5.3 EXERCISES

Determine the number of terms or factors in each expression. See Example 1.

1. $(x + 1)(x - 3)$

2. $4(y - 5)$

3. $2x + 3y$

4. $4a + 3$

5. $7x^2 + 3x + 2$

6. $(5a - 2)(7a^2 + 3a + 2)$

7. $(4y + 3)(y^2 - y - 5)$

8. $(b + 1)(b - 2)(b - 3)$

9. $(x - 1)(x + 2) - (x + 1)(x - 2)$

10. $x(2x - 3) + x(3x + 1)$

Add or subtract as indicated. See Examples 4 and 5.

11. $(2x + 3y) + (x - y)$

12. $(3a + 2b) - (a - b)$

13. $(3x - 1) - (2 - y)$

14. $(7a + 3) + (b - 2)$

15. $(x^2 + 4x - 1) + (3x^2 - 5x + 2)$

16. $(2y^2 + 5y - 3) - (2y^2 - 4y + 1)$

17. $(3a^2 + 7a - 4) - (2a - 1)$

18. $(3x + 4y - z) + (2x - y + w)$

19. $(2x - 1) - (x - 4)$

20. $(3z^2 + 2z - 1) - (z^2 - z - 4)$

Multiply. See Examples 6–10.

21. $2x(x - 1)$

22. $3a(2a + 5)$

23. $3x(x^2 - x + 4)$

24. $y^3(2y^2 + y - 3)$

25. $(x + 1)(2x - 3)$

26. $(2a + 3b)(4a + 3b)$

27. $(3y - 4)(5y - 1)$

28. $(x + 1)(x - 1)$

29. $(x - 5y)(x - 2y)$

30. $(4a + b)(4a + b)$

31. $(2x - 3)(3x^2 - 2x + 1)$

32. $(z - 1)(z^2 + z + 1)$

33. $(x + 3)(x^2 - 3x + 9)$

34. $(2y + 1)(4y^2 + 4y + 1)$

35. $(a - 2)(3a^2 + 4a - 5)$

36. $(x^2 + x + 1)(x^2 - x + 1)$

37. $(y^2 - 2y + 1)(y^2 + 3y - 2)$

38. $(2b^2 - b + 3)(b^2 + 3b - 2)$

39. $(2x + 1)^3$

40. $(3x - 2y)^3$

Simplify. See Example 11.

41. $x(x - 2) + (3x - 1)$

42. $(2x + 3) - 3(x - 2)$

43. $(x^2 + 2x - 1) - 3(x + 2)$

44. $(x^2 + 2x - 1) - x(x + 2)$

45. $(x + 1)(x - 1) - 3$

46. $(x + 2)(x + 1) - x$

47. $(2x + 1)(3x - 2) + (x - 4)$

48. $(2x - 1)(2x - 3) - x^2$

5.4   Factoring

In the problems of the last section, we began with expressions containing *factors* and rewrote them as expressions containing *terms*. In fact, the product often turned out to be a special type of algebraic expression called a *polynomial*. We will consider only polynomials in one variable.

> **Definition**
> A *polynomial* in one variable is an expression of the form $a_n x^n + a_{n-1} x^{n-1} + a_{n-2} x^{n-2} + \ldots + a_0$ where $a_n, a_{n-1}, a_{n-2}, \ldots a_0 \in R$ and $n \in I$, $n \geq 0$. We say that n is the *degree* of the polynomial and the values $a_n, a_{n-1}, a_{n-2}, \ldots a_0$ are the *coefficients*. The number a_n is called the *leading* coefficient.

In this section, we will begin with polynomials (i.e., expressions containing *terms*) and rewrite them as expressions containing *factors*. This process is called *factoring*. All of the polynomials will have integral coefficients, and the factors must have integral coefficients.

The first step in any factoring problem is to look for a common factor. If there is a factor common to *all* of the terms of the polynomial, we use the distributive property. In the polynomial $2x + 6$, we can think of 6 as $2(3)$. Then we can see that each term has a factor of 2, so we say that 2 is a common factor. That is, $2x + 6 = \mathbf{2}(x) + \mathbf{2}(3) = \mathbf{2}(x + 3)$ by the distributive property.

EXAMPLE 1

Factor out the common factor in each polynomial.

(a) $3x + 9$

(b) $2x^2 + 4x + 4$

(c) $6x^3 + 3x^2 - 9x$

Solution

(a) $3x + 9 = \mathbf{3}(x) + \mathbf{3}(3) = \mathbf{3}(x + 3)$

(b) $2x^2 + 4x + 4 = \mathbf{2}(x^2) + \mathbf{2}(2x) + \mathbf{2}(2) = \mathbf{2}(x^2 + 2x + 2)$

(c) $6x^3 + 3x^2 - 9x = \mathbf{3x}(2x^2) + \mathbf{3x}(x) + \mathbf{3x}(-3) = \mathbf{3x}(2x^2 + x - 3)$ ■

In the last section, the polynomial products were often the result of more than one application of the distributive property. In particular, *trinomials* (polynomials having three terms) resulted from two applications of the distributive property. The factors were *binomials* (polynomials with two terms). For example, $(x + 3)(x - 2) = x^2 + x - 6$.

Suppose our problem is to start with $x^2 + x - 6$ and factor it. There is no common factor, so we factor it as a product of two binomials. It is easy to see how we might factor the first term, x^2, as $x(x)$. It is also easy to see how we might factor the last term, -6, as $3(-2)$. Therefore we could "guess" that $(x + 3)$ and $(x - 2)$ are the two factors. We have not yet considered the middle term, however. If we check our guess, we see that the two factors produce not only the correct first and last terms, but also the correct middle term.

To factor a trinomial:

1. Factor out any common factor. To factor the remaining trinomial:
2. Factor the first term to get the first term of each factor.
3. Factor the last term to get the last term of each factor.
4. Insert a sign between the terms of each factor.
5. Check the middle term. If it is not correct, try again.

The process of factoring trinomials is often called "trial and error" factorization. As you gain experience with this type of problem, you will find that you make fewer false starts and often "guess" the correct factors on the first or second try. Refer to the following hints as you study the examples in this section.

Hints for Factoring Trinomials

a. Be systematic. If you aren't sure which factors to use at steps 2 or 3, choose the two with the smallest difference. If that combination does not work, then try the two with the next smallest difference, and so on.

b. Do not put two terms with a common factor in the same set of parentheses. If there is a common factor, it should have been factored out as the first step.

c. If the last term of the trinomial is positive, the two signs are alike and are the same as the sign of the middle term.

d. If the last term of the trinomial is negative, the two signs are different.

e. If the middle term is wrong only in sign, change both signs between terms of the factors.

In the following examples, we use these hints.

EXAMPLE 2

Factor completely $y^2 - 3y + 2$.

Solution   Since $y^2 = (y)(y)$, and $2 = (2)(1)$, we have $(y \quad 2)(y \quad 1)$. The signs are both negative (see hint c), so our first guess is $(y - 2)(y - 1)$, which is correct, since the middle term is $-y - 2y = -3y$. ∎

EXAMPLE 3

Factor completely $2x^2 + 9x + 4$.

Solution   We factor $2x^2$ as $(2x)(x)$ and 4 as $(1)(4)$. Although the last term has factors of 2 and 2, with a difference of 0, we know that $(2x + 2)(x + 2)$ will not work (see hint b). Our first guess is $(2x + 1)(x + 4)$, which is correct. ∎

EXAMPLE 4

Factor completely $8a^2 + 22a - 6$.

Solution   We start with $2(4a^2 + 11a - 3)$. We factor $4a^2$ as $(2a)(2a)$ and 3 as $(3)(1)$. Our first guess is $2(2a - 3)(2a + 1)$, but the middle term is $2a - 6a = -4a$, which is incorrect. There are no other factors of the last term to try. We factor $4a^2$ as $(a)(4a)$ and try $2(a + 3)(4a - 1)$. The middle term is $-a + 12a = 11a$, which is correct. ∎

EXAMPLE 5

Factor completely $6m^2 + 35m - 6$.

Solution   We factor $6m^2$ as $(2m)(3m)$ and 6 as $(3)(2)$, using factors with the smallest difference. Our first guess is $(2m + 3)(3m - 2)$, but the middle term is incorrect. It is not necessary to factor the last term as $1(6)$, since 6 cannot go in parentheses with either $2m$ or $3m$ (see hint b). Our second guess is $(6m + 1)(m - 6)$, but the middle term is $-36m + m = -35m$, which has the wrong sign. The correct answer is $(6m - 1)(m + 6)$ (see hint e). ■

Some binomials may be treated as trinomials with a zero middle term. For example, $x^2 - 9 = x^2 + 0x - 9$, and can be factored as $(x - 3)(x + 3)$.

EXAMPLE 6

Factor completely $4y^2 - 25$.

Solution Our first guess is $(2y + 5)(2y - 5)$, which is correct. ■

This binomial is called a *difference of squares* because $4y^2$ is the square of $2y$, 25 is the square of 5, and the subtraction indicates a difference. Any binomial which is a difference of squares will follow the pattern

$$a^2 - b^2 = (a - b)(a + b).$$

EXAMPLE 7

Factor completely $4p^2 - 100$.

Solution We start with $4(p^2 - 25)$. We try $4(p - 5)(p + 5)$, which is correct. ■

There are some polynomials that cannot be factored using integers as coefficients.

EXAMPLE 8

Factor completely, if possible, $x^2 - x + 1$.

Solution We try $(x - 1)(x - 1)$, but the middle term is incorrect. There are no other possibilities to try. We say that $x^2 - x + 1$ is *prime*, because there is neither a common integral factor (other than ± 1) nor are there nonconstant factors (other than $\pm(x^2 - x + 1)$). ■

EXAMPLE 9

Factor completely, if possible, $3x^2 + 2x - 6$.

Solution All possibilities are listed systematically. None of them are correct.
$(3x - 2)(x + 3)$, but the middle term is $7x$.
(It is not necessary to try $(3x + 2)(x - 3)$. See hint e.)
$(3x - 1)(x + 6)$, but the middle term is $17x$.
(It is not necessary to try $(3x + 1)(x - 6)$. See hint e.)
We conclude that $3x^2 + 2x - 6$ is prime. ■

5.4 EXERCISES

Factor completely, if possible. See Examples 1–9.

1. $2x^2 + 2x$

2. $y^4 + 3y^3$

3. $3m^3 - 3m^2 + 6m$

4. $5p^6 - 15p^3$

5. $x^2 + 5x + 6$

6. $a^2 - 7a + 12$

7. $z^2 - z - 6$

8. $x^2 + x - 12$

9. $p^2 + 10p + 24$

10. $m^2 - 2m - 24$

11. $y^2 - 11y + 24$

12. $x^2 + 9x + 7$

13. $6z^2 - 13z + 6$

14. $6a^2 - 13a - 5$

15. $5x^2 + 26x + 5$

16. $3m^2 - 5m + 2$

17. $12p^2 + 7p - 12$

18. $12y^2 + 18y - 12$

19. $3x^2 - 12x + 12$

20. $5a^2 + 10a + 5$

21. $z^2 + 5z - 24$

22. $16y^2 - 25$

23. $4m^2 - 16$

24. $x^2 + 9$

25. $25p^2 + 36$

26. $3a^3 + 9a$

27. $4y^3 + 8y^2$

28. $21x^5 + 7x^3$

29. $p^2 + 9p + 20$

30. $z^2 - z - 20$

31. $m^2 - 8m + 15$

32. $m^2 + 2m - 15$

33. $2a^2 - 24a + 40$

34. $3x^2 - 48x + 45$

35. $6a^2 - 7a - 5$

36. $3y^2 + 10y + 8$

37. $2m^2 - 3m - 9$

38. $6x^2 - 11x + 4$

39. $m^2 + 3m + 1$

40. $y^2 - 5y - 12$

41. $9x^2 - 4$

42. $4x^2 + 9$

✳**43.** Walk through the following algorithm using the polynomials in problems 10, 13, and 16.

  1. Factor out any common factor.
  To factor the remaining trinomial:
  2. Use the coefficient of the leading term as the leading coefficient of *each* of the binomial factors.
  3. Multiply the last term of the trinomial by the leading coefficient. Choose two factors of this product that have the middle term as their sum. Use these factors as the final terms of the two binomial factors.
  4. Remove and discard any common factors from the binomial factors—e.g., $8a^2 + 22a - 6 = 2(4a^2 + 11a - 3)$. Since $4(-3) = -12$, we are looking for factors of -12 that have a sum of $+11$. We use $12(-1)$. But $2(4a + 12)(4a - 1)$ is not correct. Removing the common factor of 4, we have $2(a + 3)(4a - 1)$.

5.5   Rational Expressions

In Chapter 2, a rational *number* was defined as any number that can be written as the ratio of two integers, as long as the denominator is not zero. A rational *expression* is defined similarly.

Definition
A *rational expression* is a ratio of two polynomials, such that the denominator is not equal to zero.

Rational expressions have much in common with rational numbers in fraction form. To reduce a fraction (or rational expression), we divide both numerator and denominator by all of the factors they have in common. When a fraction (or rational expression) cannot be reduced, we say it is in *lowest terms*.

First we examine these concepts as they apply to fractions. One way to reduce fractions is to write both numerator and denominator as products of prime numbers. A *prime number* is an integer greater than 1 such that the only positive integers that divide it evenly are 1 and the number itself. When the numbers involved are relatively small, it is probably unnecessary to go through these steps. But when the numbers are large, the method is a convenient means of finding all of the factors that divide both numerator and denominator.

EXAMPLE 1   Reduce $\dfrac{143}{195}$.

Solution   Factor the numerator and denominator: $143 = 11(13)$ and $195 = 3(65)$, but 65 is not a prime number. Replacing 65 by $5(13)$, we have $195 = 3(5)(13)$. We divide numerator and denominator by 13. That is $13 \div 13 = 1$. Thus,

$$\frac{143}{195} = \frac{11 \cdot 13}{3 \cdot 65} = \frac{11 \cdot \overset{1}{\cancel{13}}}{3 \cdot 5 \cdot \underset{1}{\cancel{13}}} = \frac{11}{15}. \quad \blacksquare$$

EXAMPLE 2   Reduce $\dfrac{240}{200}$.

Solution

$$\frac{240}{200} = \frac{10 \cdot 24}{10 \cdot 20} = \frac{2 \cdot 5 \cdot 3 \cdot 8}{2 \cdot 5 \cdot 4 \cdot 5} = \frac{2 \cdot 5 \cdot 3 \cdot 2 \cdot 4}{2 \cdot 5 \cdot 2 \cdot 2 \cdot 5} = \frac{\overset{1}{\cancel{2}}\,\overset{1}{\cancel{5}}\,3\,\overset{1}{\cancel{2}}\,\overset{1}{\cancel{2}}\,2}{\underset{1}{\cancel{2}}\,\underset{1}{\cancel{5}}\,\underset{1}{\cancel{2}}\,\underset{1}{\cancel{2}}\,5} = \frac{6}{5}. \quad \blacksquare$$

The process of cancellation used in the preceding examples is based on the fact that $a/a = 1$ if $a \in R$, and $a \neq 0$. Since 1 is the identity for multiplication, it has no effect on the remaining factors. That is, by cancelling identical factors in numerator and denominator, we are, in effect, removing a factor of 1 from the fraction. It is essential that you cancel *factors* of the numerator and denominator rather than *terms*, since 1 is *not* the identity for addition. The same method is used to reduce rational expressions to lowest terms.

> **To reduce a rational expression:**
> 1. Factor the numerator and denominator completely.
> 2. Divide out *factors* that are identical in both the numerator and denominator.
> 3. Multiply the remaining factors of the numerator to get the numerator of the reduced expression. Multiply the remaining factors of the denominator to get the denominator of the reduced expression.
> 4. In this chapter, you may assume that the denominators are not equal to zero.

EXAMPLE 3

Reduce $\dfrac{x^2 - 6x + 9}{x^2 - 9}$.

Solution   Factor numerator and denominator: $\dfrac{(x - 3)(x - 3)}{(x - 3)(x + 3)}$.

Divide out identical factors: $\dfrac{\cancel{(x - 3)}^1(x - 3)}{\cancel{(x - 3)}_1(x + 3)} = \dfrac{(x - 3)}{(x + 3)}$. ∎

EXAMPLE 4

Reduce $\dfrac{y^2 - 2y}{2y - 4}$.

Solution   Factor both numerator and denominator: $\dfrac{y(y - 2)}{2(y - 2)}$.

Divide out identical factors: $\dfrac{y\cancel{(y - 2)}^1}{2\cancel{(y - 2)}_1} = \dfrac{y}{2}$. ∎

EXAMPLE 5

Reduce $\dfrac{6x^2}{3x}$.

Solution $6x^2 = 2(3)(x)(x)$, and 3 and x are factors.

Divide out identical factors: $\dfrac{6x^2}{3x} = \dfrac{\cancel{3}^1 \cdot 2 \cdot \cancel{x}^1 \cdot x}{\cancel{3}_1 \cdot \cancel{x}_1} = 2x$. ∎

EXAMPLE 6

Reduce $\dfrac{z^2 + 6}{z + 3}$.

Solution $z^2 + 6$ is prime, and $z + 3$ is prime. Since the numerator and denominator each consist of two *terms,* we cannot reduce. That is, $\dfrac{z^2 + 6}{z + 3}$ is in lowest terms. ■

When multiplying rational numbers, you may multiply and then reduce; or you may reduce and then multiply. The second procedure is usually easier, because the numbers are smaller.

EXAMPLE 7

Multiply $\dfrac{3}{4}\left(\dfrac{8}{9}\right)$.

Solution   If we multiply, then reduce, we have $\dfrac{24}{36} = \dfrac{2}{3}$.

If we reduce, then multiply, we have $\dfrac{\overset{1}{\cancel{3}}}{\underset{1}{\cancel{4}}}\left(\dfrac{\overset{2}{\cancel{8}}}{\underset{3}{\cancel{9}}}\right) = \dfrac{2}{3}$. ■

EXAMPLE 8

Multiply $\dfrac{3}{4}\left(\dfrac{4}{5}\right)\left(\dfrac{5}{6}\right)\left(\dfrac{6}{7}\right)$.

Solution   If we multiply, then reduce, we have $\dfrac{360}{840} = \dfrac{3}{7}$.

If we reduce, then multiply, we have $\dfrac{3}{\underset{1}{\cancel{4}}}\left(\dfrac{\overset{1}{\cancel{4}}}{\underset{1}{\cancel{5}}}\right)\left(\dfrac{\overset{1}{\cancel{5}}}{\underset{1}{\cancel{6}}}\right)\left(\dfrac{\overset{1}{\cancel{6}}}{7}\right) = \dfrac{3}{7}$. ■

When rational expressions are multiplied, the difference in difficulty between the procedures is even more pronounced. It is much easier to factor both numerator and denominator completely, reduce, and then multiply.

> **To multiply rational expressions:**
> 1. Factor all the numerators and denominators completely.
> 2. Divide out any *factors* that are identical in both a numerator and a denominator. (They need not be in the same fraction.)
> 3. Multiply the remaining factors of the numerators to get the numerator of the product; multiply the remaining factors of the denominators to get the denominator of the product.

EXAMPLE 9

Multiply $\dfrac{x + 1}{x - 2} \left(\dfrac{x^2 - 4}{x^2 + 2x + 1} \right)$.

Solution Factor numerators and denominators:

$$\frac{(x + 1)}{(x - 2)} \frac{(x - 2)(x + 2)}{(x + 1)(x + 1)}.$$

Divide out factors that are identical in both a numerator and a denominator:

$$\frac{\overset{1}{\cancel{(x + 1)}}}{\underset{1}{\cancel{(x - 2)}}} \frac{\overset{1}{\cancel{(x - 2)}}(x + 2)}{\underset{1}{\cancel{(x + 1)}}(x + 1)} = \frac{x + 2}{x + 1}. \blacksquare$$

EXAMPLE 10

Multiply $\dfrac{m^2 - m}{2} \left(\dfrac{m + 1}{m} \right)$.

Solution Factor and reduce:

$$\frac{m(m - 1)}{2} \frac{(m + 1)}{m} = \frac{\overset{1}{\cancel{m}}(m - 1)}{2} \frac{(m + 1)}{\underset{1}{\cancel{m}}} = \frac{(m - 1)(m + 1)}{2} =$$

$$\frac{m^2 - 1}{2} \blacksquare$$

EXAMPLE 11

Multiply $\dfrac{6a^2}{4a^2 - 1} \left(\dfrac{2a - 1}{3a} \right)$.

Solution Factor and reduce:

$$\frac{2a(3a)}{(2a - 1)(2a + 1)} \frac{2a - 1}{3a} = \frac{2a\overset{1}{\cancel{(3a)}}}{\underset{1}{\cancel{(2a - 1)}}(2a + 1)} \frac{\overset{1}{\cancel{(2a - 1)}}}{\underset{1}{\cancel{3a}}} = \frac{2a}{(2a + 1)} \blacksquare$$

When we divide rational *numbers,* we use the rule, "invert the divisor and multiply." The rule also works for rational *expressions.*

> **To divide rational expressions:**
> 1. Invert the divisor.
> 2. Multiply as rational expressions.

EXAMPLE 12   Divide: $\dfrac{1}{2x + 3} \div \dfrac{5}{4x^2 - 9}$.

Solution   Invert the divisor to get: $\dfrac{1}{2x + 3}\left(\dfrac{4x^2 - 9}{5}\right)$.

Factor and reduce: $\dfrac{1}{\cancel{(2x + 3)}} \dfrac{(2x - 3)\cancel{(2x + 3)}}{5} = \dfrac{2x - 3}{5}$ ∎

EXAMPLE 13   Divide: $\dfrac{2p - 3}{3p + 2} \div \dfrac{2p + 3}{3p - 2}$.

Solution   Invert the divisor to get: $\dfrac{(2p - 3)}{(3p + 2)} \dfrac{(3p - 2)}{(2p + 3)}$.

The numerators and denominators are completely factored, and there are no factors which are identical in both a numerator and a denominator. Therefore the fraction cannot be reduced. The answer is

$$\frac{6p^2 - 4p - 9p + 6}{6p^2 + 9p + 4p + 6} = \frac{6p^2 - 13p + 6}{6p^2 + 13p + 6}.$$ ∎

5.5 EXERCISES

Reduce to lowest terms. See Examples 1 and 2.

1. $\dfrac{140}{210}$ **2.** $\dfrac{252}{315}$ **3.** $\dfrac{60}{225}$ **4.** $\dfrac{80}{96}$ **5.** $\dfrac{108}{324}$

Reduce to lowest terms. See Examples 3–6.

6. $\dfrac{x^2 + x - 2}{x^2 + 2x - 3}$ **7.** $\dfrac{y^2 + 5y + 6}{y^2 - y - 6}$ **8.** $\dfrac{m^2 - m}{m^2 + m}$

9. $\dfrac{x^3 - 4x}{x^3 + x^2 - 6x}$ **10.** $\dfrac{p^2 - 8p + 7}{p^2 - 49}$ **11.** $\dfrac{2z^2 - 5z - 12}{2z^2 - 7z - 15}$

12. $\dfrac{2x^2 - 3x - 2}{2x^2 - 5x - 3}$ **13.** $\dfrac{3y^2 + 4y - 4}{3y^2 + y - 2}$ **14.** $\dfrac{a^2 - 1}{a^2 + 1}$

Multiply. See Examples 7 and 8.

15. $\dfrac{15}{14}\left(\dfrac{12}{35}\right)$ **16.** $\dfrac{15}{14}\left(\dfrac{26}{33}\right)$ **17.** $\dfrac{3}{7}\left(\dfrac{5}{9}\right)\left(\dfrac{7}{10}\right)$

18. $\dfrac{2}{15}\left(\dfrac{27}{14}\right)\left(\dfrac{35}{54}\right)$ **19.** $\dfrac{6}{35}\left(\dfrac{7}{36}\right)\left(30\right)$

Multiply. See Examples 9–11.

20. $\dfrac{x-1}{x+2}\left(\dfrac{x^2-4}{x^2-1}\right)$

21. $\dfrac{y^2-y-2}{y^2+2y-3}\left(\dfrac{y^2+y-2}{y^2+3y+2}\right)$

22. $\dfrac{z-5}{4z^2-1}\left(\dfrac{2z+1}{2z^2-10z}\right)$

23. $\dfrac{m+2}{m^2-m-6}\left(\dfrac{m^2+6m+5}{m+1}\right)$

24. $\dfrac{x^2-5x+6}{x}\left(\dfrac{x}{x^2-3x+2}\right)$

25. $\dfrac{a^2-3a}{a-2}\left(\dfrac{a^2-3a+2}{a^3+4a^2+3a}\right)$

26. $\dfrac{x^2+5x+6}{x^2+5x+4}\left(\dfrac{x+1}{x+2}\right)$

27. $\dfrac{p+3}{p+1}\left(\dfrac{p^2-2p-3}{p^2-9}\right)$

28. $\dfrac{4y^2-1}{y+2}\left(\dfrac{y^2+3y+2}{4y^2+4y+1}\right)$

29. $\dfrac{2x^2-5x-3}{x+3}\left(\dfrac{x+3}{2x^2+9x+4}\right)$

Divide. See Examples 12 and 13.

30. $\dfrac{x+1}{x^2-49}\div\dfrac{x^2-1}{x-7}$

31. $\dfrac{2z+1}{3}\div\dfrac{8z+4}{9}$

32. $\dfrac{9y^2-3y}{2y+2}\div\dfrac{18y-6}{y^2+2y}$

33. $\dfrac{m^2-2m}{m+1}\div\dfrac{m-2}{m^2+m}$

34. $\dfrac{a}{a^2-4}\div\dfrac{a+2}{a^2-4a+4}$

35. $\dfrac{x-2}{x+3}\div\dfrac{x^2-4x+4}{x^2+4x+3}$

36. $\dfrac{x+1}{x-2}\div\dfrac{x^2-1}{x^2-3x+2}$

✳**37.** Consider the fraction $\dfrac{64}{16}$. A student who does not understand the process of cancellation writes $\dfrac{6\!\!\!/4}{1\!\!\!/6}=4$, which is the correct answer.

   (a) Give two examples that illustrate that this method does not, in general, give the correct answer.

   (b) Find another example for which it does give the correct answer.

✳**38.** In the mid-17th century, the English mathematician John Wallis discovered that
$$\frac{\pi}{2}=\frac{2}{1}\cdot\frac{2}{3}\cdot\frac{4}{3}\cdot\frac{4}{5}\cdot\frac{6}{5}\cdot\frac{6}{7}\cdot\frac{8}{7}\cdot\frac{8}{9}\cdots.$$
Write a formula for the *n*th fraction in this product.

▰▶**39.** Use pseudocode or a flowchart to show an algorithm for computing π using the first 100 factors of the formula in problem 38.

5.6 Addition and Subtraction of Rational Expressions

Before looking at addition and subtraction of rational expressions, we will review addition and subtraction of rational numbers. We will find the lowest common denominator (abbreviated LCD) of two or more fractions in order to add or subtract them.

> **Definition**
> The lowest common denominator of two or more fractions is the smallest positive number that is evenly divisible by each of their denominators.

For the fractions $\dfrac{2}{3}$ and $\dfrac{3}{4}$, the LCD is 12. We have $\dfrac{2}{3} + \dfrac{3}{4} = \dfrac{2}{3}\left(\dfrac{\mathbf{4}}{\mathbf{4}}\right) +$

$\dfrac{3}{4}\left(\dfrac{\mathbf{3}}{\mathbf{3}}\right) = \dfrac{8}{12} + \dfrac{9}{12} = \dfrac{17}{12}$. When the numbers involved are larger, however, it

is not as easy to see immediately what the LCD should be. Once again, factoring will help, and this procedure is the one we will use with rational expressions as well.

Consider the problem $\dfrac{5}{6} + \dfrac{2}{15} + \dfrac{4}{35}$. If we factor the denominators, we

have $\dfrac{5}{2 \cdot 3} + \dfrac{2}{3 \cdot 5} + \dfrac{4}{5 \cdot 7}$. The LCD must contain at least the factors $2 \cdot 3$

for 6 to divide it evenly. Furthermore, it must contain at least the factors $3 \cdot 5$ for 15 to divide it. But we don't need $2 \cdot 3 \cdot 3 \cdot 5$, since $2 \cdot 3 \cdot 5$ is divisible by both $2 \cdot 3$ and $3 \cdot 5$. That is, we do not have to repeat the factor of 3, since it is contained in two different denominators. One factor of 3 can be used for both. Finally, the LCD must contain at least the factors $5 \cdot 7$ for 35 to divide it. We have already determined that the LCD must have a factor of 5; therefore we need only introduce a factor of 7 from the last denominator. The LCD, then, would be $2 \cdot 3 \cdot 5 \cdot 7$, or 210.

The next step is to change each of the fractions to a fraction having 210 as the denominator. We have:

$$\frac{5}{2 \cdot 3} + \frac{2}{3 \cdot 5} + \frac{4}{5 \cdot 7} = \frac{?}{2 \cdot 3 \cdot 5 \cdot 7} + \frac{?}{2 \cdot 3 \cdot 5 \cdot 7} + \frac{?}{2 \cdot 3 \cdot 5 \cdot 7}.$$

It is convenient to work with the factored form, because it is immediately obvious what numbers are needed as factors in order to change the form of each fraction.

In the first fraction, we want to change the denominator from 2 · 3 to 2 · 3 · **5** · **7**. Therefore we multiply both numerator and denominator by **5** · **7**. We have:

$$\frac{5}{2\cdot3}\frac{(5\cdot7)}{(5\cdot7)} = \frac{175}{2\cdot3\cdot5\cdot7}.$$

In the second fraction, we want to change the denominator from 3 · 5 to 2 · 3 · 5 · 7. Therefore we multiply both numerator and denominator by **2** · **7**. We have:

$$\frac{2}{3\cdot5}\frac{(2\cdot7)}{(2\cdot7)} = \frac{28}{2\cdot3\cdot5\cdot7}.$$

In the third fraction, we want to change the denominator from 5 · 7 to 2 · 3 · 5 · 7. Therefore we multiply both numerator and denominator by **2** · **3**. We have:

$$\frac{4}{5\cdot7}\frac{(2\cdot3)}{(2\cdot3)} = \frac{24}{2\cdot3\cdot5\cdot7}.$$

Thus, $\frac{5}{6} + \frac{2}{15} + \frac{4}{35} = \frac{175}{210} + \frac{28}{210} + \frac{24}{210} = \frac{227}{210}$. Only one task remains: to reduce to lowest terms if possible. The factored form is again useful, because we know that if the fraction will reduce, it is because one or more of the factors 2, 3, 5, and 7 will divide out. We need only ask if 227 is divisible by any of these factors. It is not. The answer, then, is $\frac{227}{210}$ or $1\frac{17}{210}$.

You should examine the next example to be sure that you understand the principles of adding rational *numbers* before we consider adding rational *expressions*.

EXAMPLE 1

Add $\frac{1}{4} - \frac{5}{6} + \frac{7}{9}$.

Solution   Factor the denominators: $\frac{1}{2\cdot2} - \frac{5}{2\cdot3} + \frac{7}{3\cdot3}$.

Find the LCD:

From the first fraction: We need the factors 2 · 2 for the LCD.

From the second fraction: We need only the 3, since we already have a 2 in the LCD. That is, 3 is the only "new" factor.

From the third fraction: We already have one 3, but the other 3 is a "new" factor, so it goes in the LCD. Thus, the LCD is 2 · 2 · 3 · 3 = 36.

Change the fractions so that they all have 36 as the denominator.

$$\frac{1}{2 \cdot 2} - \frac{5}{2 \cdot 3} + \frac{7}{3 \cdot 3} = \frac{?}{2 \cdot 2 \cdot 3 \cdot 3} - \frac{?}{2 \cdot 2 \cdot 3 \cdot 3} + \frac{?}{2 \cdot 2 \cdot 3 \cdot 3}$$

$$= \frac{1 \cdot 3 \cdot 3}{2 \cdot 2 \cdot 3 \cdot 3} - \frac{5 \cdot 2 \cdot 3}{2 \cdot 3 \cdot 2 \cdot 3} + \frac{7 \cdot 2 \cdot 2}{3 \cdot 3 \cdot 2 \cdot 2} = \frac{9 - 30 + 28}{2 \cdot 2 \cdot 3 \cdot 3} = \frac{7}{36}. \blacksquare$$

We now apply the same principles to rational expressions.

To add or subtract rational expressions:

1. Factor each denominator.
2. Find the LCD.
   a. Use all factors from the denominator of the first fraction.
   b. Use any *new* factors from the other denominators.
3. Multiply the numerator and denominator of each fraction by the factors of the LCD that were not originally in its denominator.
4. For the answer, the numerator is the sum (or difference) of the numerators; the denominator is the LCD.
5. Factor the numerator and reduce, if possible.

EXAMPLE 2

Add $\dfrac{a}{bc} + \dfrac{b}{ac} + \dfrac{c}{ab}$.

Solution

1. Each denominator already contains factors.
2. The LCD has factors of b and c from the first fraction; a from the second. The LCD is abc.
3. We have $\dfrac{a\,\boldsymbol{a}}{bc\,\boldsymbol{a}} + \dfrac{b\,\boldsymbol{b}}{ac\,\boldsymbol{b}} + \dfrac{c\,\boldsymbol{c}}{ab\,\boldsymbol{c}}$.
4. Thus the sum is $\dfrac{a^2 + b^2 + c^2}{abc}$.
5. Since the numerator is prime, the fraction cannot be reduced. $\blacksquare$

EXAMPLE 3

Add $\dfrac{x + 2}{x - 3} + \dfrac{x + 3}{x - 2}$.

Solution

1. The denominators cannot be factored.
2. The LCD is $(x - 3)(x - 2)$.
3. We have $\dfrac{(x + 2)(\boldsymbol{x - 2})}{(x - 3)(\boldsymbol{x - 2})} + \dfrac{(x + 3)(\boldsymbol{x - 3})}{(x - 2)(\boldsymbol{x - 3})}$

   or $\dfrac{x^2 - 4}{(x - 3)(x - 2)} + \dfrac{x^2 - 9}{(x - 2)(x - 3)}$.

4. The sum is $\dfrac{2x^2 - 13}{(x - 3)(x - 2)}$.

5. Since $2x^2 - 13$ is prime, the fraction will not reduce.

   The answer is $\dfrac{2x^2 - 13}{x^2 - 5x + 6}$. ∎

EXAMPLE 4

Subtract $\dfrac{1}{x^2 - 9} - \dfrac{x}{x^2 + 6x + 9}$.

Solution   1. Factor the denominators:

$$\frac{1}{(x - 3)(x + 3)} - \frac{x}{(x + 3)(x + 3)}.$$

2. The LCD is $(x - 3)(x + 3)\ (x + 3)$

$$\text{from first} \qquad \text{from second}$$

3. We have: $\dfrac{1 \qquad (x + 3)}{(x - 3)(x + 3)(x + 3)} - \dfrac{x \qquad (x - 3)}{(x + 3)(x + 3)(x - 3)}.$

4. The result is $\dfrac{x + 3 - x^2 + 3x}{(x - 3)(x + 3)(x + 3)} = \dfrac{-x^2 + 4x + 3}{(x - 3)(x + 3)(x + 3)}.$

5. The numerator is prime, so the fraction will not reduce. ∎

EXAMPLE 5

Add $\dfrac{1}{y + 1} + \dfrac{1}{y^2} + \dfrac{1}{y^2 + y}$.

Solution

1. Factor the denominators:

$$\frac{1}{y + 1} + \frac{1}{y \cdot y} + \frac{1}{y(y + 1)}.$$

2. The LCD is $(y + 1)(y)(y)$ or $y^2(y + 1)$.
3. We have:

$$\frac{1 \quad (y^2)}{(y + 1)\,(y^2)} + \frac{1 \quad (y + 1)}{y \cdot y\,(y + 1)} + \frac{1 \quad (y)}{y(y + 1)\,(y)} =$$

$$\frac{y^2}{y^2(y + 1)} + \frac{(y + 1)}{y^2(y + 1)} + \frac{y}{y^2(y + 1)}$$

4. The sum is $\dfrac{y^2 + y + 1 + y}{y^2(y + 1)} = \dfrac{y^2 + 2y + 1}{y^2(y + 1)}.$

5. Factor the numerator and reduce: $\dfrac{(y + 1)(y + 1)}{y^2(y + 1)} = \dfrac{(y + 1)}{y^2}.$ ∎

5.6 EXERCISES

Add or subtract as indicated. See Examples 2–5.

1. $\dfrac{a}{b} + \dfrac{b}{a}$

2. $\dfrac{c}{d} + \dfrac{1}{c}$

3. $\dfrac{p}{q} + \dfrac{q}{p^2}$

4. $\dfrac{x}{x+1} + \dfrac{2x}{x-1}$

5. $\dfrac{y}{y^2-1} + \dfrac{1}{y^2-2y+1}$

6. $\dfrac{z}{z^2-3z+2} + \dfrac{3}{z^2-4z+3}$

7. $\dfrac{3}{2x^2-3x} - \dfrac{x}{4x^2-9}$

8. $\dfrac{7}{6a^2-7a-3} - \dfrac{2}{3a^2+a}$

9. $\dfrac{y-1}{y+1} + \dfrac{y+1}{y-1}$

10. $\dfrac{2z+3}{z-3} + \dfrac{2z-3}{z+3}$

11. $\dfrac{p}{2p^2-2p} - \dfrac{1}{2p^2+2p}$

12. $\dfrac{7}{x^2-6x+9} + \dfrac{x}{x^2-x-6}$

13. $\dfrac{5m}{m^2-4} + \dfrac{3}{m-2}$

14. $\dfrac{5}{t^2-25} - \dfrac{t}{t^2+10t+25}$

15. $\dfrac{2x}{9x^2+6x+1} - \dfrac{1}{3x+1}$

16. $\dfrac{2}{y} + \dfrac{3}{y^2} - \dfrac{1}{2}$

17. $\dfrac{5}{z} - \dfrac{7}{z+1} + \dfrac{z}{z^2+z}$

18. $\dfrac{2}{m} + \dfrac{3}{2m+2} - \dfrac{m}{2m^2+2m}$

19. $\dfrac{a}{a+1} + \dfrac{2}{a-1} + \dfrac{3a}{a^2-1}$

20. $\dfrac{x}{x-3} + \dfrac{1}{x+1} - \dfrac{2x}{x^2-2x-3}$

21. $\dfrac{5}{m-4} - \dfrac{m}{m+2} + \dfrac{3}{m^2-2m-8}$

22. $\dfrac{y+2}{2y-1} + \dfrac{y-2}{2y+1} - \dfrac{5}{4y^2-1}$

23. $\dfrac{3}{z} + \dfrac{2z}{2z-1} - \dfrac{5z-2}{2z^2-z}$

24. $\dfrac{x}{3x+2} + \dfrac{1}{2x-3} - \dfrac{x+5}{6x^2-5x-6}$

✳**25.** Simplify: $\dfrac{x+y}{x-y} \left(\dfrac{y^{-1}-x^{-1}}{x^{-1}+y^{-1}} \right)$

✳**26.** Consider $1 + \left(1 + \dfrac{1}{2} + \dfrac{1}{6} + \dfrac{1}{24} + \dfrac{1}{120} + \dfrac{1}{720} + \dots\right)$.

  (a) Find a formula for the nth term in parentheses.

  (b) Find the sum (through the 4th term in parentheses) by adding the fractions using the lowest common denominator. Express the sum as a decimal rounded to seven places.

  (c) Express each fraction (through the 4th term in parentheses) as a decimal rounded to seven places, and add the decimals.

  (d) What is the difference between the answers to (a) and (b)?

✳**27.** 159/46 can be expressed as a "continued fraction" by writing $159/46 =$

$$3 + 21/46 = 3 + \cfrac{1}{\cfrac{46}{21}} = 3 + \cfrac{1}{2 + \cfrac{4}{21}} = 3 + \cfrac{1}{2 + \cfrac{1}{\cfrac{21}{4}}} = 3 + \cfrac{1}{2 + \cfrac{1}{5 + \cfrac{1}{4}}}.$$

Express 149/68 as a continued fraction.

5.7   Polynomials in Nested Form

Throughout this chapter we have been working with polynomials. In Pascal, there is no symbol for exponentiation. Even in languages such as FORTRAN and BASIC that do have exponentiation symbols, polynomials are often written without using them. You learned in Section 2.6 that it is better programming practice to write repeated multiplications rather than to use exponents, because the results are more accurate. Thus, we might write a polynomial like $3x^2 - 2x + 1$ as 3*X*X − 2*X + 1. We will call this form the *expanded* form.

Expressions written in this manner, however, are awkward. A more common procedure is to write the polynomial in *nested* form. For example, consider $3x^2 - 2x + 1$ again. There is no common factor, but the first two terms have a common factor of x. If we factor just the first part of the expression, we have $3x^2 - 2x + 1 = x(3x - 2) + 1$. (You should realize that we cannot say we have factored the polynomial.) Using computer notation, we write X*(X*3 − 2) + 1.

Consider a slightly longer example. The polynomial $3x^3 - 2x^2 + x - 1$ could be written as $x[3x^2 - 2x + 1] - 1$. But the expression within brackets can be rewritten as it was in the previous paragraph. Thus, we have $x[x(3x - 2) + 1] - 1$, or using computer notation, X*(X*(X*3 − 2) + 1) − 1.

The procedure for writing a polynomial in nested form can be generalized as an algorithm.

> **To write a polynomial with *n* terms in nested form:**
> 1. Factor the common factor out of the first $n - 1$ terms.
> 2. Ignore the last term in parentheses, and factor the common factor from the other terms in parentheses.
> 3. Repeat step 2 until there are only two terms in the innermost set of parentheses.

EXAMPLE 1

Write $x^2 + 2x - 4$ in nested form using computer notation.

Solution   We factor the first part: $x^2 + 2x - 4 = x(x + 2) - 4$. The expression within parentheses contains just two terms. Thus the nested form of $x^2 + 2x - 4$ is X*(X + 2) − 4. ■

EXAMPLE 2

Write $2x^3 + x^2 - 3x + 4$ in nested form using computer notation.

Solution   Factor the first part: $2x^3 + x^2 - 3x + 4 = x[2x^2 + x - 3] + 4$. Factor the first part inside brackets: $x[x(2x + 1) - 3] + 4$. Since the expression in parentheses contains just two terms, the polynomial is in nested form. Writing it in computer notation, we have X*(X*(X*2 + 1) − 3) + 4. ■

EXAMPLE 3

Write $x^4 + x^3 - x^2 + x - 1$ in nested form using computer notation.

Solution Factor the first part:
$x^4 + x^3 - x^2 + x - 1 = x\{x^3 + x^2 - x + 1\} - 1$. Factor the first part inside braces: $x\{x[x^2 + x - 1] + 1\} - 1$. Factor the first part inside brackets: $x\{x[x(x + 1) - 1] + 1\} - 1$. Since the expression in parentheses contains just two terms, the polynomial is in nested form. Writing it in computer notation, we have X*(X*(X*(X + 1) - 1) + 1) - 1. ■

The advantage of writing polynomials in nested form is that there are fewer multiplications to perform. When the number of operations in a computer problem is used as a measure of efficiency, generally only multiplications and divisions are counted. Additions and subtractions are done so much faster than multiplications and divisions that their effect can usually be ignored when comparing two procedures. Consider the polynomial $3x^2 - 2x + 1$. To simplify the *expanded* form 3*X*X - 2*X + 1 requires three multiplications, but to simplify the *nested* form, X*(X*3 - 2) + 1, requires only two multiplications. Using the nested form is more efficient.

EXAMPLE 4

Compare the number of operations required to evaluate $3x^4 - 2x^2 + x - 1$ in expanded form and in nested form.

Solution
The expanded form 3*X*X*X*X - 2*X*X + X - 1 requires six multiplications. The nested form X*(X*(X*(X*3) - 2) + 1) - 1 requires only four multiplications. ■

It is not necessary to write the expanded form in order to count the multiplications. For example, since x^3 is written as X*X*X, there are two multiplications. In general, an exponent of n indicates $n - 1$ multiplications. Thus $3x^3 - 2x^2 + x - 1$ has three multiplications in the first term (two for x^3 and one for the factor 3), and two multiplications in the second term (one for x^2 and one for the factor -2). There are no multiplications in the third and fourth terms. The polynomial requires a total of five multiplications. When the polynomial is written in the nested form X*(X*(X*3 - 2) + 1) - 1, however, only three multiplications are required. That is, the number of multiplications is the same as the number of multiplications in the first term alone.

To count the number of multiplications in a polynomial:
1. In *expanded* form: Count the number of multiplications required in *all* of the terms. An exponent of n indicates $n - 1$ multiplications.
2. In *nested* form: Count the number of multiplications required in the *first* term.

EXAMPLE 5

Compare the number of operations required to evaluate $3x^4 - 2x^2 + 5$ in expanded form and in nested form.

Solution The expanded form requires four multiplications for the first term (three for x^4 and one for the factor 3) and two multiplications for the second term (one for x^2 and one for the factor -2), for a total of six multiplications. The nested form requires only four multiplications (the same number as the first term alone). ∎

By writing a base two number in a nested form, you can see why the algorithms of Section 4.2 are valid. Although we consider a single example, the arguments can be generalized to cover any integer in any base. Consider the binary number 101101. We have

$$
\begin{aligned}
101101_2 &= \mathbf{1} \cdot 32 + \mathbf{0} \cdot 16 + \mathbf{1} \cdot 8 + \mathbf{1} \cdot 4 + \mathbf{0} \cdot 2 + \mathbf{1} \\
&= \mathbf{1} \cdot 2^5 + \mathbf{0} \cdot 2^4 + \mathbf{1} \cdot 2^3 + \mathbf{1} \cdot 2^2 + \mathbf{0} \cdot 2 + \mathbf{1} \\
&= 2\{\mathbf{1} \cdot 2^4 + \mathbf{0} \cdot 2^3 + \mathbf{1} \cdot 2^2 + \mathbf{1} \cdot 2 + \mathbf{0}\} + \mathbf{1} \\
&= 2\{2[\mathbf{1} \cdot 2^3 + \mathbf{0}.2^2 + \mathbf{1} \cdot 2 + \mathbf{1}] + \mathbf{0}\} + \mathbf{1} \\
&= 2\{2[2(\mathbf{1} \cdot 2^2 + \mathbf{0} \cdot 2 + \mathbf{1}) + \mathbf{1}] + \mathbf{0}\} + \mathbf{1} \\
&= 2\{2[2(2\{\mathbf{1} \cdot 2 + \mathbf{0}\} + \mathbf{1}) + \mathbf{1}] + \mathbf{0}\} + \mathbf{1}.
\end{aligned}
$$

The digits 101101 of the binary number are shown in boldface type. If we perform the arithmetic operations in the normal order (grouping symbols first, from the inside out), we will be doing the steps described in the rule for converting from base 2 to base 10.

On the other hand, if we divide the base ten number (still written in nested form) by two, 2 is not a factor of the entire expression. There is a remainder of 1. If we "remove" this remainder, and divide by two again, the remainder will be 0. If we continue the process, the series of remainders will be the digits of the binary number 101101. That is, we will be doing the steps described in the rule for converting from base 10 to base 2.

As an example of conversion from a decimal fraction to a binary fraction, we consider the binary number 0.1111.

$$
0.1111_2 = \mathbf{1}\left(\frac{1}{2}\right) + \mathbf{1}\left(\frac{1}{4}\right) + \mathbf{1}\left(\frac{1}{8}\right) + \mathbf{1}\left(\frac{1}{16}\right)
$$

We can write this number in nested form as

$$
\frac{1}{2}\{\mathbf{1} + \mathbf{1}\left(\frac{1}{2}\right) + \mathbf{1}\left(\frac{1}{4}\right) + \mathbf{1}\left(\frac{1}{8}\right)\}
$$

$$
\frac{1}{2}\{\mathbf{1} + \frac{1}{2}[\mathbf{1} + \mathbf{1}\left(\frac{1}{2}\right) + \mathbf{1}\left(\frac{1}{4}\right)]\}
$$

$$
\frac{1}{2}\{\mathbf{1} + \frac{1}{2}[\mathbf{1} + \frac{1}{2}(\mathbf{1} + \mathbf{1}\left(\frac{1}{2}\right))]\}
$$

$$
\frac{1}{2}\{\mathbf{1} + \frac{1}{2}[\mathbf{1} + \frac{1}{2}(\mathbf{1} + \frac{1}{2} \cdot \mathbf{1})]\}.
$$

Multiplication of the decimal fraction by two removes the first fraction of 1/2 and produces (as an integer) the leftmost bit of the binary fraction. Multiplication of the remaining fraction by two removes the second fraction of 1/2 and produces (as an integer) the next bit, and so forth. The argument could also be used for bases 8 and 16.

5.7 EXERCISES

Write each polynomial in nested form, using computer notation. See Examples 1–3.

1. $x^2 - 7x + 2$

2. $2x^2 + 5x - 3$

3. $4x^2 + 3x - 1$

4. $3x^2 - x + 2$

5. $x^3 + 4x^2 - 3x + 2$

6. $2x^3 - 5x^2 + 7x - 1$

7. $8x^3 - 6x^2 + 4x - 3$

8. $2x^3 - 4x^2 + 1$

9. $x^3 - 2x^2 + 3$

10. $x^3 - x^2 + x - 1$

11. $x^4 + 3x^3 + 2x^2 - x - 1$

12. $2x^4 - 4x^3 + 5x^2 - 6x + 4$

13. $3x^4 + 2x^3 - x^2 + 7x - 5$

14. $5x^4 - 4x^3 + 3x^2 - 2x + 1$

15. $x^4 - x^3 + x^2 - x + 1$

Compare the number of multiplications required to evaluate each polynomial in expanded form and in nested form. See Examples 4 and 5.

16. $x^2 - 7x + 2$

17. $2x^2 + 5x - 3$

18. $4x^2 + 3x - 1$

19. $3x^2 - x + 2$

20. $x^3 + 4x^2 - 3x + 2$

21. $2x^3 - 5x^2 + 7x - 1$

22. $8x^3 - 6x^2 + 4x - 3$

23. $2x^3 - 4x^2 + 1$

24. $x^3 - 2x^2 + 3$

25. $x^3 - x^2 + x - 1$

26. $x^4 + 3x^3 + 2x^2 - x - 1$

27. $2x^4 - 4x^3 + 5x^2 - 6x + 4$

28. $3x^4 + 2x^3 - x^2 + 7x - 5$

29. $5x^4 - 4x^3 + 3x^2 - 2x + 1$

30. $x^4 - x^3 + x^2 - x + 1$

✳**31.** Write the following octal numbers in nested form, and convert them to base 10 to show how the conversion algorithm works.

    a. 127_8

    b. 231_8

✳**32.** Using the nested form of the base 2 number 0.1111, develop an algorithm for conversion from a binary fraction to a decimal fraction.

CHAPTER REVIEW

An exponent is a symbol used to indicate a repeated _____.

In the expression x^p, the base is _____, and the exponent is

_____.

Algebraic expressions that are to be added are called _____.

Algebraic expressions that are to be multiplied are called _____.

A _____ in one variable is an expression of the form $a_n x^n + a_{n-1} x^{n-1} + a_{n-2} x^{n-2} + \ldots + a_0$ where $a_n, a_{n-1}, a_{n-2}, \ldots, a_0 \in R$ and $n \in I$, $n \geq 0$.

The degree of the polynomial $a_n x^n + a_{n-1} x^{n-1} + a_{n-2} x^{n-2} + \ldots + a_0$ is _____, and the values $a_n, a_{n-1}, a_{n-2}, \ldots, a_0$ are called _____.

A _____ is a polynomial with exactly two terms.

A _____ is a polynomial with exactly three terms.

A _____ expression is a ratio of two polynomials such that the denominator is not equal to zero.

A _____ number is an integer greater than 1, such that the only positive integers that divide it evenly are 1 and the number itself.

The LCD, or _____ _____ _____, of two or more rational numbers is the smallest positive integer that is evenly divisible by each of their denominators.

CHAPTER TEST

Simplify each expression. There should be no zero or negative exponents in the final answer.

1. $5x^3y(7xy^2)$ **2.** $5a^{-3}b(7ab^{-2})$ **3.** $4z^2(4z^{-2})$ **4.** $\dfrac{36m^2n}{4m}$ **5.** $\dfrac{5x^3y^{-1}}{3^{-1}z^{-2}w^2}$

6. $(2ab^2)^3$ **7.** $3(a^3b^{-2})^2$ **8.** $\left(\dfrac{2x}{3}\right)^3$ **9.** $\left(\dfrac{2z^{-1}}{3}\right)^{-2}$ **10.** $(125)^{2/3}$

Perform the indicated operations:

11. $(3x^2 + 4x - 1) + (2x^2 - 3x + 4)$ **12.** $(2y^2 - 3y + 1) - (y^2 + y - 3)$

13. $(3m + 1)(2m - 3)$ **14.** $(2p - 1)(4p^2 + 2p - 1)$

15. $(z - 2)(z - 3) + 5$

Factor completely, if possible:

16. $5x^2 - 10x$ **17.** $z^2 - z - 12$ **18.** $a^2 + 11a + 28$

19. $6x^2 + 7x - 3$ **20.** $9y^3 - y$

Reduce to lowest terms:

21. $\dfrac{70}{105}$ **22.** $\dfrac{2x^2 - x - 1}{x^2 + x - 2}$

Perform the indicated operation:

23. $\dfrac{26}{35} \cdot \dfrac{14}{39}$ **24.** $\dfrac{m^2 - 5m + 6}{m^2 - 4} \cdot \dfrac{m^2 + 3m + 2}{m^2 - 9}$

25. $\dfrac{2p - 5}{p + 3} \div \dfrac{2p^2 - 11p + 15}{p^2 + 4p + 3}$ **26.** $\dfrac{x}{y} + \dfrac{y}{x^2}$

27. $\dfrac{a}{a+2} + \dfrac{2a}{a-2}$

28. $\dfrac{z}{2z+3} - \dfrac{1}{2z^2+z-3}$

29. $\dfrac{p}{p^2-3p} + \dfrac{1}{p+2}$

30. $\dfrac{2}{z} + \dfrac{3}{z^2} - \dfrac{1}{3}$

Write each polynomial in nested form, using computer notation:

31. $3x^2 + 2x - 1$ **32.** $3x^3 - 2x^2 + 5x - 4$ **33.** $x^4 + 2x^3 - 3x^2 + 4x - 5$

Compare the number of multiplications required to evaluate each polynomial in expanded form and in nested form:

34. $3x^4 - 2x^2 + 1$ **35.** $2x^3 + 4x^2 - 3x + 5$

➡36. The flowchart in the following figure shows an algorithm for computing $1^3 + 2^3 + 3^3 + \dots n^3$ for values of n from 1 through 100. Modify it so that the sum is displayed only if $1^3 + 2^3 + \dots n^3 = (1 + 2 + 3 + \dots + n)^2$. It might surprise you to learn that these two expressions are equal for every whole number n.

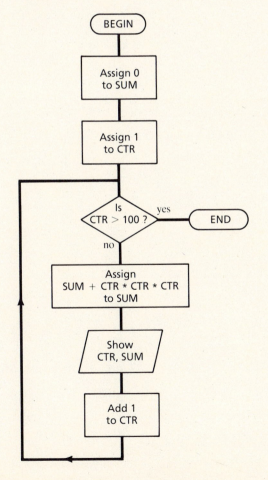

FOR FURTHER READING

Autrey, Melanie Ann and Joe Dan Austin. "A Novel Way to Factor Quadratic Polynomials," *The Mathematics Teacher,* Vol. 72 No. 2 (February 1979), 127–128. (An algorithm for factoring trinomials is given.)

Crombie, A. C. "Descartes," *Scientific American,* Vol. 201 No. 4 (October 1959), 160–173.

Harris, Whitney, Jr. "Right Answer—Wrong Method," *Mathematics and Computer Education,* Vol. 19 No. 1 (Winter 1985), 49–51. (A BASIC program to find fractions in which a digit may be cancelled is given.)

Kalman, Dan and Warren Page. "Nested Polynomials and Efficient Exponentiation Algorithms for Calculators," *The College Mathematics Journal,* Vol. 16 No. 1 (January 1985), 57–60.

Biography   René Descartes

René Descartes was born March 31, 1596 in France. His mother died within months of his birth. His father was a lawyer whose sons were expected to pursue careers in the army or the church, as did most gentlemen of the time.

When he was about eight, René was sent to a Jesuit school. He was a sickly child, and the rector suggested that he stay in bed as long as he wished in the mornings. Long after he had regained his health, Descartes continued to meditate in bed in the mornings.

Later he studied law and mathematics. At the age of twenty-two, he joined the Dutch army as an unpaid officer. There is a legend that as he was walking down a street in Holland, he saw a poster written in Flemish. He asked a passerby to translate it. The poster carried a challenge to solve a difficult mathematics problem, which he did within a few hours.

Descartes left the Dutch army, travelled, and joined the Bavarian army. On the night of November 10, 1619, he had three consecutive dreams, which he interpreted as a sign that he should systematically doubt everything and deductively construct a universal science. It is believed that he began to develop his analytic geometry at about the same time.

He continued to study mathematics in his leisure time. The notation using positive integers as exponents is one of his contributions. He settled in Holland in 1628. He spent twenty years studying and writing, interrupted only by his grief at the death of his illegitimate five-year-old daughter.

In 1649, he reluctantly accepted Queen Christina's invitation to Sweden. She had him tutor her in philosophy at 5 o'clock each morning. It was one of Sweden's coldest winters, and Descartes died February 11, 1650 of pneumonia.

Chapter Six
Equations

An equation is solved by removing the numbers that conceal the value of the variable, much like a child removes the wrapping that conceals the contents of a package.

6.1 Linear Equations and Inequalities

The word *algebra* is derived from the name of a book written about 825 by Mohammed ibn musa al-Khowârizmî. The book was called *Hisâb al-jabr w'al Muqabâlah*. The title refers to the science of solving equations. Though the word *algebra* today refers to a larger body of knowledge, solving equations is an important topic of elementary algebra.

To solve an equation means to find the values that, when substituted for the variables, make the equation a true statement. In this section, we will solve linear equations in one variable. In a linear equation, the largest exponent of the variable is 1, and there are no fractional or negative exponents.

> **Definition**
> A linear equation in the variable x is an equation that can be put in the form $ax + b = 0$ where a and $b \in R$, $a \neq 0$.

The value of the variable that makes the equation true is said to *satisfy* the equation, and it is called the *solution* of the equation. Two equations with the same solution are said to be *equivalent*.

First, let us explore some ideas that are important in solving an equation. Inverse operations are two operations such that one operation ''undoes'' the other, as do addition and subtraction. Suppose we begin with 2. If we first add 3, then subtract 3, we have $2 + 3 = 5$, and $5 - 3 = 2$. The subtraction ''undoes'' the addition, and we have 2 again. Multiplication and division are also inverse operations.

There is an analogy between unwrapping a package and solving an equation. To unwrap a package, the ribbon comes off first, then the paper, and finally the box. That is, the package is unwrapped in the *opposite* order from which it is wrapped. Also notice that tying and untying the ribbon, wrapping and unwrapping the paper, and closing and opening the box, are examples of inverse operations. Likewise, an equation is put together using addition, subtraction, multiplication, and division with the standard order of operations. To ''take it apart,'' we undo it in the opposite order, and we use inverse operations.

Finally, if we add or subtract the same number on both sides of the equation, or if we multiply or divide both sides of the equation by the same nonzero number, an equivalent equation is obtained. That is, the solution of the new equation is the same as the solution of the original equation.

To solve a linear equation, there are three steps:

1. Simplify each side.
2. Remove extra terms.
3. Remove extra factors or divisors.

Let's examine these three steps in more detail.

1. Simplify each side. Use the distributive property, if necessary, and combine like terms. For example, $2(x + 3) - 5 = 6$ should be rewritten as

$$2x + 6 - 5 = 6$$
$$2x + 1 \qquad = 6.$$

2. Remove extra terms. We want to isolate the x on one side of the equation. In the previous example, the "1" is a term that we do not want. Since the "1" has been added to $2x$, we can undo the addition with a subtraction. Subtracting 1 from both sides of the equation gives

$$2x + 1 - \mathbf{1} = 6 - \mathbf{1} \text{ or}$$
$$2x \qquad\quad = 5.$$

3. Remove extra factors or divisors. We only want x, not $2x$. We need to remove the factor 2. Since multiplication can be "undone" by division, we divide both sides of the equation by 2.

$$\frac{2x}{\mathbf{2}} = \frac{5}{\mathbf{2}} \text{ or } x = 2\frac{1}{2}.$$

To solve a linear equation in one variable:

1. Simplify each side.
   a. Use the distributive property, if necessary.
   b. Combine like terms.
2. Using inverse operations, remove extra terms (i.e., add or subtract the same number on both sides of the equation).
3. Using inverse operations, remove extra factors or divisors (i.e., multiply or divide both sides of the equation by the same nonzero number).

EXAMPLE 1   Solve $\dfrac{x}{3} - 1 = 4$.

Solution
1. Neither side can be simplified.

2. We want to remove the 1. Since 1 was subtracted, we add 1 to both sides of the equation: $\dfrac{x}{3} - 1 + 1 = 4 + 1$ or

$$\frac{x}{3} = 5.$$

3. We want to remove the 3. Since 3 was divided into x, we multiply both sides of the equation by 3: $(3)\,\dfrac{x}{3} = (3)\,5$ or

$$x = 15. \ \blacksquare$$

EXAMPLE 2

Solve $2y - 1 = 3y + 2$.

Solution

1. Neither side can be simplified.
2. The final answer should have only one y term. Therefore, we begin by removing the extra y term. To remove $2y$, we would subtract $2y$ from both sides. (The equation could also be solved by subtracting $3y$ from both sides.) We have $2y - 1 - 2y = 3y + 2 - 2y$ or

$$-1 = y + 2.$$

Now we remove the 2 by subtracting 2 from both sides.

$$-1 - 2 = y + 2 - 2 \text{ or } -3 = y \ \blacksquare$$

EXAMPLE 3

Solve $2z - (z + 1) = 3(z - 1) + 4$.

Solution

1. Simplify each side: $2z - (z + 1) = 3(z - 1) + 4$

$$2z - z - 1 = 3z - 3 + 4$$
$$z - 1 = 3z + 1.$$

2. Remove extra terms: $z - 1 - z = 3z + 1 - z$

$$-1 = 2z + 1$$
$$-1 - 1 = 2z + 1 - 1$$
$$-2 = 2z.$$

3. Remove extra factors or divisors: $-2/2 = 2z/2$

$$-1 = z. \ \blacksquare$$

EXAMPLE 4

Solve $5x - 1 = x - 2(x + 1)$.

Solution

$$5x - 1 \quad = x - 2(x + 1)$$

$$5x - 1 \quad = x - 2x - 2$$

$$5x - 1 \quad = -x - 2$$

$$5x - 1 + x = -x - 2 + x$$

$$6x - 1 \quad = -2$$

$$6x - 1 + 1 = -2 + 1$$

$$6x \quad\quad = -1$$

$$6x/6 \quad = -1/6 \text{ or } x = -\frac{1}{6}. \quad\blacksquare$$

A statement that two quantities are not equal is called an *inequality*.

Inequality Symbols

$\neq$ means *is not equal to*

$>$ means *is greater than*

$<$ means *is less than*

$\geq$ means *is greater than or equal to*

$\leq$ means *is less than or equal to*

Whenever two numbers are compared, the inequality symbol points toward the smaller number. Thus we can say $2 < 3$ or $3 > 2$. Both inequalities express the same relationship between 2 and 3. Solving a linear inequality is almost like solving a linear equation. There is one additional rule: If you multiply or divide both sides of an inequality by a negative number, you must reverse the direction of the inequality symbol.

To solve a linear inequality in one variable:

1. Simplify each side.
   a. Use the distributive property, if necessary.
   b. Combine like terms.
2. Using inverse operations, remove extra terms (i.e., add or subtract the same number on both sides of the inequality).
3. Using inverse operations, remove extra factors or divisors (i.e., multiply or divide both sides of the inequality by the same nonzero number).
4. If you multiply or divide both sides by a *negative* number, reverse the direction of the inequality symbol.

EXAMPLE 5

Solve $3m + 1 \geq 2$.

Solution

$3m + 1 - \mathbf{1} \geq 2 - \mathbf{1}$

$3m \qquad\qquad \geq 1$

$3m/3 \qquad\quad \geq 1/3$

$m \qquad\qquad \geq \dfrac{1}{3}.$ ■

We did not multiply or divide by a *negative* number, so the inequality continued to "point" in the same direction throughout the problem.

EXAMPLE 6

Solve $-3p + 1 < 2$.

Solution $\quad -3p + 1 - \mathbf{1} < 2 - \mathbf{1}$

$-3p \qquad\qquad < 1$

$-3p/(\mathbf{-3}) \qquad > 1/(\mathbf{-3})$

$p \qquad\qquad > -\dfrac{1}{3}.$ ■

Division by -3 changed the $<$ to a $>$.

EXAMPLE 7

Solve $-4x + 3 > 2x + 1$.

Solution $\quad -4x + 3 - \mathbf{2x} > 2x + 1 - \mathbf{2x}$

$-6x + 3 \qquad\quad > \qquad 1$

$-6x + 3 - \mathbf{3} > \qquad 1 - \mathbf{3}$

$-6x \qquad\qquad > \qquad -2$

$-6x/(\mathbf{-6}) \qquad < \qquad -2/(\mathbf{-6})$

$x \qquad\qquad < \qquad \dfrac{1}{3}.$ ■

6.1 EXERCISES

Solve each equation. See Examples 1–4.

1. $5x + 3 = 7$

2. $4y + 7 = 2$

3. $6z + 1 = -3$

4. $2x + 4 = -7$

5. $\dfrac{x}{5} + 2 = 7$

6. $\dfrac{3y}{2} - 1 = 3$

7. $\dfrac{m}{3} - 2 = 1$

8. $\dfrac{p}{4} + 3 = -1$

9. $\dfrac{2x}{3} + 1 = 4$

10. $3m + 2 = 5m - 1$

11. $2p - 6 = 7p + 8$

12. $5x + 1 = 1 - x$

13. $4x + 3 = 2 - 5x$

14. $n - 2 = 4n + 1$

15. $3(x + 1) = 2x + 3$

16. $4(y - 2) = y - 1$

17. $5(1 - p) = 3p - 2$

18. $-1(2 - x) = 5x + 4$

19. $2(m + 1) = 3m - 7$

20. $3(z - 4) + 2 = z$

21. $-(x - 2) + 2x = 0$

22. $2(n - 1) + 4 = 3n + 2(n - 1)$

23. $-3(x + 2) - 1 = 2x - 4(x + 3)$

24. $-1(y - 3) - 4 = 7y - (y - 7)$

25. $2(2z + 3) - 5 = 2z - (3z - 11)$

Solve each inequality. See Examples 5–7.

26. $3x - 5 > 7$

27. $2y - 4 \leq 3$

28. $z - 6 \geq 3$

29. $4x - 7 < 2$

30. $\dfrac{m}{2} - 5 < 2$

31. $\dfrac{2p}{3} - 3 > 1$

32. $3x - 1 \geq x + 5$

33. $2y + 8 \leq 7y + 6$

34. $2(z + 3) \leq 3z - 1$

35. $m + 2 > 4(m + 1)$

36. $3(1 - x) < 5x + 2$

Replace each letter by a single digit:

✳37. TWO + TWO = FOUR

✳38. FOUR + ONE = FIVE

✳39. Solve for x: $8^{(x-1)} = 4^{-(x-1)}$

✳40. Solve for x: $2(3(4(5(x - 1) + 6) - 5) + 4) - 2 = 0$

6.2   Formulas

A formula is an expression relating two or more variables by means of an equation. One variable usually appears by itself while the other variables are all on the other side of the equation. A formula is said to be *solved for* the variable that appears alone on one side of the equation. For example, $E = mc^2$ is a formula relating the variables E, m, and c. We say it is solved for E. Sometimes it is necessary to rewrite a formula so that it is solved for a *different* one of the variables. We could write $m = E/c^2$ and the formula would be solved for m.

Solving a formula is just like solving an equation in one variable if we have a clear idea of which variable we are trying to isolate. To leave the desired variable on one side of the equation by itself, we simplify both sides of the equation, remove extra terms, then remove extra factors or divisors.

EXAMPLE 1

Solve for v_0: $v = v_0 + at$.

Solution
1. Each side is in simplest form.
2. Since we are solving for v_0, we want to remove the term at by subtracting at from both sides of the equation.

$$v - at = v_0 + at - at$$

$$v - at = v_0 \blacksquare$$

EXAMPLE 2

Solve for t: $v = v_0 + at$.

Solution
1. Each side is in simplest form.
2. Since we are solving for t, we subtract v_0 from both sides, leaving only those terms that contain t.

$$v - v_0 = v_0 + at - v_0$$

$$v - v_0 = at$$

3. Since we are solving for t, we remove the factor a by dividing both sides of the equation by a.

$$\frac{v - v_0}{a} = \frac{at}{a}$$

$$\frac{v - v_0}{a} = t \blacksquare$$

EXAMPLE 3

Solve for k: $Fr^2 = kQ_1Q_2$.

Solution
1. Each side is in simplest form.
2. There are no extra terms.
3. We remove the extra factors of Q_1 and Q_2 by dividing both sides of the equation by Q_1Q_2.

$$\frac{Fr^2}{Q_1Q_2} = \frac{kQ_1Q_2}{Q_1Q_2}$$

$$\frac{Fr^2}{Q_1Q_2} = k \blacksquare$$

Two problems arise in solving formulas. First, fractions can be difficult to work with. However, if you multiply both sides of the equation by the lowest common denominator of all the fractions in the problem, the denominators will divide out. Second, even though the same variable may appear several times, it may not appear in like terms, so the terms cannot be combined. An expression, however, may be factored so that the variable appears only once.

> **To solve a formula for a particular variable:**
> 1. If the formula contains denominators, multiply both sides of the equation by the lowest common denominator.
> 2. Simplify each side.
> 3. Using inverse operations, remove extra terms. (Add or subtract the same expression on both sides of the equation.)
> 4. If the variable being solved for appears more than once, factor it out as a common factor.
> 5. Using inverse operations, remove extra factors or divisors. (Multiply or divide both sides of the equation by the same nonzero factor. You may assume that denominators are not equal to zero in this chapter.)

EXAMPLE 4

Solve for R: $\dfrac{1}{o} + \dfrac{1}{i} = \dfrac{2}{R}$.

Solution

1. Multiply both sides by oiR, the LCD of the three fractions:

$$oiR \left(\frac{1}{o} + \frac{1}{i} \right) = oiR \left(\frac{2}{R} \right)$$

$$iR + oR = 2oi.$$

2. Neither side has like terms, so neither side can be simplified.

3. Both terms on the left contain the variable R, which is the one we are to isolate, so there are no extra terms.

4. Since the variable R appears twice on the left, we factor it out as a common factor: $R(i + o) = 2oi$.

5. Since we are solving for R, we want to remove the factor $(i + o)$:

$$\frac{R(i + o)}{(i + o)} = \frac{2oi}{(i + o)}$$

$$R = \frac{2oi}{i + o}. \quad \blacksquare$$

EXAMPLE 5

Solve for a: $S = \dfrac{n}{2}(a + l)$.

Solution

1. Multiply both sides by 2, the lowest common denominator.

$$2(S) = 2\left[\frac{n}{2}(a + l) \right]$$

$$2S = n(a + l)$$

2. Simplify both sides: $2S = na + nl$.

3. Remove the term nl: $2S - \mathbf{nl} = na + nl - \mathbf{nl}$

$$2S - nl = na.$$

4. The variable a appears only once.

5. Remove the factor n: $\dfrac{2S - nl}{n} = \dfrac{na}{n}$

$$\frac{2S - nl}{n} = a. \ \blacksquare$$

6.2 EXERCISES

Solve each formula for the indicated variable. See Examples 1–3.

1. $F = ma$ for m

2. $F = ma$ for a

3. $\omega = nmgh$ for g

4. $\omega = nmgh$ for m

5. $P = 2L + 2W$ for L

6. $P = 2L + 2W$ for W

7. $v^2 = v_0^2 + 2a(x - x_0)$ for a

8. $v^2 = v_0^2 + 2a(x - x_0)$ for x

9. $E = F/q$ for F

10. $D = M/V$ for M

11. $h = \dfrac{V}{\pi r^2}$ for V

12. $A = \dfrac{\pi r^2 h}{2}$ for h

13. $F = \dfrac{GM_1 M_2}{r^2}$ for G

14. $F = \dfrac{GM_1 M_2}{r^2}$ for M_1

Solve each formula for the indicated variable. See Examples 4 and 5.

15. $E = \dfrac{F}{q}$ for q

16. $D = \dfrac{M}{V}$ for V

17. $x = x_0 + v_0 t + \dfrac{1}{2} at^2$ for x_0

18. $x = x_0 + v_0 t + \dfrac{1}{2} at^2$ for v_0

19. $v = \dfrac{s_2 - s_1}{t_2 - t_1}$ for t_2

20. $v = \dfrac{s_2 - s_1}{t_2 - t_1}$ for s_2

21. $M = \dfrac{Itm}{ze}$ for e

22. $\dfrac{1}{s} = \dfrac{1}{p} - \dfrac{1}{E}$ for E

23. $\dfrac{1}{s} = \dfrac{1}{p} - \dfrac{1}{E}$ for s

24. $v^2 = k^2 m \left(\dfrac{2}{r} + \dfrac{1}{a} \right)$ for a

25. $v^2 = k^2 m \left(\dfrac{2}{r} + \dfrac{1}{a} \right)$ for r

26. $\dfrac{1}{f} = \dfrac{1}{o} + \dfrac{1}{i}$ for f

27. $\dfrac{1}{f} = \dfrac{1}{o} + \dfrac{1}{i}$ for i

28. $\dfrac{p^2}{a^3} = \dfrac{4\pi^2}{k^2(m_1 + m_2)}$ for m_1

29. $\dfrac{p^2}{a^3} = \dfrac{4\pi^2}{k^2(m_1 + m_2)}$ for m_2

30. $\dfrac{p^2}{a^3} = \dfrac{4\pi^2}{k^2(m_1 + m_2)}$ for k^2

✱**31.** Bode's law provides a convenient formula for approximating the distances in astronomical units (1 a. u. is approximately 93,000,000 miles) of the planets from the sun. For Mercury, the distance is 4/10. Thereafter the distance (d) of the nth planet is given by $d = (3 \cdot 2^{n-2} + 4)/10$. The planets, in order, are Mercury, Venus, Earth, Mars, the asteroid belt, Jupiter, Saturn, and Uranus. (Bode's law does not work for Neptune and Pluto.)

(a) Use Bode's law to approximate the distance from the sun to Saturn.

(b) Use Bode's law to approximate the distance from Earth to Mars.

(c) How much closer are the asteroids to Mars than to Jupiter?

➠**32.** The formula in Example 5 gives the sum S of a series of n numbers obtained by taking a as the first number and adding d to get each successive number. The last number is l. Use pseudocode or a flowchart to show an algorithm for adding the numbers 1 through 100:

(a) Using this formula.

(b) Using an accumulator.

(c) Which algorithm do you think would be executed faster on a computer?

✱**33.** A closed plane figure with straight edges is called a *polygon*. A line segment joining two nonadjacent vertices (corners) is called a *diagonal*. Find a formula for the number of distinct diagonals that can be drawn for a polygon with n sides.

✱**34.** A solid that is bounded by planar surfaces is called a *polyhedron*. It is said to be convex if every line segment joining two of its points contains only points inside the polyhedron. Given the information in the table, determine a formula relating the number of faces (F), vertices (V), and edges (E) of a convex polyhedron.

Polyhedron	Number of faces	Number of vertices	Number of edges
tetrahedron	4	4	6
octahedron	8	6	12
icosahedron	20	12	30
hexahedron	6	8	12
dodecahedron	12	20	30

6.3 Word Problems

In solving verbal problems, there are no hard and fast rules. Each problem may be different from the preceding one. There are, however, a few guidelines that may help to get you started.

    1. Read the problem quickly. Don't worry about the numbers in the problem. If it says, "A bank teller has $250 in tens and twenties," think to yourself, "A bank teller has *some money* in tens and twenties." Try to picture the situation. Watch for key words like "sum" and "product," which indicate that addition and multiplication are involved.

2. Look for the question. Let the variable represent the quantity you're trying to find. This isn't always the best thing to do, but it *usually* is. If the question is How long did it take?, write, "Let t = time." If the question is, How far did he travel?, write, "Let d = distance traveled." If the problem asks for more than one thing, write down expressions to represent all of the quantities.

3. Write down and label relevant information. You will need to reread the problem more carefully at this stage. It helps to have all the bits and pieces of the problem in one place and written in a concise form. If the problem mentions perimeter of a rectangle, you would write down anything that you know about length and width, because perimeter is based on length and width. You may want to draw a picture or write down a formula that seems relevant.

4. Write an equation. The most common mistake in working problems is to try to do step 4 before step 1. It is tempting to think that the first step in solving a word problem is to write an equation. Don't fall into this trap. If you take the time to do steps 1, 2, and 3, step 4 (writing the equation) is not hard. For this step, you will probably want to reread the problem.

5. Solve the equation. This step is the easiest one.

6. Check your answer. In any type of problem, however, it is important to check your answer in the *original* problem. If, for example, you write the wrong equation, you may solve that equation correctly, but have the wrong answer for the word problem. Checking the equation (instead of the original problem) would not show you that you had an error. Reread the problem, substituting your answer to see if it meets the conditions described.

These steps are summarized as follows.

To solve a word problem:
1. Read the problem for understanding.
2. Identify the variable. Usually it will be the quantity you are trying to find.
3. Write down relevant information about all of the quantities involved in the problem. Draw a picture or write a formula.
4. Write the equation. (It may be necessary to reread the problem to see how the various quantities are related.)
5. Solve the equation.
6. Check the answer by substituting your answer into the *original* problem.

EXAMPLE 1

A plane leaves the airport flying west at 300 mph. One hour later another plane leaves the same airport flying east at 350 mph. For how long will each travel before they are 1730 miles apart?

Solution

1. Visualize the scene. Don't worry about numbers yet. One plane takes off. Later, another plane takes off headed in the opposite direction. It is going faster than the first plane.

2. Identify the variables. The question is For how long will *each* travel?, so we let the variable represent the time that one of the planes travels. We write, "Let t = time of the first plane." But we should also specify the time that the second plane travels, since we were asked to find that time as well. Since it took off one hour later, its travel time is one hour *less*. We write, "$t - 1$ = time of the second plane."

3. Write down other relevant information. Rereading the problem, we see that we are given the rates of the two planes. We write down 300 = rate of first, 350 = rate of second. We know the distance between the two planes is 1730 miles.

   Summarizing this information, we have:

   Let t = time of first plane

   $t - 1$ = time of second plane

   300 mph = rate of first plane

   350 mph = rate of second plane

   1730 miles = distance between planes

   We might want to use the formula $D = RT$.

   A diagram of the plane routes might also be useful.

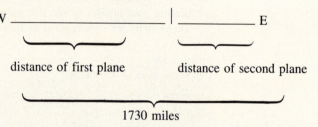

4. Now we are ready to write the equation. By looking at the information we have summarized in step 3, we can see how to use the formula $D = RT$ to find each individual distance. The distance of the first plane would be its rate multiplied by its time or $300t$. The distance of the second plane would be its rate multiplied by its time or $350(t - 1)$. The diagram shows that the distance 1730 is obtained by adding the two individual distances, so $300t + 350(t - 1) = 1730$.

5. Solve: $300t + 350t - 350 = 1730$

$$650t - 350 = 1730$$
$$650t - 350 + 350 = 1730 + 350$$
$$650t = 2080$$
$$\frac{650t}{650} = \frac{2080}{650}$$
$$t = 3.2$$

Since t was the time of the first plane, and $t - 1$ was the time of the second, we have 3.2 hours (or 3 hours and 12 minutes) for the first, and 2.2 hours (or 2 hours and 12 minutes) for the second.

6. Check. We read the original problem as "A plane leaves the airport flying west at 300 mph. One hour later another plane leaves the same airport traveling east at 350 mph. When the second plane has travelled for 2.2 hours, the two planes are 1730 miles apart." ∎

EXAMPLE 2

A chemical company has 220 gallons of a solution that has been diluted to 40 percent strength, but needs to be stronger. How much of the chemical should be added to bring the solution up to 45 percent strength?

Solution

1. Try to visualize a vat of this solution. Part of the solution is the chemical; part is a diluting agent. The chemist needs to pour some more of the chemical into it.

2. Since the question is, How much chemical should be added?, let g (for gallons) = amount of chemical to be added.

3. Write down other relevant information. Summarizing, we have:
    Let g = amount of chemical (100 percent chemical)
    220 = amount of solution (40 percent chemical)
    $g + 220$ = total amount of mixture (45 percent chemical).

4. The amount of chemical is the same whether it is in two containers (as it was at the start of the problem) or whether it is all mixed together. Consider the amount of chemical: 100 percent of g is chemical; 40 percent of 220 gallons is chemical; and 45 percent of $g + 220$ gallons is chemical.
    Therefore we have:

$$g + .40(220) = .45(g + 220)$$
$$g + 88 = .45g + 99$$
$$g + 88 - .45g = .45g + 99 - .45g$$
$$.55g + 88 = 99$$
$$.55g + 88 - 88 = 99 - 88$$
$$.55g = 11$$
$$.55g/.55 = 11/.55$$
$$g = 20 \text{ gallons}$$

6. Check: A chemical company has 220 gallons of a solution that has been diluted to 40 percent strength (so 40 percent of 220 gallons, or 88 gallons, would be chemical). Twenty gallons of chemical are added (so there would be 220 + 20 or 240 gallons of solution, and 88 + 20 or 108 gallons would be chemical). The strength would be 108/240 = 45 percent. ∎

EXAMPLE 3

An estimate for ideal body weight for a man is 106 pounds for the first five feet of height and six pounds for each additional inch. If a basketball player weighs 250 pounds and that is his ideal body weight, how tall is he?

Solution

Let h = height in inches

$h - 60$ = inches over 5′

250 = ideal weight

Ideal weight = 106 + 6 · (number of inches over 5′)

$$250 = 106 + 6(h - 60)$$
$$250 = 106 + 6h - 360$$
$$250 = 6h - 254$$
$$250 + 254 = 6h - 254 + 254$$
$$504 = 6h$$
$$504/6 = 6h/6$$
$$84 = h$$

Therefore, his height is 84 inches, or 7 feet.

Alternate Solution:

Let i = number of inches over 5′

Height = 5′ + i

$6i$ = weight over 106 pounds

$6i$ = 250 − 106

$6i$ = 144

i = 24

Therefore, his height is 5′ + 24 inches or 5′ + 2′ = 7′.

Check: If he is 7′ tall, he should weigh 106 pounds for the first 5′ and 6 × 24 or 144 pounds for the last 24″. That is, he would weigh 106 + 144 or 250 pounds, which agrees with the information in this problem. ∎

Programmers are sometimes faced with problems that can be solved by equations.

EXAMPLE 4

A Pascal programmer wants to print a list of employee names and social security numbers. She specifies that each name is to be printed in the first 20 positions of a line. We say the "field width" is 20. The social security number also has a field width of 20. When the program is run, the names are all right-justified (or aligned so that the last letter of each is in the 20th position). She can left-justify the names by specifying that each is printed with a field width equal to the length (number of characters) of the name itself. Let L represent that length. Find a formula for the field width of the social security number that will keep these numbers in the same positions as before.

Solution

Let W = field width of social security numbers

$\quad\ \ L$ = length of name

$\quad 40$ = total number of positions to be used

$L + W = 40$

$\qquad W = 40 - L$ ▰

EXAMPLE 5

If the second field of Example 4 were street address, the programmer would want to left-justify the values in the second column also. Devise a formula for the second field width.

Solution　To left-justify the street address, we need to know its exact length, just as we used the length of the name in the previous example.

Let W = width of address field

$\quad\ \ L$ = length of name

$\quad\ \ A$ = length of address

The first 20 positions are to be occupied by the name and some blank spaces. The next A positions are to be occupied by the address. The total length of the printed line, then, is $20 + A$, but this is also the sum of the field widths. We have $L + W = 20 + A$

$\qquad W = 20 + A - L.$ ▰

6.3 EXERCISES

Use an equation to solve each problem. See Examples 1–5.

1. A student has grades of 85, 89, 92, and 90. What grade must he make on the fifth test in order to have an average of 91?

2. The formula for determining the BTU rating of an air conditioner is: BTU = area × exposure factor × climate factor. If an air conditioner with a rating of 1520 BTU is used with an exposure factor of 20 and a climate factor of 0.95, how large a room can be cooled?

3. When an object is dropped, the distance that it falls in t seconds is $s = \frac{1}{2} gt^2$.

   Calculate the gravitational constant g, if you know that in two seconds, an object falls 64 feet.

4. The formula for converting Fahrenheit temperature (F) to Celsius temperature (C) is $F = \frac{9}{5} (C) + 32$. At what temperature are the Fahrenheit and Celsius readings numerically equal?

5. The Rule of 72 states that an estimate for years required to double an investment at R percent annually can be obtained by dividing 72 by R. What interest rate is required to double an investment in 4.8 years?

6. If doubling the current interest rate means an investment would double in 12 years, use the Rule of 72 (see problem 4 above) to find the current rate.

7. Ten liters of a salt and water solution is 90 percent salt. How much water should be added to dilute it to 75 percent salt?

8. Two liters of a salt and water solution is 70 percent salt. How much salt should be added to bring it up to an 80 percent solution?

9. An executive wants to contribute to an IRA account. She wants to split $2,000 so that part is invested at a fixed rate of 8 percent and the remainder goes into a variable rate account paying 9 percent for the current year. If she wants to earn $175 for the year, how much should she invest at each rate?

10. A specialty shop sells unusual blends of coffee. In order to close out the stock of two blends, the owner wants to mix them. There are 20 pounds of coffee worth $4.00 a pound and 15 pounds of coffee worth $4.50 a pound. For how much should the mixture sell?

11. A real estate broker is paid a 2 percent commission on each sale that she makes and a 1 percent commission on each home that she lists. She estimates that she will list homes worth twice the value of her sales. If she wishes to have an annual income of $40,000, find the total sales she must make and the total value of the real estate she must list.

12. A real estate broker is paid a certain rate of commission on each home that he lists, and for each home that he sells, he is paid at twice that rate. If he lists $300,000 and sells $600,000 for a commission of $45,000, what percent is the commission on each listing and each sale?

13. Two cars leave point A at the same time, traveling the same direction. The faster travels at 55 mph while the slower travels at 48 mph. How long will it take for the faster car to get ahead by 15.4 miles?

14. Two trains leave from the same place heading in opposite directions. The faster train travels at 110 mph while the slower train travels at 96 mph. How long will it take for the trains to be 309 miles apart?

15. Two people begin walking toward each other from points A and B, which are 13 miles apart. One person walks at a rate 0.5 mph faster than the other. If they meet after two hours, how fast was each walking?

16. A merchant wishes to make a 15 percent profit on the cost of an item for which he paid $102. What should the selling price be?

17. A merchant wishes to make a 15 percent profit on the selling price of an item that cost $102. What should the selling price be?

18. A poll reports that 975 people, or 65 percent of the sample, think that high school students should be required to pass a standardized test in order to graduate. How many people were polled?

19. A poll reports that 86 percent of the respondents favor a certain piece of legislation and 4 percent are undecided. If 1500 people were polled, how many were opposed?

20. A woman's ideal body weight can be estimated by taking 100 pounds for the first five feet of height and five pounds for each additional inch. However, 10 percent should be deducted for a small frame. How tall is a small-framed woman if her ideal body weight is 108 pounds?

21. A 5′6″ man of average frame weighs 143 pounds. By how many pounds is he over the ideal body weight? (See Example 3 for the formula.)

22. A 5′6″ woman of average frame weighs 143 pounds. What is her ideal body weight and by what percent is she overweight? (See problem 20 for the formula.)

23. A man can burn 350 calories per hour playing tennis and 110 calories per hour washing dishes. How many hours would he have to wash dishes to burn as many calories as he would playing tennis for $1\frac{1}{2}$ hours?

24. A woman can burn 250 calories per hour golfing and 125 calories per hour ironing. If she golfs twice as long as she irons and burns a total of 750 calories, for how long does she iron?

25. A man can burn 250 calories per hour bowling and 80 calories per hour watching television. If he bowls for two hours, then watches tv for two hours, how many ounces of beer can he drink (14 calories per ounce) to replace the burned calories?

✳26. If four cats can catch four mice in four hours, how many mice can one cat catch in one hour?

✳27. If you got a 10 percent raise last year and a 10 percent cut in salary this year and are now making $19,800, what was your salary before the raise?

✳28. A bicycle leaves point A and a car leaves point B at the same time. They travel toward each other, and the car travels at twice the rate of the bicycle. If the car left ten minutes later than the bicycle, they would meet three miles closer to point B. Find their rates.

6.4   Quadratic Equations

In Section 6.1, we solved linear equations in one variable. In a linear equation, the highest exponent of the variable is 1. In this section, we will solve quadratic equations in one variable. In a quadratic equation, the highest exponent of the variable is 2.

Definition

A quadratic equation in x is an equation that can be put in the form $ax^2 + bx + c = 0$ where a, b, and c are real numbers, $a \neq 0$.

The form $ax^2 + bx + c = 0$ is called the *standard* form. There are two important characteristics of this form of the equation. First, one side of the equation must equal zero. Second, the terms must appear in descending order of their exponents. To solve a quadratic equation, we will factor and use the following special property of zero.

Special Property of Zero

If a and b are real numbers such that $ab = 0$, then $a = 0$ or $b = 0$.

This property says that if the product of two numbers is 0, then one of the two numbers must be zero. Zero is the only number that has this property. For contrast, consider $ab = 5$. We could have 5 as one of the factors since $1(5) = 5$, but we could also have $\frac{1}{2}(10) = 5$ or $\frac{1}{3}(15) = 5$, and so on.

It is important to have a quadratic equation in standard form so that we will be able to use the special property of zero. The standard form of a quadratic equation, however, involves terms, and the special property of zero requires factors. Factoring will be easier if the terms are arranged with descending exponents.

Consider $6x^2 + x - 2 = 0$. It is a quadratic equation, and it is in standard form where $a = 6$, $b = 1$, and $c = -2$. (In order for the equation to follow the pattern $ax^2 + bx + c = 0$, we must include the negative sign as part of the number c.) We factor the expression on the lefthand side to obtain $(2x - 1)(3x + 2) = 0$. Since the product of the two factors is zero, the special property of zero tells us that one of the factors must be zero. That is, either $(2x - 1) = 0$ or $(3x + 2) = 0$. But now we have two *linear* equations instead of one *quadratic* equation. If $2x - 1 = 0$, we have $x = \frac{1}{2}$. If $3x + 2 = 0$, we have $x = -\frac{2}{3}$. Thus, the solution of the equation $6x^2 + x - 2 = 0$ is

$$x = \frac{1}{2} \text{ or } x = -\frac{2}{3}.$$

> **To solve a quadratic equation:**
> 1. If the equation is not in standard form $ax^2 + bx + c = 0$, write an equivalent equation that is in standard form (i.e., add or subtract the same term(s) on both sides of the equation in order to make one side equal zero, and arrange the terms in descending order of their exponents).
> 2. Factor the polynomial.
> 3. Use the special property of zero to set each factor equal to zero.
> 4. Solve the resulting linear equations.

EXAMPLE 1

Solve $x^2 = 9$.

Solution

1. The equation is not in standard form. Subtract 9 from both sides.

$$x^2 - 9 = 9 - 9$$
$$x^2 - 9 = 0$$

2. Factor: $(x - 3)(x + 3) = 0$.
3. Set each factor equal to zero: $x - 3 = 0$ or $x + 3 = 0$.
4. Solve the linear equations:

$$
\begin{aligned}
x - 3 \quad\quad &= 0 &\quad \text{or} \quad & x + 3 \quad\quad = 0 \\
x - 3 + 3 &= 0 + 3 &\quad \text{or} \quad & x + 3 - 3 = 0 - 3 \\
x \quad\quad &= 3 &\quad \text{or} \quad & x \quad\quad = -3. \ \blacksquare
\end{aligned}
$$

EXAMPLE 2

Solve $y^2 - 5y = 0$.

Solution

1. The equation is in standard form.
2. Factor: $y(y - 5) = 0$.
3. Set each factor equal to zero: $y = 0$ or $y - 5 = 0$.
4. Solve each liner equation:

$$
\begin{aligned}
y = 0 \quad &\text{or} \quad y - 5 + 5 = 0 + 5 \\
y = 0 \quad &\text{or} \quad y \quad\quad\quad = 5. \ \blacksquare
\end{aligned}
$$

EXAMPLE 3

Solve $z^2 + 6 = 5z$.

Solution

1. The equation is not in standard form. Subtract $5z$ from both sides.

$$z^2 - 5z + 6 = 5z - 5z$$
$$z^2 - 5z + 6 = 0$$

2. Factor: $(z - 2)(z - 3) = 0$

3. Set each factor equal to zero: $z - 2 = 0$ or $z - 3 = 0$

4. Solve each linear equation:

$$z - 2 \quad = 0 \qquad \text{or} \quad z - 3 \quad = 0$$
$$z - 2 + 2 = 0 + 2 \quad \text{or} \quad z - 3 + 3 = 0 + 3$$
$$z \qquad\quad = 2 \qquad \text{or} \quad z \qquad\quad = 3. \; \blacksquare$$

EXAMPLE 4

Solve $6x^2 - 15x + 9 = 0$.

Solution   The equation is in standard form, so we factor:

$$3(2x^2 - 5x + 3) = 0$$
$$3(2x - 3)(x - 1) = 0.$$

One of the factors must be zero. It is impossible to have $3 = 0$, so we have:

$$2x - 3 = 0 \quad \text{or } x - 1 = 0$$

$$2x \qquad = 3 \quad \text{or } x \qquad = 1$$

$$x \qquad = 3/2 \text{ or } x \qquad = 1. \; \blacksquare$$

Factoring to solve quadratic equations is the fastest and easiest method to use on many problems. The method has one serious disadvantage, however. Not every quadratic equation can be solved by factoring. For example, $x^2 - 3x + 1 = 0$ cannot be done by this method. In the next section, we will learn how to solve this type of problem.

6.4 EXERCISES

Solve each quadratic equation. See Examples 1–4.

1. $x^2 - 5x + 4 = 0$
2. $y^2 + 7y + 12 = 0$
3. $z^2 + z - 6 = 0$
4. $x^2 - 10x + 24 = 0$
5. $m^2 + 2m - 24 = 0$
6. $p^2 - 5p - 24 = 0$
7. $x^2 - x = 2$
8. $y^2 - 2y = 8$
9. $z^2 - 9z + 8 = 0$
10. $x^2 + 6x + 8 = 0$
11. $6m^2 - 13m + 6 = 0$
12. $3p^2 - 5p + 2 = 0$
13. $4x^2 + 4x = 3$
14. $6y^2 + 5y + 1 = 0$
15. $9z^2 - 6z + 1 = 0$
16. $6x^2 - 7x = 3$
17. $2m^2 + 2m = 0$
18. $3v^2 + 6v = 0$
19. $x^2 - 1 = 0$
20. $y^2 - 16 = 0$
21. $r^2 = 16$
22. $s^2 = 25$
23. $25x^2 = 36$
24. $25y^2 = 36y$
25. $6z^2 - 13z = 5$
26. $2x^2 + 5x = 3$
27. $6p^2 = 3 - 7p$
28. $m^2 = 11m - 24$
29. $r^2 - 10 = 3r$
30. $-s^2 + 21 = 4s$
31. $x^2 - 8x + 15 = 0$
32. $2y^2 - 7y = 15$
33. $3z^2 + 13z = 10$
34. $3x^2 + 17x + 10 = 0$
35. $2y^2 = 8$
36. $2z^2 = 8z$

✳**37.** An object falls according to the formula $s = -16t^2 + 144$ where s represents the distance of the object above the ground at any time t in seconds. How long will it take an object to fall 80 feet?

✳**38.** A 25-foot ladder is placed so that it touches a wall 15 feet above ground. If the bottom of the ladder is pulled out four feet, how far down the wall will the top of the ladder slip? (Hint: Use the Pythagorean theorem.)

✳**39.** A boy was sitting 600 yards from the front of a train. As the train started to move, he saw his dog run from beside his window to the front of the train and back to his window. How far did the dog run if the train went 800 yards during the same time and each moved at a uniform rate of speed?

6.5   The Quadratic Formula

If a quadratic equation is in standard form $ax^2 + bx + c = 0$, it differs from every other quadratic equation only in the values of a, b, and c. Hence the solution of the equation depends on the values of a, b, and c. The formula known as the *quadratic formula* gives the solution of a quadratic equation in standard form. It gives the solution of quadratic equations that cannot be factored, as well as those that can be.

The Quadratic Formula
For a quadratic equation in standard form $ax^2 + bx + c = 0$, the solutions are given by

$$x = \frac{-b \pm \sqrt{b^2 - 4ac}}{2a}$$

EXAMPLE 1

Solve $x^2 - 3x + 1 = 0$.

Solution   In this equation $a = 1$, $b = -3$, and $c = 1$. Substituting these values into the formula, we have:

$$x = \frac{\overset{-b}{-(-3)} \pm \sqrt{\overset{b^2}{(-3)^2} \overset{-4\,a\,c}{- 4(1)(1)}}}{\underset{2a}{2(1)}}$$

$$x = \frac{3 \pm \sqrt{9 - 4}}{2}$$

$$x = \frac{3 \pm \sqrt{5}}{2}. \blacksquare$$

The $\pm$ sign indicates that there are two solutions. One solution contains the plus sign and the other contains the minus sign. That is,

$$x = \frac{3 + \sqrt{5}}{2} \text{ is one answer, and } x = \frac{3 - \sqrt{5}}{2} \text{ is the other.}$$

The expression $b^2 - 4ac$, which appears in the formula, is called the *discriminant*. If $b^2 - 4ac < 0$, you need not complete the computation, for the square root of a negative number is not a real number. A computer program written to use the quadratic formula should include a check to determine whether the discriminant is negative before doing the computation.

Unlike factoring, the quadratic formula will *always* produce a solution to a quadratic equation when a solution exists. The method is not, however, as efficient as factoring for many problems. The method is summarized as follows.

To solve a quadratic equation using the quadratic formula:
1. If the equation is not in standard form $ax^2 + bx + c = 0$, write an equivalent equation that is in standard form (i.e., add or subtract the same term(s) on both sides of the equation in order to make one side equal zero, and arrange the terms in descending order of exponents).
2. Identify the values of a, b, and c.
3. Substitute the values of a, b, and c into the formula

$$x = \frac{-b \pm \sqrt{b^2 - 4ac}}{2a}$$

4. a. Using the $+$ sign, simplify the expression to obtain one solution.
   b. Using the $-$ sign, simplify the expression to obtain the other solution.

EXAMPLE 2

Use the quadratic formula to solve $x^2 - 5x + 6 = 0$.

Solution
1. The equation is in standard form.
2. $a = 1$, $b = -5$, and $c = 6$.
3. $x = \dfrac{-(-5) \pm \sqrt{(-5)^2 - 4(1)(6)}}{2(1)}$
4. $x = \dfrac{5 \pm \sqrt{25 - 24}}{2}$

   $x = \dfrac{5 \pm \sqrt{1}}{2}$

   $x = \dfrac{5 \pm 1}{2}$

$$x = \frac{5 + 1}{2} = \frac{6}{2} = 3 \quad \text{or} \quad x = \frac{5 - 1}{2} = \frac{4}{2} = 2 \ \blacksquare$$

EXAMPLE 3

Use the quadratic formula to solve $2x^2 + 3x = -1$.

Solution

1. The equation is not in standard form, so we add 1 to both sides.

$$2x^2 + 3x + 1 = -1 + 1$$
$$2x^2 + 3x + 1 = 0$$

2. $a = 2$, $b = 3$, and $c = 1$

3. $$x = \frac{-3 \pm \sqrt{3^2 - 4(2)(1)}}{2(2)}$$

4. $$x = \frac{-3 \pm \sqrt{9 - 8}}{4}$$

$$x = \frac{-3 \pm \sqrt{1}}{4}$$

$$x = \frac{-3 \pm 1}{4}$$

$$x = \frac{-3 + 1}{4} = \frac{-2}{4} = \frac{-1}{2} \quad \text{or} \quad x = \frac{-3 - 1}{4} = \frac{-4}{4} = -1 \ \blacksquare$$

EXAMPLE 4

Use the quadratic formula to solve $x^2 - 5x = 0$.

Solution The equation is in standard form. $a = 1$, $b = -5$, and $c = 0$.

$$x = \frac{-(-5) \pm \sqrt{(-5)^2 - 4(1)(0)}}{2(1)} = \frac{5 \pm \sqrt{25 - 0}}{2} = \frac{5 \pm \sqrt{25}}{2} =$$

$$\frac{5 \pm 5}{2} = 5 \text{ or } 0 \ \blacksquare$$

Even though the square root that occurs in the formula may not be an integer, it may be possible to write the square root in a simplified form. Because $\sqrt{xy} = (xy)^{1/2} = x^{1/2}(y^{1/2}) = \sqrt{x}\sqrt{y}$, it may be possible to factor the number under the square root symbol so that the square root of one of the factors is an integer. For example,

$$\sqrt{18} = \sqrt{9(2)} = \sqrt{9}\sqrt{2} = 3\sqrt{2}.$$

EXAMPLE 5   Use the quadratic formula to solve $x^2 - 5 = 0$.

Solution   The equation is in standard form. $a = 1$, $b = 0$, and $c = -5$.

$$x = \frac{-0 \pm \sqrt{0^2 - 4(1)(-5)}}{2(1)} = \frac{-0 \pm \sqrt{0 + 20}}{2} = \frac{\pm\sqrt{20}}{2} =$$

$$\frac{\pm\sqrt{4(5)}}{2} = \frac{\pm\sqrt{4}\sqrt{5}}{2} = \frac{\pm 2\sqrt{5}}{2} = \pm\sqrt{5} \ ■$$

Now that you know how to use the formula, you might wonder how we know it will always produce a solution when a solution exists. We begin with the general equation $ax^2 + bx + c = 0$. Our objective is to solve this equation for x.

1. We multiply both sides of the equation by $4a$. The reason that we use $4a$ is to make the expression on the lefthand side easier to factor. We have: $4a^2x^2 + 4abx + 4ac = 0$.

2. We subtract $4ac$ from both sides of the equation. This step leaves the two terms with x in them on the lefthand side. We have $4a^2x^2 + 4abx = -4ac$.

3. We add b^2 to both sides of the equation. We could add *any* number, as long as we add it on both sides of the equation. It just happens that b^2 will make the expression on the lefthand side of the equation easier to factor. We have $4a^2x^2 + 4abx + \boldsymbol{b^2} = \boldsymbol{b^2} - 4ac$.

4. Factor the expression on the lefthand side. We have

$$(2ax + b)(2ax + b) = b^2 - 4ac$$
$$(2ax + b)^2 = b^2 - 4ac.$$

5. Since we are solving for x, we need to remove the square on the lefthand side for the next step. Finding the square root is the inverse operation of finding the square. Therefore we take the square root on both sides of the equation. The radical sign, which restricts the square root to a positive value, did not appear in the original problem. That is, we do not want to restrict the answer to the positive square root only, so we introduce the $\pm$ sign to show both square roots. We have

$$\sqrt{(2ax + b)^2} = \pm\sqrt{b^2 - 4ac},$$

or since the square root "undoes" the square, we have

$$2ax + b = \pm\sqrt{b^2 - 4ac}.$$

6. The next step will be to remove the b term.

$$2ax + b - b = -b \pm \sqrt{b^2 - 4ac}$$
$$\text{or } 2ax = -b \pm \sqrt{b^2 - 4ac}$$

7. Finally, we remove the factor $2a$.

$$\frac{2ax}{2a} = \frac{-b \pm \sqrt{b^2 - 4ac}}{2a}$$

$$\text{or } x = \frac{-b \pm \sqrt{b^2 - 4ac}}{2a}$$

You can see from this derivation that from a theoretical standpoint, the quadratic formula is completely accurate. From a practical standpoint, however, the limitations of computer arithmetic may lead to inaccurate results. For example, if $b > 0$ and is very large compared to a and c, $\sqrt{b^2 - 4ac}$ is very close to $\sqrt{b^2} = b$. From the formula, we would have as one solution,

$$x = \frac{-b \pm \sqrt{b^2 - 4ac}}{2a}$$

and x is very close to

$$\frac{-b + b}{2a} = 0.$$

An alternate solution was devised by William Kahan.[†]

The quadratic formula for a computer solution:

If $b \geq 0$ $x = \dfrac{-b - \sqrt{b^2 - 4ac}}{2a}$ and $x = \dfrac{2c}{-b - \sqrt{b^2 - 4ac}}$

If $b < 0$ $x = \dfrac{-b + \sqrt{b^2 - 4ac}}{2a}$ and $x = \dfrac{2c}{-b + \sqrt{b^2 - 4ac}}$

The solutions of the equation $0.5x^2 + 200,000x + 0.5 = 0$ are given on a computer as approximately 0.00021 and $-400,000$ by the standard quadratic formula. They are given as -0.0000025 and $-400,000$ by the alternate formula. The alternate formula is presented here only to bring to your attention the

[†]Forsythe, George E. "Solving a Quadratic Equation On a Computer," *The Mathematical Sciences*, edited by COSRIMS, The MIT Press, 1969.

fact that considerations for solving a problem mathematically and solving it on a computer are often different. Writing a computer program that always produces a correct solution for a particular type of problem requires more than simply coding an algebraic formula or procedure for the computer.

6.5 EXERCISES

Use the quadratic formula to solve each equation. See Examples 1–5.

1. $x^2 - 4x + 2 = 0$ **2.** $y^2 - 3y + 1 = 0$

3. $z^2 - 5z + 3 = 0$ **4.** $2x^2 - 3x - 1 = 0$

5. $2y^2 + 5y = 1$ **6.** $2p^2 + 3p = 7$

7. $q^2 - 21 = 0$ **8.** $r^2 - 43 = 0$

9. $2x^2 - 7x = 0$ **10.** $3y^2 + 2y - 3 = 0$

11. $3z^2 - z - 1 = 0$ **12.** $x^2 - 5x = 7$

13. $p^2 - 6p = 7$ **14.** $10q^2 + q - 1 = 0$

15. $r^2 + 8r + 1 = 0$ **16.** $x^2 + 3x + 1 = 0$

17. $2y^2 + 4y = 1$ **18.** $3z^2 + 4z = 1$

19. $x^2 + 5x + 2 = 0$ **20.** $y^2 - 5y = 2$

21. $3z^2 + z = 1$ **22.** $x^2 = 3$

23. $y^2 = 3y$ **24.** $z^2 - 7z = 8$

25. Use pseudocode or a flowchart to show an algorithm for solving a quadratic equation (with $c \neq 0$) by computer. If $b^2 - 4ac < 0$, take the appropriate action.

26. Solve problems 1–4 of Section 6.4 using the quadratic formula for a computer to verify that the solution is correct.

27. Assuming $x \neq 0$, multiply the equation $ax^2 + bx + c = 0$ by $1/x^2$.

(a) Use the quadratic formula to solve the equation for $1/x$, and invert the solution to obtain an expression for x.

(b) If $b \geq 0$, we replaced the solution $\dfrac{-b + \sqrt{b^2 - 4ac}}{2a}$ by $\dfrac{2c}{-b - \sqrt{b^2 - 4ac}}$ to avoid $\dfrac{-b + b}{2a}$. If $b < 0$, why is $\dfrac{-b - \sqrt{b^2 - 4ac}}{2a}$ replaced by $\dfrac{2c}{-b + \sqrt{b^2 - 4ac}}$?

28. A cubical box that measures $3' \times 3' \times 3'$ is pushed against a wall. A ladder $15'$ long is placed so that it rests against the edge of the box and the wall. Use quadratic equations to determine how far from the box the foot of the ladder is.

6.6 Higher Degree Equations

Ancient Babylonian mathematicians knew how to solve quadratic equations. A general method for solving cubic equations (equations of the form $ax^3 + bx^2 + cx + d = 0$, $a, b, c, d \in R$, $a \neq 0$), however, was not found until the sixteenth century. A method for solving quartic equations (equations of the form $ax^4 + bx^3 + cx^2 + dx + e = 0$, $a, b, c, d, e \in R$, $a \neq 0$) was found soon after that. Success with the cubic and quartic led to a 300-year search for a method to solve higher degree equations. In the early 1800s, a young Norwegian mathematician, Neils Henrik Abel, proved that there can be no formula for solving quintic equations (equations of the form $ax^5 + bx^4 + cx^3 + dx^2 + ex + f = 0$, $a, b, c, d, e, f \in R$, $a \neq 0$). Much mathematics evolved from the search, however. In particular, the work of Evariste Galois in France led to important concepts in modern algebra.

Even though there is a method for solving the general cubic equation $ax^3 + bx^2 + cx + d = 0$, it is a long and complicated procedure. Some cubics (and higher degree equations) are factorable, but they are special cases. For example, $x^3 + 5x^2 + 6x = 0$ can be written as

$$x(x^2 + 5x + 6) = 0 \text{ or}$$

$$x(x + 2)(x + 3) = 0.$$

Thus, we have $x = 0$, $x + 2 = 0$, or $x + 3 = 0$.
That is, $x = 0$, $x = -2$, or $x = -3$.

To solve a factorable equation of degree 3 or more:
1. If it is not in standard form, write an equivalent equation that is in standard form (i.e., add or subtract the same term(s) on both sides of the equation to make one side equal zero, and arrange the terms in descending order of exponents).
2. Factor completely.
3. Set each factor equal to 0.
4. Solve each linear or quadratic equation.

EXAMPLE 1

Solve $x^4 - 13x^2 + 36 = 0$.

Solution
1. The equation is in standard form.
2. Factor: $(x^2 - 4)(x^2 - 9) = 0$, which can be factored further $(x - 2)(x + 2)(x - 3)(x + 3) = 0$.
3. $x - 2 = 0$, $x + 2 = 0$, $x - 3 = 0$, or $x + 3 = 0$
4. $x = 2$, $x = -2$, $x = 3$, or $x = -3$ ∎

| **EXAMPLE 2** | Solve $x^3 + 3x^2 + x = 0$. |

Solution

1. The equation is in standard form.
2. Factor: $x(x^2 + 3x + 1) = 0$. The second factor is prime, so we have
3. $x = 0$ or $x^2 + 3x + 1 = 0$.
4. $x = 0$, or by the quadratic formula $x = \dfrac{-3 \pm \sqrt{5}}{2}$. ∎

The computer is a useful tool for finding *approximate* solutions to those equations that do not lend themselves to finding *exact* solutions easily.

As an example of a *numerical* (as opposed to an *algebraic*) method, we will examine the bisection method for finding the solution to higher degree polynomial equations. (A polynomial equation has the form $a_n x^n + a_{n-1} x^{n-1} + a_{n-2} x^{n-2} + \ldots + a_0 = 0$ where $a_i \in R$.) The bisection method is just one of many available, but it illustrates the overall nature of an *iterative* process. The approach is to take a guess, use that guess to obtain a better one, and so on, until we have the answer to the desired degree of precision.

Consider the equation $x^3 + x - 3 = 0$. Suppose we want the solution rounded to hundredths. We might say that we are trying to find the value for x which makes the polynomial $x^3 + x - 3$ equal to 0. By making a few preliminary calculations, we discover that when $x = 1$, $x^3 + x - 3 = -1$ and when $x = 2$, $x^3 + x - 3 = 7$. That is, as x is allowed to take on values between 1 and 2, the polynomial $x^3 + x - 3$ takes on values between -1 and 7. Since 0 falls in the interval between -1 and 7, we know that the value we are seeking must fall within the interval from 1 (which produced the -1) to 2 (which produced the 7). For our first guess, we bisect this interval and guess 1.5, which is the average of 1 and 2. When $x = 1.5$, the polynomial is equal to $(1.5)^3 + 1.5 - 3 = 1.875$.

Since the polynomial must take on a value of zero somewhere between -1 and 1.875 (i.e., between a negative value and a positive value), the x value we seek must be between 1 (which produced the -1) and 1.5 (which produced the 1.875). We bisect the interval from 1 to 1.5 and guess 1.25, which is the average of 1 and 1.5, for our second guess.

We must continue this process until when x is rounded to the nearest hundredth, we get the same guess twice in a row. The repetition is a signal that further computation will not produce a better guess. (Computer programs may be written to terminate execution when the difference between two x values is less than some predetermined amount, such as 0.005.) It is easier to see the steps involved if we put the results in table form (see Table 6.1).

Each guess comes from realizing that $x^3 + x - 3$ will take on both positive and negative values, but zero must be somewhere between them. Each time we add a new row to the table, we find the last entry in the column under $x^3 + x$

Table 6.1

Iteration no.	x	$x^3 + x - 3$	Rounded value of x
0	1.00000	−1.00	1.00
0	2.00000	7.00	2.00
1	1.50000	1.88	1.50
2	1.25000	0.20	1.25
3	1.12500	−0.45	1.13
4	1.18750	−0.14	1.19
5	1.21875	0.03	1.22
6	1.203125	−0.06	1.20
7	1.2109375	−0.01	1.21
8	1.21484375	0.01	1.21

− 3 that has the *opposite* sign. Taking the average of the two corresponding unrounded x values gives the next x value. Notice that as the x values approach the solution 1.21, the values of $x^3 + x - 3$ approach 0.

While we are interested in a final answer rounded to hundredths, the computations for x in the example are not rounded. Rounding all of the calculations to two places would introduce unacceptably large round-off error; therefore, the rounding step is last. The polynomial values, however, are rounded, since we are interested only in their signs. The procedure can be generalized as an algorithm.

To solve a polynomial equation in x using the bisection method:

1. Find the vicinity of the solution.
   a. Find a value for x that makes the polynomial positive.
   b. Find a value for x that makes the polynomial negative.

2. For the first iteration:
   a. Use the average of the two x values from step 1.
   b. Calculate the value of the polynomial.
   c. Round off the x value to the desired precision.

3. For the next iteration:
   a. Find the last iteration for which the value of the polynomial has the opposite sign from the value calculated for the polynomial in the previous step.
   b. Use the average of the corresponding x values for the next approximation. Use as many decimal places as your calculator or computer allows.
   c. Calculate the value of the polynomial.
   d. Round off the x value to the desired precision.

4. Repeat step 3 until the same x value appears twice consecutively in step 3 as the rounded value for x.

EXAMPLE 3

Solve $x^3 + x - 3 = 0$ for x rounded to thousandths.

Solution   The following table gives a computer-generated solution. Each x value is shown rounded to thousandths, although *more digits were used in the calculations*.

Iteration no.	Rounded value of x	Sign of $x^3 + x - 3$
0	1.000	−
0	2.000	+
1	1.500	+
2	1.250	+
3	1.125	−
4	1.188	−
5	1.219	+
6	1.203	−
7	1.211	−
8	1.215	+
9	1.213	−
10	1.214	+
11	1.213	−
12	1.214	+
13	1.214	+

After the 13th iteration, the rounded value of x appears as 1.214 twice, so that number is the solution rounded to thousandths. Notice that increasing the desired precision from hundredths to thousandths increased the number of iterations required to find a solution. ■

If you are using a calculator that does not have an exponent key or memory, it is very tedious to enter numbers having five or six decimal places. But the purpose of working these problems is to learn the procedure, and the procedure is the same whether you use three decimal places or eight. For the next example, then, we will use three decimal places for *each* calculation. Using only one or two decimal places beyond the desired precision, however, may introduce a round-off error into the final solution.

EXAMPLE 4

Solve $x^5 - 27 = 0$ for x rounded to hundredths.

Solution
1.   When $x = 1$, $x^5 - 27 = -26$.
     When $x = 2$, $x^5 - 27 = 5$.

2. For the first iteration, we use the average of 1 and 2, or 1.5. When $x = 1.5$, the polynomial has the value $(1.5)^5 - 27 = -19.406$.

3. For the second iteration, we use the x values corresponding to -19.406 and 5. That is, we average 2 and 1.5 to get 1.75. The polynomial has a value of $(1.75)^5 - 27 = -10.587$.

4. For the third iteration, we use the x values corresponding to -10.587 and 5. That is, we average 1.75 and 2 to get 1.875. The value of the polynomial is $(1.875)^5 - 27 = -3.826$.

5. For the fourth iteration, we use the x values corresponding to -3.826 and 5. That is, we average 1.875 and 2 to get 1.938. The value of the polynomial is 0.338.

6. For the fifth iteration, we use the x values corresponding to 0.338 and -3.826. That is, we average 1.938 and 1.875 to get 1.907. The value of the polynomial is -1.780.

7. Iterations 6–10 are shown in the following table.

Iteration no.	x	(Rounded)	$x^5 - 27$
0	1	(1.00)	-26
0	2	(2.00)	5
1	1.5	(1.50)	-19.406
2	1.75	(1.75)	-10.587
3	1.875	(1.88)	-3.826
4	1.938	(1.94)	0.338
5	1.907	(1.91)	-1.780
6	1.923	(1.92)	-0.704
7	1.931	(1.93)	-0.152
8	1.935	(1.94)	0.127
9	1.933	(1.93)	-0.013
10	1.934	(1.93)	0.057

Since the rounded value 1.93 appears twice consecutively, we say that 1.93 is the solution rounded to hundredths. ■

6.6 EXERCISES

Solve each equation by factoring. See Examples 1 and 2.

1. $x^3 - 5x^2 + 6x = 0$

2. $x^3 - 6x^2 + 9x = 0$

3. $x^4 - 10x^2 + 9 = 0$

4. $x^4 - 13x^2 + 36 = 0$

5. $x^4 - 5x^2 + 4 = 0$

6. $x^4 - 17x^2 + 16 = 0$

7. $x^4 - 2x^2 + 1 = 0$

8. $4x^5 - 9x^3 = 0$

9. $2x^3 - 3x^2 + x = 0$

10. $6x^4 - 13x^3 + 6x^2 = 0$

Use the bisection method to find a solution (rounded to hundredths) between 1 and 2 for each equation. See Examples 3 and 4.

11. $x^2 - 3 = 0$ **12.** $x^3 - x^2 - 1 = 0$

13. $x^2 - x - 1 = 0$ **14.** $x^3 - x^2 - x = 0$

15. $2x^3 - 5x - 3 = 0$

Use the bisection method to find a solution (rounded to tenths) between 2 and 3 for each equation. See Examples 3 and 4.

16. $x^2 - 6 = 0$ **17.** $x^3 - 15 = 0$

18. $3x^2 - 4x - 6 = 0$ **19.** $x^3 - x^2 - 7 = 0$

20. $2x^3 - 5x - 8 = 0$

Use the bisection method to find a solution (rounded to hundredths) between 4 and 5 for each equation. See Examples 3 and 4.

21. $x^3 - 100 = 0$ **22.** $x^4 - 500 = 0$

23. $x^3 - x^2 - 50 = 0$ **24.** $x^3 - 2x^2 - 10x = 0$

➠**25.** Use pseudocode or a flowchart to show an algorithm for solving a polynomial equation by the bisection method.

6.7   Equations with Rational Expressions

In Section 6.2 we learned that the denominator can be eliminated from a formula by multiplying by the LCD. This technique can also be applied to equations involving rational expressions. For example, to solve $\dfrac{2}{2x - 1} = \dfrac{x}{x + 1}$, multiply both sides of the equation by $(2x - 1)(x + 1)$, which is the LCD. We have

$$(2x - 1)(x + 1)\left(\frac{2}{2x - 1}\right) = (2x - 1)(x + 1)\left(\frac{x}{x + 1}\right).$$

The denominators divide out, leaving

$$(x + 1)(2) = x(2x - 1)$$
$$2x + 2 = 2x^2 - x.$$

Writing the equation in standard form, we have

$$0 = 2x^2 - 3x - 2$$
$$0 = (2x + 1)(x - 2).$$

Thus,

$$2x + 1 = 0 \text{ or } x - 2 = 0$$
$$x = -\frac{1}{2} \text{ or } x = 2.$$

To check this solution, we first replace x by $-\dfrac{1}{2}$:

$$\frac{2}{-1-1} = \frac{-\dfrac{1}{2}}{\dfrac{1}{2}}$$

$$\frac{2}{-2} = -1.$$

Then we replace x by 2:

$$\frac{2}{4-1} = \frac{2}{2+1}$$

$$\frac{2}{3} = \frac{2}{3}.$$

Up to this point, when denominators have included variables, we have said that you may assume there are no zero denominators. We no longer make that assumption. Solving the equation that results from eliminating the denominators may produce a value for which the denominator of one of the fractions in the original equation is zero. Such a value is not a solution of the original equation.

To solve an equation with rational expressions:
1. Multiply both sides of the equation by the LCD of all the rational expressions in the equation.
2. Solve the resulting equation.
3. Eliminate from the solution any value that produces a zero denominator in the original equation.

EXAMPLE 1

Solve for x: $\dfrac{x}{x-3} = \dfrac{3}{x-3} + 3$.

Solution   Multiply both sides of the equation by $(x-3)$.

$$(x-3)\left(\frac{x}{x-3}\right) = (x-3)\left(\frac{3}{x-3} + 3\right)$$

$$\cancel{(x-3)}\left(\frac{x}{\cancel{x-3}}\right) = \cancel{(x-3)}\left(\frac{3}{\cancel{x-3}}\right) + (x-3)(3)$$

$$x = 3 + (x-3)(3) \text{ since the denominators divide out.}$$

$$x = 3 + 3x - 9 \text{ from the distributive property.}$$

$$x = -6 + 3x \text{ when the righthand side is simplified.}$$

$$-2x = -6 \text{ after } 3x \text{ is subtracted from both sides.}$$

$$x = 3 \text{ after both sides are divided by } -2.$$

But if $x = 3$, both denominators in the original equation are zero. Hence, there is no solution to this problem. ∎

EXAMPLE 2

Solve $\dfrac{1}{3} + \dfrac{1}{x + 1} = \dfrac{-2}{(x + 1)(x - 3)}$.

Solution Multiply both sides by $3(x + 1)(x - 3)$, the LCD.

$$3(x + 1)(x - 3)\left(\frac{1}{3} + \frac{1}{x + 1}\right) = 3(x + 1)(x - 3)\left(\frac{-2}{(x + 1)(x - 3)}\right)$$

$$3(x + 1)(x - 3)\frac{1}{3} + 3(x + 1)(x - 3)\frac{1}{x + 1}$$

$$= 3(x + 1)(x - 3)\frac{-2}{(x + 1)(x - 3)}$$

$(x + 1)(x - 3) + 3(x - 3) = 3(-2)$ since the denominators divide out.

$x^2 - 3x + x - 3 + 3x - 9 = -6$ from the distributive property.

$x^2 + x - 12 = -6$ when the lefthand side is simplified.

$x^2 + x - 6 = 0$ after adding 6 to both sides.

$(x + 3)(x - 2) = 0$ when the lefthand side is factored.

$x + 3 = 0$ or $x - 2 = 0$

$x = -3$ or $x = 2$

Both answers are valid solutions.

When $x = -3$, we have $\dfrac{1}{3} + \dfrac{1}{-2} = \dfrac{-2}{(-2)(-6)}$

$$\frac{2}{6} - \frac{3}{6} = \frac{-2}{12} \quad \text{or} \quad -\frac{1}{6} = -\frac{1}{6}.$$

When $x = 2$, $\dfrac{1}{3} + \dfrac{1}{3} = \dfrac{-2}{3(-1)}$ or $\dfrac{2}{3} = \dfrac{2}{3}$. ∎

EXAMPLE 3

Solve $\dfrac{1}{x + 1} + \dfrac{x}{x - 2} = \dfrac{7x - 8}{x^2 - x - 2}$.

Solution $\dfrac{1}{x + 1} + \dfrac{x}{x - 2} = \dfrac{7x - 8}{(x + 1)(x - 2)}$

Multiply both sides of the equation by the LCD, $(x + 1)(x - 2)$.

$$(x + 1)(x - 2)\left(\frac{1}{x + 1} + \frac{x}{x - 2}\right) = (x + 1)(x - 2)\left(\frac{7x - 8}{(x + 1)(x - 2)}\right)$$

$$(x + 1)(x - 2)\frac{1}{x + 1} + (x + 1)(x - 2)\frac{x}{x - 2}$$

$$= (x + 1)(x - 2)\frac{7x - 8}{(x + 1)(x - 2)}$$

$x - 2 + x(x + 1) = 7x - 8$ since the denominators divide out.

$x - 2 + x^2 + x = 7x - 8$ from the distributive property.

$x^2 + 2x - 2 = 7x - 8$ after the lefthand side is simplified.

$x^2 - 5x + 6 = 0$ after $7x$ is subtracted and 8 added to both sides.

$(x - 2)(x - 3) = 0$ when the lefthand side is factored.

Thus, $x - 2 = 0$ or $x - 3 = 0$.

We have $x = 2$ or $x = 3$.

But $x = 2$ would produce a denominator of zero in the original problem. We conclude that $x = 3$ is the only solution. When $x = 3$, we have

$$\frac{1}{4} + 3 = \frac{21 - 8}{9 - 3 - 2}$$

$$\frac{1}{4} + \frac{12}{4} = \frac{13}{4}. \quad \blacksquare$$

EXAMPLE 4

Solve $\dfrac{1}{x - 1} + \dfrac{x}{x - 3} = \dfrac{4x - 6}{x^2 - 4x + 3}$.

Solution $\dfrac{1}{x - 1} + \dfrac{x}{x - 3} = \dfrac{4x - 6}{(x - 1)(x - 3)}$

Multiply both sides of the equation by the LCD, $(x - 1)(x - 3)$.

$$(x - 1)(x - 3)\left(\frac{1}{x - 1} + \frac{x}{x - 3}\right) = (x - 1)(x - 3)\left(\frac{4x - 6}{(x - 1)(x - 3)}\right)$$

$$(x - 1)(x - 3)\,\frac{1}{x - 1} + (x - 1)(x - 3)\,\frac{x}{x - 3} =$$

$$(x - 1)(x - 3)\,\frac{4x - 6}{(x - 1)(x - 3)}$$

$x - 3 + (x - 1)x = 4x - 6$

$x - 3 + x^2 - x = 4x - 6$

$x^2 - 3 = 4x - 6$

$x^2 - 4x + 3 = 0$

$(x - 3)(x - 1) = 0$

$x - 3 = 0$ or $x - 1 = 0$

$x = 3$ or $x = 1$

Both values would produce zero denominators in the original equation. We conclude that this problem has no solution. $\blacksquare$

From Examples 2, 3, and 4, you can see that eliminating denominators may lead to a quadratic equation. It is possible that of the two solutions, both, only one, or neither is a solution of the original equation involving rational expressions.

6.7 EXERCISES

Solve each equation. See Examples 1–4.

1. $\dfrac{x}{3} + \dfrac{4}{5} = \dfrac{1}{2}$

2. $\dfrac{y-1}{4} + \dfrac{5}{3} = \dfrac{2y+7}{6}$

3. $\dfrac{2x+5}{3x} + \dfrac{7}{x} = 5$

4. $\dfrac{y}{4} + \dfrac{y-1}{6} = \dfrac{y+2}{3}$

5. $\dfrac{3z-2}{z} + \dfrac{7}{2z} = 9$

6. $\dfrac{x+5}{2x} - 3 = \dfrac{10}{x}$

7. $\dfrac{2a+3}{3a} + \dfrac{5}{3} = \dfrac{1}{a}$

8. $\dfrac{2}{y-1} + \dfrac{1}{3} = \dfrac{5}{3y-3}$

9. $\dfrac{2}{x-1} + \dfrac{3}{5} = \dfrac{5x-1}{4x-4}$

10. $\dfrac{4}{z} + \dfrac{2}{3} = \dfrac{2z}{3}$

11. $\dfrac{x}{5} - \dfrac{5}{x} = \dfrac{2x-1}{30}$

12. $\dfrac{2y}{y+2} - \dfrac{3y}{y-2} = \dfrac{25}{y^2-4}$

13. $\dfrac{a+1}{a-3} - \dfrac{2a+2}{a-2} = \dfrac{6}{a^2-5a+6}$

14. $\dfrac{1}{x-1} + \dfrac{1}{2} = \dfrac{2}{x^2-1}$

15. $\dfrac{2}{y-1} + \dfrac{3y}{y+2} = \dfrac{y+5}{y^2+y-2}$

16. $\dfrac{2z}{z-3} + \dfrac{1}{z+3} = \dfrac{2z}{z^2-9}$

17. $\dfrac{a-1}{a+1} - \dfrac{a-2}{a+2} = \dfrac{a^2+2a-1}{a^2+3a+2}$

18. $\dfrac{m+1}{m-2} - \dfrac{m+2}{m-3} = \dfrac{-m^2+4}{m^2-5m+6}$

19. $\dfrac{x}{2x+1} - \dfrac{1}{x} = \dfrac{-x^2-3x-1}{2x^2+x}$

20. $\dfrac{3y}{3y-1} - \dfrac{2}{y+2} = \dfrac{-5y+4}{3y^2+5y-2}$

21. $\dfrac{1}{4p-2} + \dfrac{p}{p+3} = \dfrac{p^2-3p+3}{2p^2+5p-3}$

22. $\dfrac{a-2}{a+3} + \dfrac{a}{a-1} = \dfrac{a^2-2a+5}{a^2+2a-3}$

23. $\dfrac{x-1}{x} + \dfrac{x}{x+1} = \dfrac{-3x-1}{2x^2+2x}$

24. $\dfrac{y-2}{y} + \dfrac{y}{2y-1} = \dfrac{y^2-4y+2}{2y^2-y}$

✳**25.** Two children can paddle a boat at the rate of 12 mph in still water. If they can go 12 miles upstream in the same length of time they go 20 miles downstream, what is the rate of the current?

✳**26.** If the rate of the current is 4 mph and a trip 12 miles upstream and back takes 1 hour and 52 minutes, what is the rate of the boat in still water?

✳**27.** If you travel 50 miles at 50 mph, but make the return trip at 60 mph, what is the average rate of speed for the round trip?

EXAMPLE 4

Determine the final value of each variable in the program segment.

A = 2

B = 3

C = A

A = B

B = C

Solution

A	B	C
2̸ 3	3̸ 2	2

Thus, we have A = 3, B = 2, and C = 2. ∎

In Section 3.6, the IF. . .THEN statement was introduced. Remember that the single alternative decision structure takes the form IF (condition) THEN (action) while the double alternative decision structure takes the form IF (condition) THEN (action 1) ELSE (action 2). The assigning of a value to a variable is one possible action that may occur in a decision structure.

EXAMPLE 5

Determine the final value of each variable in the program segment.

A = 1

B = 2

IF (A + B = 3) THEN A = 2

Solution   The first statement assigns the value 1 to A. The second statement assigns the value 2 to B. Since A + B = 1 + 2 = 3, the condition is true, so the action is performed. That is, the value 2 is assigned to A. We would write:

A	B
1̸ 2	2

Thus, we have A = 2 and B = 2. ∎

EXAMPLE 6

Determine the final value of each variable in the program segment.

A = 1

B = 2

IF (A + B = 5) THEN B = 5

Solution   The first statement assigns the value 1 to A. The second statement assigns the value 2 to B. Since A + B = 1 + 2 ≠ 5, the action is not performed. That is, the value 5 is *not* assigned to B. B retains its current value. We would write:

A	B
1	2

Thus, we have A = 1 and B = 2. ∎

6.8 EXERCISES

Determine the final value of each variable in the program segments. See Examples 1 and 2.

1. X = 2
   X = X + 3

2. Y = 5
   Y = Y − 1

3. TC = 0
   TC = TC + 10

4. SUM = 0
   SUM = SUM + 5

Determine the final value of each variable in the program segments. See Examples 3 and 4.

5. X = 4
   Y = X
   X = Y

6. A = 1
   B = 2
   A = A + B

7. A = 2
   B = 3
   B = A + B

8. X = 1
   Y = 2
   X = Y
   Y = X

9. X = 3
   Y = 2
   X = Y
   Y = X

10. X = 1
   Y = 2
   Z = X
   X = Y
   Y = Z

11. A = 4
   B = 5
   C = A
   A = B
   B = C

Determine the final value of each variable in the program segments. See Examples 5 and 6.

12. X = 2
   Y = 3
   IF (X + Y = 5) THEN X = 3

13. A = 2
   B = 3
   IF (A + B = 4) THEN A = 3

14. X = 1
   Y = 2
   IF (X = Y) THEN Y = X
   X = Y

15. X = −1
   Y = 4
   IF (X + Y = 2) THEN X = 3

16. P = 3
   Q = P
   IF (Q = 3) THEN P = 5

17. A = 5
   B = 3
   C = A + B
   IF (C = 2) THEN A = 2
   B = 4

18. X = −6
   Y = −3
   Z = X − Y
   IF (Z = −3) THEN X = 6
   Y = 3

CHAPTER REVIEW

A linear equation in the variable x is an equation that can be put in the form
_____ where a and $b \in R$, $a \neq 0$.

The value of the variable that makes the equation a true statement is said to
_____ the equation, and it is called the _____ of
the equation.

A quadratic equation in the variable x is an equation that can be put in the form
_____ where a, b, and c are real numbers, $a \neq 0$. This form of
the equation is called the _____ form.

The Special Property of Zero says that if a and b are real numbers such that
$ab = 0$, then _____ or _____.

For a quadratic equation in standard form, the solutions are given by $x =$
_____.

An equation of the form $ax^3 + bx^2 + cx + d = 0$, a, b, c, and $d \in R$,
$a \neq 0$, is called a _____ equation.

An equation of the form $ax^4 + bx^3 + cx^2 + dx + e = 0$, a, b, c, d, and
$e \in R$, $a \neq 0$, is called a _____ equation.

CHAPTER TEST

Solve each equation.

1. $3x - 2 = 7$

2. $\dfrac{P}{5} + 1 = 11$

3. $2(z + 3) + 1 = -z + 4$

4. $3x - (x + 1) = 4(x - 1) + 1$

Solve each inequality.

5. $2y - 1 \leq 5$

6. $3x + 1 > 5x + 3$

Solve each formula for the indicated variable.

7. $D = R \cdot T$ for R

8. $A = L \cdot W$ for W

9. $F = \dfrac{9C}{5} + 32$ for C

10. $L = a + (n - 1)d$ for a

11. $L = a + (n - 1)d$ for d

12. $S = \dfrac{a}{1 - r}$ for a

13. If the interest rate is 9 percent per year, how much should you deposit now in
order to have a total (principal + interest) of \$32,700 in one year?

14. Twelve liters of an acid and water solution is 4 percent acid. How many liters of
a 1 percent acid solution should be added to make a 3 percent solution?

15. Two cars 210 miles apart are headed toward each other and meet after two hours.
If it is known that one travels five mph faster than the other, find their rates.

Solve each quadratic equation by factoring.

16. $x^2 - 15x + 14 = 0$ **17.** $y^2 - 10y + 21 = 0$

18. $2x^2 + 7x + 5 = 0$ **19.** $x^2 + 4x = 5$

20. $z^2 = 2z + 15$

Use the quadratic formula to solve each equation.

21. $x^2 + 5x + 3 = 0$ **22.** $y^2 + 2y - 1 = 0$

23. $2z^2 - 3z - 2 = 0$ **24.** $2x^2 = 4x - 1$

25. $3p^2 - 2p = 2$

Solve each equation by factoring.

26. $x^3 - 3x^2 + 2x = 0$ **27.** $2x^3 + 2x^2 - 4x = 0$ **28.** $x^5 - 5x^3 + 4x = 0$

Use the bisection method to find a solution (rounded to hundredths) for each equation.

29. $x^3 - 25 = 0$ between 2 and 3   **30.** $x^3 - 29 = 0$ between 3 and 4

Solve each equation.

31. $\dfrac{1}{x} + \dfrac{2}{3x} = 1$ **32.** $\dfrac{2}{a} + \dfrac{3}{2} = \dfrac{5}{6}$

33. $\dfrac{1}{x - 2} + \dfrac{2}{x + 3} = \dfrac{2x + 1}{x^2 + x - 6}$ **34.** $\dfrac{1}{x - 2} - \dfrac{2}{x + 3} = \dfrac{x^2 + 3x - 5}{x^2 + x - 6}$

35. $\dfrac{1}{x + 3} + \dfrac{2}{x - 3} = \dfrac{x^2 + 3x - 6}{x^2 - 9}$

Determine the final value of each variable in the program segments.

36. A = 2
B = 3*A

37. B = 6
B = B + 3

38. A = 2
B = 3*A
A = 3*B

39. X = 1
Y = X + 3
Z = X*Y

40. X = 2
Y = 2*X
IF (Y = X*X) THEN (X = Y*Y)

▐▶**41.** A geometric series is a sum in which each term after the first (a) is obtained by multiplying the previous term by a fixed constant (r). The following flowchart shows an algorithm for computing the sum of a geometric series with n terms. Modify it so that the sum is displayed only if $a + ar + ar^2 + \ldots + ar^{n-1} = (a - ar^n)/(1 - r)$.

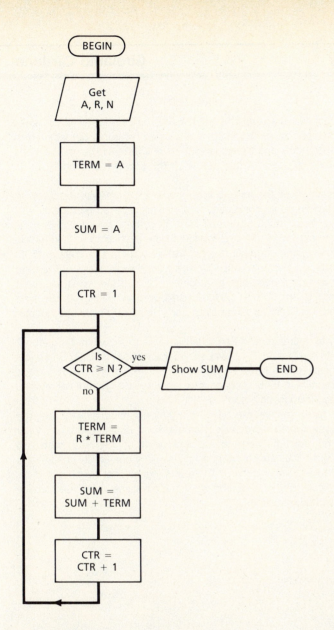

FOR FURTHER READING

Kimberling, Clark. "Roots: Half-Interval Search," *The Mathematics Teacher*, Vol. 78 No. 2 (February 1985), 120–123. (A BASIC program to compute roots of polynomial functions is developed.)

Kroopnick, Allen. "How Good is the 'Rule of 72'?," *The Two-Year College Mathematics Journal*, Vol. 10 No. 4 (September 1979), 279–280.

Rothman, Tony. "Genius and Biographers: The Fictionalization of Evariste Galois," *The American Mathematical Monthly*, Vol. 89, No. 2 (February 1982), 84–106.

Todd, Aaron R. "The Quadratic Equation and Inexactness of Computation," *The MATYC Journal*, Vol. 10 No. 3 (Fall 1976), 186–187.

Biography   Girolamo Cardano

Girolamo Cardano was born September 24, 1501 in Italy. His father was a lawyer and a friend of Leonardo da Vinci. Cardano studied medicine, but lectured in mathematics, astronomy, and Greek. In 1531, he married Lucia Bandareni. He is said to have had such a temper that he cut off the ears of his youngest son in anger.

It was customary at the time to keep mathematical discoveries secret in order to use them in public contests whose winners received prestige as well as prize money. About 1515, Scipio del Ferro discovered how to solve cubic equations of the form $x^3 + bx = c$. He taught the method to his son-in-law, Annibale della Nave, and his pupil, Antonio Maria Fior, before he died. In 1535, Fior heard that Nicolo Fontana (called Tartaglia) could solve cubics of the form $x^3 + bx^2 = c$ and challenged him to a contest. Each of them was to submit 30 problems for the other to solve. After 30 days, whoever had solved the most problems would win. Tartaglia solved all of Fior's problems. Fior solved none of Tartaglia's.

Cardano was impressed and persuaded Tartaglia to show him his method, promising not to reveal it. In 1542, when Cardano was given Scipio del Ferro's method by Annibale della Nave, he no longer felt bound by his promise. He published the solution in 1545. Cardano gave credit to Tartaglia, but the method is still known as "Cardan's method."

Cardano's oldest son was executed in 1560 for poisoning his wife after she boasted that none of her children was related to her husband. Cardano's work suffered, and he took up gambling. He later wrote on the mathematics of chance. In 1570, he was arrested for casting a horoscope of Christ's life and was forbidden to lecture or publish. There is a legend that he predicted his death on a particular day in 1576, and when the day came, committed suicide in order to fulfill the prediction.

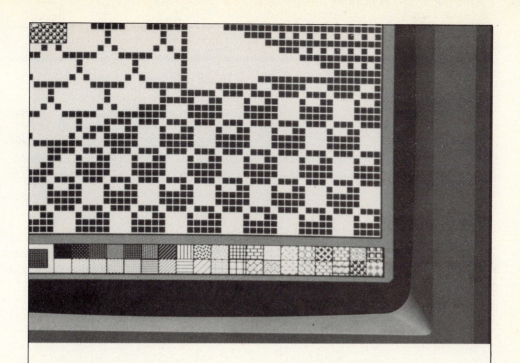

Chapter Seven
Functions

A computer monitor displays pictures composed of small dots called pixels. A coordinate system similar to the one for graphing mathematical functions is used to identify individual pixels.

7.1   Functions

The concept of a function is one of the most basic in mathematics.

> **Definition**
> A *function* is a correspondence between two sets such that there is associated with each element of the first set exactly one element of the second set. The first set is called the *domain*, and the second set is called the *range*.

One way to indicate a function is to set up a table. The elements of the domain are listed in a column on the left, and the corresponding elements of the range in a column on the right. For example, consider the sets $A = \{1, 3, 5, 7\}$ and $B = \{2, 4, 6, 8\}$.

The table below shows a function.

A	B
1	2
3	4
5	6
7	8

In this example, set A is the domain and set B is the range. For each element in set A, there is exactly one element in set B that corresponds to it.

EXAMPLE 1

For the two sets below, determine whether the correspondence in the table is a function. If it is a function, state the domain and the range.

A	B
2	1
4	3
6	5
8	7

Solution   The correspondence is a function. For each element in set A, there is exactly one element in B that corresponds to it. The domain is $A = \{2, 4, 6, 8\}$ and the range is $B = \{1, 3, 5, 7\}$. ■

EXAMPLE 2

For the two sets given, determine whether the correspondence in the table is a function. If it is a function, state the domain and range.

A	B
1	−1
0	0
1	1
4	2

Solution   The table does *not* show a function, because if we choose the element 1 from set A, both -1 and 1 in set B are associated with it. The definition of a function requires that *each* element in set A correspond to *exactly* one element in set B. ■

EXAMPLE 3

For the two sets given, determine whether the correspondence in the table is a function. If it is a function, state the domain and range.

X	Y
−1	1
0	0
1	1
2	4

Solution   The table shows a function, because for each element in set X, there is exactly one element in set Y that is associated with it. For example, if we examine -1, the only number associated with it is 1. If we examine 1, the only number associated with it is 1, and so on. The domain is $X = \{-1, 0, 1, 2\}$ and the range is $Y = \{0, 1, 4\}$. ■

Sometimes there are so many values in the sets that it is not convenient, or even possible, to list them all. Then we use an equation. The symbol $f(x)$ is read "f of x." The letter f represents the function, and the number in parentheses is called the *argument* of the function. Beginning algebra students sometimes think that $f(x)$ is a variable f multiplied by a variable x, but it should be clear from the context of a problem whether $f(x)$ is a multiplication problem or a function value.

The statement $f(x) = x^2 + 1$ specifies a function relating a domain (whose elements we may substitute for x) and a range (whose elements will replace $f(x)$). If the domain is $\{0, 1, 2, \ldots, 100\}$, then we will have

$$f(0) = 0^2 + 1 = 1$$
$$f(1) = 1^2 + 1 = 2$$
$$f(2) = 2^2 + 1 = 5$$

and so on.

> **Definition**
> $f(x)$ is the element of the range associated with the element x of the domain under the function f. To find $f(a)$ for a real number a, substitute a for x and evaluate the resulting expression.

EXAMPLE 4

Let $f(x) = x^2$ and $g(x) = 2x$. Then evaluate each expression:

(a) $f(3)$

(b) $g(3)$

(c) $f(-3)$

Solution

(a) $f(3) = 3^2 = 9$

(b) $g(3) = 2(3) = 6$

(c) $f(-3) = (-3)^2 = 9$ ∎

EXAMPLE 5

Let $f(x) = (x - 1)^2$ and let $g(x) = x^2 - 1$. Then evaluate each expression below:

(a) $f(-2)$

(b) $-f(2)$

(c) $g(-2)$

Solution

(a) $f(-2) = (-2 - 1)^2 = (-3)^2 = 9$

(b) $-f(2) = -(2 - 1)^2 = -1^2 = -1$

(c) $g(-2) = (-2)^2 - 1 = 4 - 1 = 3$ ∎

EXAMPLE 6

Let $f(x) = 2x + 3$ and $g(x) = 3x + 2$. Then evaluate each expression:

(a) $f(2) + g(2)$

(b) $f(2) + f(3)$

(c) $f(2) + g(3)$

Solution

(a) $f(2) = 2(2) + 3 = 4 + 3 = 7$, and $g(2) = 3(2) + 2 = 6 + 2 = 8$
   Therefore, $f(2) + g(2) = 7 + 8 = 15$.

(b) $f(2) = 7$ and $f(3) = 2(3) + 3 = 6 + 3 = 9$
   Therefore, $f(2) + f(3) = 7 + 9 = 16$.

(c) $f(2) = 7$ and $g(3) = 3(3) + 2 = 9 + 2 = 11$
   Therefore, $f(2) + g(3) = 7 + 11 = 18$. ∎

When the domain is not specified, we assume that the domain is the set of all real numbers for which the function is defined. Remember that an expression that has a zero denominator or the square root of a negative number is not defined in the real number system.

To determine the domain of a function:

1. The domain is the set of real numbers for which the function is defined. Start with the set of real numbers.
2. If the function has a denominator, eliminate values for which the denominator is zero.
3. If the function has a square root (or other even root), eliminate values for which the expression under the radical sign is negative.

EXAMPLE 7

Determine whether the equation $g(x) = 1/(2x + 3)$ represents a function, and if it does, state the domain.

Solution $g(x) = 1/(2x + 3)$ is a function, because each time we substitute a value for x, the equation yields a single value. The domain is the set of real numbers, with the exception of any value for which the denominator is zero. Because $2x + 3 = 0$ when $x = -3/2$, the function is undefined at $x = -3/2$. We write this domain in set notation as $\{x|\ x \in R,\ x \neq -3/2\}$. ■

EXAMPLE 8

Determine whether the equation $k(x) = x^3$ represents a function, and if it does, state the domain.

Solution $k(x) = x^3$ is a function, because each time we substitute a value for x, the equation yields a single value. The domain is the set of real numbers, R. ■

EXAMPLE 9

Determine whether the equation $f(x) = \sqrt{x - 1}$ represents a function, and if it does, state the domain.

Solution $f(x) = \sqrt{x - 1}$ is a function, because each time we substitute a value for x, the equation yields a single value. Remember that $\sqrt{}$ indicates only the positive square root. The domain is $\{x|\ x \in R,\ x \geq 1\}$, since the expression under the radical sign, $x - 1$, must be greater than or equal to zero. ■

Since x can take on any value in the domain, x is often called the *independent variable* of the function f defined by $y = f(x)$. The value $f(x)$ depends on the value of x, so $f(x)$ is called the *dependent* variable. In the formula for the perimeter of a rectangle ($P = 2l + 2w$), perimeter depends on both length and width. We might say that "perimeter is a function of length and width" and write $P = f(l, w) = 2l + 2w$. In this chapter, however, we will confine our work to functions of a single variable.

7.1 EXERCISES

Determine whether the correspondence in each table is a function. See Examples 1–3.

1.	X	Y
	-2	6
	3	-9
	-4	12
	5	-15

2.	X	Y
	2	3
	2	1
	4	5
	4	3

3.	X	Y
	-2	3
	-3	3
	-4	3
	-5	3

4.	X	Y
	2	2
	3	3
	-1	-1
	0	0

5.	X	Y
	2	-2
	3	-3
	4	-4
	5	-5

6.	X	Y
	1	1
	2	4
	3	9

7.	X	Y
	1	1
	4	2
	9	3

8.	X	Y
	4	2
	4	-2
	9	3
	9	-3

9.	X	Y
	1	4
	1	3
	2	2
	3	1

10.	X	Y
	4	1
	3	1
	2	2
	1	3

Evaluate each expression if $f(x) = x^2 - 1$ and $g(x) = 3x + 2$. See Examples 4–6.

11. $f(1)$　　　　　　　　**12.** $f(0)$　　　　　　　　**13.** $f(-1)$

14. $f(-2)$　　　　　　　**15.** $g(1)$　　　　　　　　**16.** $g(0)$

17. $g(-1)$　　　　　　　**18.** $g(-2)$　　　　　　　**19.** $f(3) + g(3)$

20. $f(4) - g(4)$　　　　　**21.** $f(-3) + g(-3)$　　　　**22.** $f(5) - f(3)$

23. $g(4) - g(2)$　　　　　**24.** $f(1) + g(-1)$　　　　**25.** $f(-1) + g(1)$

For each equation, determine whether the correspondence is a function. If it is a function, specify the domain. See Examples 7–9.

26. $f(x) = 2x + 1$

27. $f(x) = 3x \pm 2$

28. $f(x) = x^2 - 3$

29. $g(x) = \dfrac{x + 4}{x - 1}$

30. $g(x) = \dfrac{x - 3}{2x + 1}$

31. $k(x) = \sqrt{x - 4}$

32. $k(x) = \sqrt{2x - 1}$

33. $f(x) = x^2$

34. $g(x) = x^3$

35. $k(x) = x^4$

36. $h(x) = \dfrac{\pm \sqrt{x - 3}}{x^2 + 1}$

✴**37.** The composition of functions f and g is denoted $f \circ g(x)$, and $f \circ g(x) = f(g(x))$. If $f(x) = x^2$ and $g(x) = 2x + 3$, find

(a) $f \circ g(x)$

(b) $g \circ f(x)$

✴**38.** The correspondence denoted by $f^{-1}(x)$ is the inverse of a function $f(x)$ if $f \circ f^{-1}(x) = x$ for all x in the domain of f^{-1}, and $f^{-1} \circ f(x) = x$ for all x in the domain of f. For each pair of functions, determine whether $g(x) = f^{-1}(x)$.

(a) $f(x) = x + 1$ $g(x) = x - 1$

(b) $f(x) = x/2$ $g(x) = 2x$

(c) $f(x) = 2x - 1$ $g(x) = 2x + 1$

(d) $f(x) = 2x - 1$ $g(x) = (x + 1)/2$

(e) $f(x) = 2x - 1$ $g(x) = x - 2$

✴**39.** It is possible to find the inverse (when it exists) of a function by interchanging x and y in the equation. Thus, $y = 3x + 1$ becomes $x = 3y + 1$ or $y = (x - 1)/3$. Find the inverse of each function:

(a) $y = x - 4$

(b) $y = 3x$

(c) $y = \sqrt{x}$ for $x \geq 0$

✴**40.** Is the inverse of a function always a function?

✴**41.** What do the following functions have in common?

(a) $f(n) = n^2 - n + 41$ where the domain is the set of integers greater than or equal to 1 and less than 32.

(b) $g(n) = n^2 - 79n + 1601$ where the domain is the set of integers greater than or equal to 40 and less than 71.

✴**42.** The function f defined by $f(n) = 2^{(2^n)} + 1$ was once (falsely) thought to produce only prime numbers for integral values of n. Such primes are called *Fermat primes*. Find as many Fermat primes as you can.

7.2   Graphing

The relationship between the independent and dependent variables of a function is often apparent from a graph. A horizontal line, called the *x*-axis, is used to locate the independent variable. A vertical line, called the *y*-axis, is used to locate the dependent variable (see Figure 7.1). The point of intersection of the two lines is called the *origin*. For graphing, the variable $f(x)$ is frequently replaced by the variable y, so that instead of writing $f(x) = x^2 - 1$, we would write $y = x^2 - 1$.

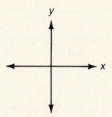

Figure 7.1

The two lines divide the plane into four regions called *quadrants*. The quadrants are usually numbered with Roman numerals, starting at the upper right with I and proceeding counterclockwise (see Figure 7.2). Such a system is called a "rectangular coordinate system" or a "Cartesian coordinate system" in honor of René Descartes. Each line is marked off in equal segments to indicate the integers, with 0 at the origin on both axes. The same scale is often used on both axes.

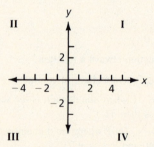

Figure 7.2

For a function, each element *x* in the domain has associated with it a unique element *y* of the range. For example, $f(2) = 3$ means that when $x = 2$, we have $f(2)$ or $y = 3$. We designate this point by using parentheses (2, 3) and call it an ordered pair. The value of the independent variable is always listed first.

These values are called the coordinates of the point, and they determine a unique point on the plane. For example, the point (2, 3) is located where a vertical line from $x = 2$ (on the x-axis) and a horizontal line from $y = 3$ (on the y-axis) intersect. To *plot* a point on a plane means to indicate its location by a dot. The point (2, 3) is plotted on the plane in Figure 7.3

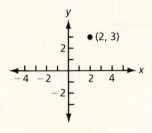

Figure 7.3

To graph a function, we plot the set of ordered pairs (x, y) where x belongs to the domain, and y is the value of the range associated with x.

To sketch the graph of a function given by $y = f(x)$:
1. Sketch and label the x- and y-axes.
2. Indicate the scale on each axis.
3. Plot all of the pairs (x, y) on the same coordinate system.

EXAMPLE 1

Graph the function indicated by the following table:

x	y
2	3
4	5
-6	6

Solution

EXAMPLE 2

Graph the function indicated by the following table:

x	y
-1	-1
0	1
1	3

Solution

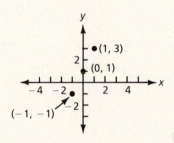

If the domain is the set of real numbers, then each value of x on the x-axis will have a y value associated with it, and the graph will be a series of points. Not every graph, however, is the graph of a function. Since the definition of a function requires that only one value of y is to be associated with each value of x, *any* vertical line on the plane will intersect the graph of a function only once. If the graph is not the graph of a function, it is possible to place a vertical line somewhere on the plane so that the line intersects the graph in more than one point.

To determine whether a graph represents a function:
1. If it is impossible to draw a vertical line on the plane that intersects the graph more than once, it represents a function.
2. If it is possible to draw a vertical line (anywhere on the plane) that intersects the graph more than once, it does not represent a function.

EXAMPLE 3

Determine whether each graph represents a function.

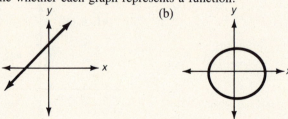

(a)

(b)

Solution

(a) The graph shown in Example 3(a) represents a function, because a vertical line drawn anywhere on the plane intersects the graph only once.

(b) The graph shown in Example 3(b) does *not* represent a function, because a vertical line can be drawn to intersect the graph in more than one point. The y-axis, for example, intersects the graph in two points. ▪

EXAMPLE 4

Determine whether each graph represents a function.

(a) 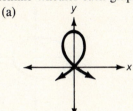   (b)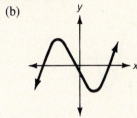

Solution

(a) The graph shown in Example 4(a) does *not* represent a function, because a vertical line can be drawn so that it intersects the graph in more than one point.

(b) The graph in Example 4(b) represents a function, because anywhere a vertical line is drawn through the graph, it intersects the graph only once. ▪

7.2 EXERCISES

Graph each function on a single plane. See Examples 1 and 2.

1.

x	y
−2	6
3	−9
−4	12

2.

x	y
−2	3
−3	3
−4	3
−5	3

3.

x	y
2	2
3	3
−1	−1
0	0

4.

x	y
2	−2
3	−3
4	−4
5	−5

5.

x	y
1	1
2	4
3	9

6.

x	y
1	1
4	2
9	3

7.

x	y
1	1
0	0
−1	−1

8.

x	y
−2	4
0	−1
2	4

9.

x	y
−2	7
−1	0
0	−1
1	0

Determine whether each graph represents a function. See Examples 3 and 4.

10.

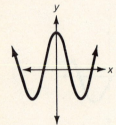

11.

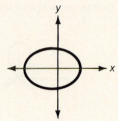

12.

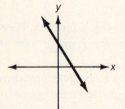

13.

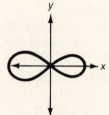

14.

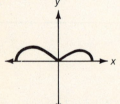

15.

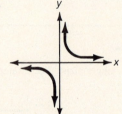

16.

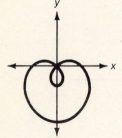

17.

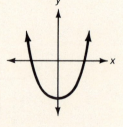

18.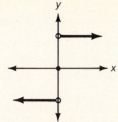

19.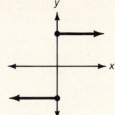

(The open circle means
that the point is not
included on the graph.)

✳**20.** Graphs of functions are often shown in newspaper and magazine articles. For example, the horizontal axis may indicate years while the vertical axis indicates the federal deficit. Find three graphs of functions in current periodicals, and identify the independent and dependent variables.

✳**21.** The ASCII character set is a set of 95 printable symbols as well as some nonprintable communication symbols used by many microcomputers. The 65th character in the character set is A, and the other letters of the alphabet follow in sequence. Consider the computer function which returns the character in the Nth position of the ASCII character set. This function is usually denoted CHR(N) in Pascal or CHR$(N) in BASIC. Find

(a) CHR(69) or CHR$(69)

(b) CHR(72) or CHR$(72)

(c) CHR(84) or CHR$(84)

(d) CHR(89) or CHR$(89)

✳**22.** The array of numbers that follows is known as *Pascal's triangle*. Each number (other than a 1) is the sum of the two above it. Write a function that associates with the number of each row, the sum of the numbers in that row (if the 1 at the top is in row 0).

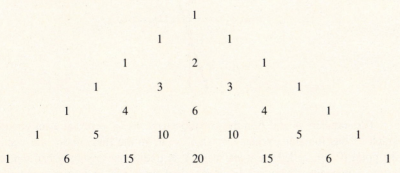

✳**23.** Can you draw slanted lines through the triangle so that the sum associated with each slanted line is one of the Fibonacci numbers {1, 1, 2, 3, 5, 8, . . .}?

7.3   Linear Functions

In the previous section, we saw some graphs of functions with just a few values in the domain. If the domain is the set of real numbers, it is impossible to list every ordered pair. If we examine a sample of values, however, a pattern emerges.

When the points fall along a straight line, we draw the line connecting these points to graph the other values of the function. This type of function is called a *linear* function.

It is easy to recognize and graph a linear function. Any function that fits the pattern $f(x) = mx + b$ (or $y = mx + b$) where m and b are real numbers is linear. Since a straight line is determined by two points, it is not necessary to plot lots of points in order to sketch the graph. It is a good idea, though, to plot at least three points. If you know the function is linear, but the points don't fall on a straight line, you know you have made a mistake and you can find and correct it.

EXAMPLE 1

Sketch the graph of $y = 3x - 1$.

Solution   We choose three values of x, say -1, 0, and 2. The corresponding values for y are $3(-1) - 1 = -4$, $3(0) - 1 = -1$, and $3(2) - 1 = 5$. That is, $(-1, -4)$, $(0, -1)$, and $(2, 5)$ are three points on the graph. We locate these three points and connect them to show all points of the function.

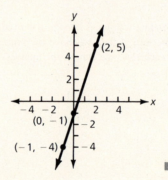

The numbers m and b in the equation $y = mx + b$ reveal a lot about the graph of this function. When $x = 0$, $y = b$, so the line crosses the y-axis at the point $(0, b)$, called the *y-intercept*. It is traditional to simply call the value b the y-intercept. The value of m is called the *slope*. It is a measure of the steepness of the line. The larger the absolute value of the slope, the more nearly vertical the line is; the closer the slope is to zero, the more nearly horizontal the line is. The slope is sometimes described as "rise/run." That is, to get from one point to another on the line, instead of moving along the line itself, we could rise and run, or move up and over. The ratio of rise to run is constant for a given line.

We can use the slope and y-intercept to sketch the graph of a linear function.

> **To sketch the graph of a linear function given by $y = mx + b$:**
> 1. Identify the slope (m) and y-intercept (b). Write the slope m as a fraction.
> 2. Plot the y-intercept at $(0, b)$. Starting there,
>   (a) Use the numerator of the slope (rise) to count up (if it is positive) or down (if it is negative).
>   (b) Use the denominator of the slope (run) to count over to the right (if it is positive) or to the left (if it is negative).
> 3. Plot a second point at the position found in step 2.
> 4. Connect the two points to graph the remaining values of the function.

EXAMPLE 2

Sketch the graph of $y = 3x - 1$.

Solution
1. The slope is 3 or 3/1 as a fraction, and the y-intercept is -1.
2. We plot the y-intercept at $(0, -1)$. From that point, count up three and over one to the right to get $(1, 2)$.
3. Plot the point $(1, 2)$.
4. Connect $(0, -1)$ and $(1, 2)$.

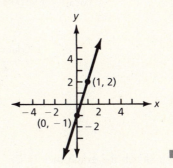

EXAMPLE 3

Sketch the graph of $y = -x - 2$.

Solution
1. The slope is -1 or $-1/1$ and the y-intercept is -2.
2. The y-intercept is at $(0, -2)$. From that point we count down one and over one to the right to get $(1, -3)$.
3. Plot the point $(1, -3)$.
4. Connect the two points.

Alternate solution:
1.   The slope is -1 and the y-intercept is -2.
2.   The negative sign in the slope may go with either numerator or denominator. If we put it with the denominator $(1/-1)$, we would count up one and to the left one to get $(-1, -1)$.
3.   Plot the point $(-1, -1)$.
4.   Connect the two points.

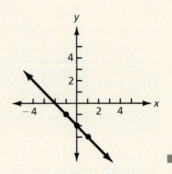

<hr>

EXAMPLE 4

Sketch the graph of $y = \dfrac{2x}{3} + \dfrac{1}{3}$.

Solution
1. $m = 2/3, \ b = 1/3$

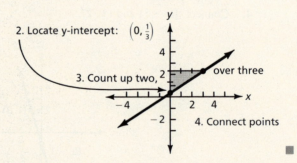

<hr>

EXAMPLE 5

Sketch the graph of $y = 2$.

Solution   Since the x term is missing, we must have $m = 0$ (i.e., $y = 0x + 2$). A slope of 0 indicates a horizontal line. The y-intercept is 2. In fact, for any value of x, the y value will be 2. See the following figure.

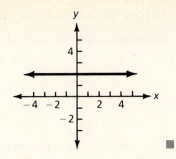

EXAMPLE 6

Sketch the graph of $x = 2$.

Solution   The equation cannot be written in the form $y = mx + b$. It is impossible to change the coefficient of y from 0 to 1. We say the slope is *undefined*. The graph will be a vertical line. Since the value $x = 2$ has an infinite number of y values associated with it, the correspondence is not a function. At $x = 2$, the entire set of real numbers is associated with x. See the following figure.

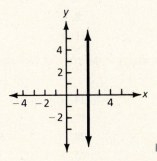

Linear functions are not always written in the form $y = mx + b$. For example, $2x + 3y = 1$ does not fit this pattern. But we can solve this equation for y, giving $3y = -2x + 1$, or $y = \dfrac{-2x + 1}{3} = \dfrac{-2x}{3} + \dfrac{1}{3}$.

EXAMPLE 7

Identify the slope and y-intercept of the graph of $x - y = 4$.

Solution   Solve the equation for y: $-y = -x + 4$

$$y = x - 4$$

Since the equation is now in the form $y = mx + b$, we can say that the slope is 1 and the y-intercept is -4. ■

EXAMPLE 8   Identify the slope and y-intercept of the graph of $x - 2y = 3$.

Solution   Solve the equation for y: $-2y = -x + 3$

$$y = \frac{-x + 3}{-2}$$

$$y = \frac{1x}{2} - \frac{3}{2}$$

Since the equation is now in the form $y = mx + b$, we can say that the slope is $\frac{1}{2}$ and the y-intercept is $-\frac{3}{2}$ or $-1\frac{1}{2}$. ■

7.3 EXERCISES

Graph each equation. See Examples 2–6.

1. $y = 2x + 3$

2. $y = -x - 2$

3. $y = \frac{1}{2}x + 3$

4. $y = \frac{2}{3}x - 1$

5. $y = -\frac{3}{5}x + 4$

6. $y = 1$

7. $y = -1$

8. $x = -1$

9. $x = 1$

10. $y = -2$

11. $y = 2x$

12. $y = -x$

Identify the slope and y-intercept of the graph of each equation below. See Examples 7 and 8.

13. $y = 3x - 2$

14. $y = x + 3$

15. $y = -2x + 5$

16. $y = \frac{1}{2}x - 1$

17. $2x - y = 4$

18. $2x + 3y = 6$

19. $x = 2y + 1$

20. $3y = 2x - 6$

21. $3x - y = 7$

22. $4x - 5y = 3$

23. $x = 2y$

24. $y = 3$

25. $y = 4$

26. $x = -y$

Graph each equation.

27. $x - y = 3$

28. $2x - 3y = 9$

29. $2y = 3x + 4$

30. $2y - x = 3$

31. $x - 2y = 4$

32. $2x - y = 3$

33. $x + y = 5$

34. $3x + y = -1$

35. $2x = 3$

36. $2y = 3$

✳**37.** **(a)** What is the relationship between lines that have the same slope?

  (b) What is the relationship between lines that have slopes whose product is -1?

❋**38.** **(a)** Graph $y = x + 1$ and its inverse function $y = x - 1$.

   (b) Graph $y = 2x$ and its inverse function $y = \dfrac{x}{2}$.

❋**39.** **(a)** Graph $y = 2x - 1$ and its inverse function.

   (b) How would you describe the graph of the inverse of a function?

❋**40.** If a function is given by $ax + by = c$, give formulas for the x- and y-intercepts of its graph.

7.4   Quadratic Functions

In Section 6.4, we studied quadratic *equations*. In this section, we study quadratic *functions*.

> **Definition**
> A *quadratic function* is a function that can be written in the form $f(x) = ax^2 + bx + c$ (or $y = ax^2 + bx + c$), where a, b, and $c \in R$, $a \neq 0$.

Consider the function given by $f(x) = x^2 - 5x + 6$. The domain is the set of real numbers. We could sketch a graph by examining a few values in the domain to find the corresponding values in the range. For example, $f(0) = 0^2 - 5(0) + 6 = 6$, $f(1) = 1^2 - 5(1) + 6 = 2$, and $f(2) = 2^2 - 5(2) + 6 = 0$.

That is, $(0, 6)$, $(1, 2)$, and $(2, 0)$ are points on the graph. See Figure 7.4.

It isn't as easy to graph the quadratic function as it is the linear function, because the points don't fall on a straight line.

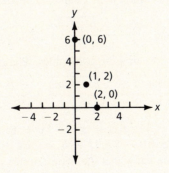

Figure 7.4

Knowing what to expect makes graphing the linear function easy. Quadratic functions do, however, follow a pattern. The graph of a quadratic function is call a *parabola*. See Figure 7.5.

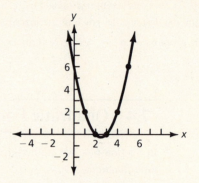

Figure 7.5

Because the lefthand side is a mirror image of the righthand side, we say that the parabola is *symmetric*. If $a > 0$ in the equation $y = ax^2 + bx + c$, the parabola will open upward and is sometimes called a *fill* parabola. If $a < 0$, the parabola will open downward and is sometimes called a *spill* parabola.

If we knew the *vertex*, the highest or lowest point of the parabola, and two other points, we could obtain a reasonably good sketch of the graph by connecting the three points to complete a parabola. The x-coordinate of the vertex is given by $-b/(2a)$. By substituting this value for x in the function, we can calculate the y-coordinate. It is easy to see why $-b/(2a)$ locates the center of the parabola. When the graph crosses the x-axis, the y-coordinate will be zero. That is, we have $ax^2 + bx + c = 0$. But this equation is the quadratic equation that we saw in Chapter 6. It has two solutions. They are $\dfrac{-b + \sqrt{b^2 - 4ac}}{2a}$ and $\dfrac{-b - \sqrt{b^2 - 4ac}}{2a}$. Since a parabola is symmetric, the center is halfway between these two values. We average them to find the midpoint:

$$\frac{1}{2}\left(\frac{-b + \sqrt{b^2 - 4ac}}{2a} + \frac{-b - \sqrt{b^2 - 4ac}}{2a}\right) = \frac{1}{2}\left(\frac{-2b}{2a}\right) = \frac{-b}{2a}.$$

For $f(x) = x^2 - 5x + 6$, the vertex occurs at $x = 5/2$. Since $f(5/2) = (5/2)^2 - 5(5/2) + 6 = 25/4 - 25/2 + 6 = 25/4 - 50/4 + 24/4 = -1/4$, we know that the vertex occurs at $(5/2, -1/4)$.

Once we know where the vertex is, we can find two points on either side of it. For the example above, we might examine $x = 1$ and $x = 4$. Since $f(1) = 1 - 5 + 6 = 2$ and $f(4) = 16 - 20 + 6 = 2$, we know that the points $(1, 2)$ and $(4, 2)$ are on the graph. (See Figure 7.5.)

The procedure for graphing a quadratic function is summarized in the following table.

> **To sketch the graph of a quadratic function given by**
> $f(x) = ax^2 + bx + c$:
> 1. Determine whether the parabola opens upward or downward.
>   a. If $a > 0$, it will open upward.
>   b. If $a < 0$, it will open downward.
> 2. Locate the vertex.
>   a. Let $x = -b/(2a)$.
>   b. Find $f(x)$.
> 3. Choose two other values for x (one on each side of the vertex) and find the corresponding function values.
> 4. Plot these three points and connect them to complete the parabola.

EXAMPLE 1

Sketch the graph of $f(x)$ or $y = -x^2 + 4$.

Solution

1. For this equation $a = -1$, $b = 0$, and $c = 4$. Since $a = -1$, the parabola will open downward.

2. The vertex will occur at $x = -0/-2 = 0$. Since $f(0) = 4$, we know the vertex is at $(0, 4)$.

3. We choose two other values for x, one on each side of the vertex. We use 2 and -2. Since $f(-2) = -(4) + 4 = 0$, we know that $(-2, 0)$ is on the graph. Since $f(2) = -(4) + 4 = 0$, we know that $(2, 0)$ is on the graph.

4. We locate these three points and connect them to complete the parabola.

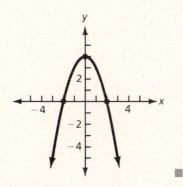

EXAMPLE 2

Sketch the graph of $f(x)$ or $y = 2x^2 + 3x + 1$.

Solution

1. For this equation $a = 2$, $b = 3$, and $c = 1$. Since $a = 2$, the parabola will open upward.

2. The vertex will be at $x = -3/(2 \cdot 2) = -3/4$. Since $f(-3/4) = 2(9/16) + 3(-3/4) + 1 = 9/8 - 9/4 + 1 = 9/8 - 18/8 + 8/8 = -1/8$, the vertex is at $(-3/4, -1/8)$.

3. Choose two other values, one on each side of $-3/4$. We use -2 and 0. $f(-2) = 2(4) + 3(-2) + 1 = 8 - 6 + 1 = 3$, so $(-2, 3)$ is on the graph, and $f(0) = 2(0) + 3(0) + 1 = 1$, so $(0, 1)$ is on the graph.

4. We plot these three points and connect them to complete the parabola.

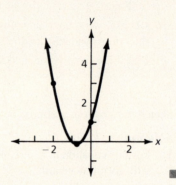

7.4 EXERCISES

For each quadratic function below, determine whether the parabola will open upward or downward, find the vertex, and sketch the graph. See Examples 1 and 2.

1. $y = x^2 + 7x + 12$

2. $y = -x^2 + 2x + 8$

3. $y = x^2 - 4x - 4$

4. $y = x^2 + 6x + 5$

5. $y = x^2 - 1$

6. $y = x^2 - 9$

7. $y = -x^2 + 4x$

8. $y = x^2 - 6x$

9. $y = 4x^2 - 4x + 1$

10. $y = 4x^2 - 4x - 3$

11. $y = 9x^2 - 9x + 2$

12. $y = 9x^2 - 3x - 2$

13. $y = x^2 + 2x + 3$

14. $y = -x^2 + 2x + 3$

15. $y = 2x^2 + 3x + 1$

16. $y = -2x^2 + 3x + 1$

17. $y = -2x^2 - 3x - 1$

18. $y = -x^2 - x + 1$

19. $y = -x^2 + x - 1$

20. $y = x^2 + x - 1$

✳21. A function f is said to be even if $f(-x) = f(x)$ for every x and odd if $f(-x) = -f(x)$ for every x. Determine whether each of the following functions is even, odd, or neither.

(a) $f(x) = x^2$

(b) $f(x) = |x|$

(c) $f(x) = x^3$

(d) $f(x) = |x^3|$

(e) $f(x) = -x^2$

(f) $f(x) = -x^3$

✳**22.** Graph the following quadratic equations:

(a) $y = x^2$

(b) $y = 2x^2$

(c) $y = (x - 1)^2$

(d) $y = x^2 + 3$

(e) $y - 3 = 2(x - 1)^2$

✳**23.** If a quadratic equation is written in the form $y - c = d(x - e)^2$, what is the significance of each number (c, d, and e)?

✳**24.** A parabola $y = ax^2 + bx + c$ may be graphed by first plotting the vertex. To locate n points on the graph, start at the vertex and successively move 1 horizontal unit and $a(2n - 1)$ vertical units where $n \in \{1, 2, 3, \ldots\}$. Because the graph is symmetric, the lefthand side can also be sketched. Verify that this procedure works on problems 3, 7, and 13.

7.5 Polynomial Functions and Computer Graphics

For higher degree polynomial functions, there are some shortcuts like those we used with linear and quadratic functions, but they involve calculus. Without using calculus, the best thing to do is plot several points and connect them. One thing you might notice, though, is that a linear function has degree one and zero bends. A quadratic function has degree two and one bend. A cubic function has degree three and two bends. When we have found both bends, we know that we have a reasonably good graph.

EXAMPLE 1

Graph $y = x^3 - x^2 - 2x + 2$.

Solution We choose several values for x, say -2, -1, 0, 1, and 2. For convenience, we list these in table form and calculate the corresponding function values.

x	x^3	$- x^2 - 2x$	$+ 2$	y
-2	$(-2)^3 -$	$(-2)^2 - 2(-2) +$	2	$-8 - 4 + 4 + 2 = -6$
-1	$(-1)^3 -$	$(-1)^2 - 2(-1) +$	2	$-1 - 1 + 2 + 2 = 2$
0	$0^3 -$	$0^2 - 2(0) +$	2	$0 - 0 + 0 + 2 = 2$
1	$1^3 -$	$1^2 - 2(1) +$	2	$1 - 1 - 2 + 2 = 0$
2	$2^3 -$	$2^2 - 2(2) +$	2	$8 - 4 - 4 + 2 = 2$

We plot these five points and connect them with a smooth curve going from left to right. See the figure on the following page.

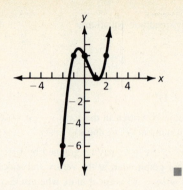

Most cubic functions will follow this pattern. They will either look like the graph above, or they will look like that graph if it were viewed from the back.

EXAMPLE 2

Sketch the graph of $y = x^3 + 1$.

Solution   We choose several values for x. Again we use -2, -1, 0, 1, and 2. We list these in table form and calculate the corresponding function values.

x	$x^3 + 1$	y
-2	$(-2)^3 + 1$	$-8 + 1 = -7$
-1	$(-1)^3 + 1$	$-1 + 1 = 0$
0	$0^3 + 1$	$0 + 1 = 1$
1	$1^3 + 1$	$1 + 1 = 2$
2	$2^3 + 1$	$8 + 1 = 9$

We plot these five points and connect them to form a smooth curve. See the following figure.

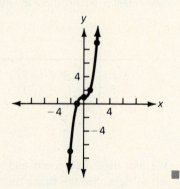

At first, Example 2 may not appear to follow the pattern described for a cubic function. But if you imagine that the graph were made of wire, two bends would be required to shape it.

When the degree of the polynomial function is higher than three, however, all we can say is that the number of bends is *at most* one less than the degree.

> **To sketch the graph of a polynomial function:**
> 1. Choose several values for x. Five values will usually provide a good start toward the graph.
> 2. Calculate the corresponding function values.
> 3. Plot the points.
> 4. Connect them with a smooth curve going from left to right.
> 5. If the number of bends in the graph is one less than the degree of the polynomial, the graph is adequate. If not, examine more points until all of the bends are located or you are convinced there are no more bends.

EXAMPLE 3

Sketch the graph of $y = x^4 - 3$.

Solution A table of values follows.

x	$x^4 - 3$	y
-2	$(-2)^4 - 3$	$16 - 3 = 13$
-1	$(-1)^4 - 3$	$1 - 3 = -2$
0	$0^4 - 3$	$0 - 3 = -3$
1	$1^4 - 3$	$1 - 3 = -2$
2	$2^4 - 3$	$16 - 3 = 13$

We plot these five points and connect them. See the following figure.

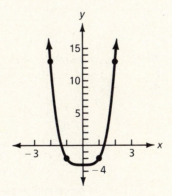

Since only one bend appears in the graph, we examine the values $x = 3$ and $x = -3$. When $x = 3$, $y = 3^4 - 3 = 81 - 3 = 78$, and when $x = -3$, $y = (-3)^4 - 3 = 81 - 3 = 78$. Thus, we conclude that the graph has only one bend. ∎

To graph equations on a microcomputer, we think of the screen as a Cartesian plane. The boxes formed by an imaginary grid on the screen are called *pixels* (for "picture elements"). Each pixel has both an x and a y value. The numbering system varies from computer to computer. The origin may be at the upper left or the lower left. For purposes of illustration, we will assume it is at the lower left. Figure 7.6 shows the lower lefthand portion of the screen with the coordinates of each pixel labeled.

0, 4	1, 4	2, 4	3, 4	4, 4	5, 4
0, 3	1, 3	2, 3	3, 3	4, 3	5, 3
0, 2	1, 2	2, 2	3, 2	4, 2	5, 2
0, 1	1, 1	2, 1	3, 1	4, 1	5, 1
0, 0	1, 0	2, 0	3, 0	4, 0	5, 0

Figure 7.6

A typical system has 280 pixels (numbered 0 to 279) in each horizontal row and 160 pixels (numbered 0 to 159) in each vertical column.

If we drew the graph of $y = x^2 - 5x + 6$, the screen would show only the first quadrant. We would prefer having the axes in the center of the screen. We could draw them by lighting up pixels (140, N) where N takes on the values from 0 to 159 and (K, 80) where K takes on the values from 0 to 279. See Figure 7.7.

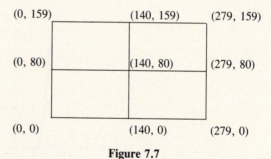

Figure 7.7

We have to be careful in plotting the coordinates of the function, however, because the lower lefthand corner of the screen still has (0, 0) as its coordinates. But (0, 0) of our graph has (140, 80) as the coordinates on the screen. That is, we are really working with two different graphs. To avoid confusion, then, let (x, y) be the actual coordinates of the function, but let (X, Y) be the coordinates we want the computer to show. That is, when $(x, y) = (0, 0)$, we have (X, Y)

$= (140, 80)$. When $x = 1$, $y = 1^2 - 5(1) + 6 = 2$. But we want the point $(1, 2)$ to show up one unit over and two units up from $(140, 80)$. When (x, y) $= (1, 2)$, $(X, Y) = (141, 82)$.

In fact, we could generalize. When our function has coordinates (x, y), we want the computer to show $(X, Y) = (x + 140, y + 80)$. Using the domain $\{x \mid x \in I, -140 \leq x < 140\}$ would make the graph show up with the axes in the middle of the screen.

The graph is very narrow if we let one pixel correspond to one unit on the graph. We would have a better graph if we let ten pixels represent one unit on the graph. Each time x or y changes by one, X or Y would change by ten pixels. That is, $X = 10(x) + 140$ and $Y = 10(y) + 80$. If ten pixels $=$ one unit, then one pixel $= 1/10$ of a unit. That is, each time x or y changes by $1/10$, X or Y would change by one pixel.

The x and y scales need not be the same. In fact, in order to make the graph fit the screen, the scales may have to be different.

EXAMPLE 4

Describe how to plot $f(x)$ or $y = x^3 + 1$ in the center of the screen using a scale of ten pixels per unit for the x-axis and a scale of five pixels per unit for the y-axis.

Solution

1. Plot the axes in the center of the screen as described above.
2. Instead of having x take on values from -140 to 140, we allow x to take on values from -14 to 14, since ten pixels represent one unit. Since each pixel represents $1/10$ of a unit, x will increase in increments of 0.1.
3. $y = f(x) = x^3 + 1$
4. $X = 10x + 140$
5. $Y = 5y + 80$
6. Plot (X, Y) ■

We generalize the procedure as an algorithm:

To plot a function given by $y = f(x)$ on a microcomputer (280 x 160 pixels):

(a) the x-axis is to be up N units with a scale of one unit $=$ A pixels.
(b) the y-axis is to be over M units with a scale of one unit $=$ B pixels.

1. The x-axis will run from $(0, N)$ to $(279, N)$.
2. The y-axis will run from $(M, 0)$ to $(M, 159)$.
3. x runs from $-M/A$ to $(280 - M)/A$ in increments of $1/A$.
4. $y = f(x)$.
5. Plot (X, Y) where $X = Ax + M$ and $Y = By + N$.

EXAMPLE 5	Describe how to plot $y = x^2 + x - 1$ with the x-axis up 30 pixels and the y-axis over 40 pixels. Use a scale of 20 pixels per unit on the x-axis and a scale of 10 pixels per unit on the y-axis.

Solution

1.   The x-axis will run from (0, 30) to (279, 30).
2.   The y-axis will run from (40, 0) to (40, 159).
3. x runs from -2 to 12 in increments of 0.05.
4. $y = x^2 + x - 1$.
5.   Plot (X, Y) where $X = 20x + 40$ and $Y = 10x + 30$. ■

7.5 EXERCISES

Sketch the graph of each function below. See Examples 1–3.

1. $y = x^3 + x^2 - 9x - 9$

2. $y = x^3 - x^2 - 9x + 9$

3. $y = x^3 - 4x^2 - 4x + 16$

4. $y = x^3 - 2x^2 - 5x + 6$

5. $y = x^3 - 2x + 1$

6. $y = x^3 - 2x^2 + 1$

7. $y = x^3 - x^2 + 4x - 4$

8. $y = x^4 - 10x^2 + 9$

9. $y = 4x^4 - 17x^2 + 4$

10. $y = 9x^4 - 13x^2 + 4$

11. $y = x^4 + 4x^2 + 3$

12. $y = x^4 - 5x^3 + 5x^2 + 5x - 6$

Assume a 280×160 coordinate system on a computer monitor for each question. Also assume that the axes are to be in the center of the screen. Describe how to graph each function using the given scales. See Examples 4 and 5.

13. $y = x^2 - 5x + 6$ with a scale of 20 pixels per unit on each axis.

14. $y = x^2 - 5x + 6$ with a scale of 20 pixels per unit on the x-axis and a scale of 10 pixels per unit on the y-axis.

15. $y = x^3 - 5x + 6$ with a scale of 10 pixels per unit on the x-axis and a scale of 20 pixels per unit on the y-axis.

16. $y = x^2 - 9$ with a scale of 70 pixels per unit on each axis.

17. $y = x^2 - 9$ with a scale of 40 pixels per unit on the x-axis and a scale of 20 pixels per unit on the y-axis.

18. $y = x^2 - 9$ with a scale of 20 pixels per unit on the x-axis and a scale of 40 pixels per unit on the y-axis.

✳19. Sketch the graph of $x^2 + y^2 = 4$.
Does the graph represent a function?

➠20. Using the data given in problem 13 above, use pseudocode or a flowchart to show an algorithm for graphing the function on a microcomputer.

✳**21.** X (MOD N) is the remainder when X is divided by N.

Let red represent 0
   blue represent 1
   yellow represent 2
   green represent 3 and
   orange represent 4.

For integral values of x and y ($0 \leq x \leq 6$ and $0 \leq y \leq 6$), determine the color associated with $x^2 + y^2$ (MOD 5). Using a coordinate system where boxes represent coordinates (like the one shown in Figure 7.6 of this section), color each box (x, y) with the color determined by $x^2 + y^2$ (MOD 5).

✳**22.** Describe how plotting a function on a computer would be different if the origin were located at the upper lefthand corner of the screen, and the values of y increase as you move down the axis.

7.6   Exponential and Logarithmic Functions

In Chapter 5, we worked with exponents. Some of the problems had a base of 2, some had a base of 3, some had a base of 10, and so on. If we restrict the use of exponents to a single base, we call the exponents *logarithms*. A commonly used base is 10. When the base is 10, logarithms are called *common logarithms*. The symbol *log* stands for the common logarithm. Thus if $a = 10^b$, we write $b = \log a$. One of the most useful bases is the number e. To eight significant digits, $e = 2.7182818$, but like π, it is an irrational number, and therefore is an infinite nonrepeating decimal.

Definition
If $a = e^b$, then b is the *natural logarithm* of a. We write $b = \ln a$.

The symbol ln stands for *natural logarithm* and indicates that the number e is the base.

EXAMPLE 1

Write each statement below using the natural logarithm notation.

(a) $e^0 = 1$

(b) $e^2 = 7.39$

(c) $e^{-1} = 0.37$

Solution

(a) Since 0 is the exponent or logarithm, we have $\ln 1 = 0$.

(b) Since 2 is the exponent or logarithm, we have $\ln 7.39 = 2$.

(c) Since -1 is the exponent or logarithm, $\ln 0.37 = -1$. ■

EXAMPLE 2

Write each statement below using exponent notation.

(a) $\ln 1 = 0$

(b) $\ln 2 = 0.69$

(c) $\ln 3 = 1.10$

Solution

(a) $e^0 = 1$

(b) $e^{0.69} = 2$

(c) $e^{1.10} = 3$ ∎

Tables have been computed so that for any positive number x, we can look up its natural logarithm. That is, for any positive number x, there is associated with it a unique number y such that $y = \ln x$. This correspondence is a function. It is called the logarithmic function. The domain is $\{x \mid x \in R, x > 0\}$. Remember that $y = \ln x$ means $x = e^y$. Since $e > 0$, it is impossible for x to be less than or equal to zero.

We can also say that for any real number x, we can look up the number y that has x as its natural logarithm. That is, the correspondence $y = e^x$ is also a function. It is called the exponential function. The domain is the set of real numbers.

The exponential function is especially useful in programming languages like Pascal, which do not have a special exponent symbol. Many business applications involve large exponents, and it would be very inconvenient to have to use repeated multiplication for these problems. For example, $P = (1 + r/360)^{365t}$ is the formula for computing the principal after t years if \$1.00 is invested at an annual rate of r percent (written as a decimal), compounded daily, after t years. Using the exponential function allows us to work with exponents like $365t$.

Often we need to compute values of the form a^b, where $a > 0$. The base is a, not e. But we know that there is a number y such that $y = \ln a$, and therefore $a = e^y$. But since logarithms are exponents, they must follow the rules of Chapter 5.

We have $a^b = (e^y)^b = (e^{\ln a})^b = e^{b(\ln a)}$.

> **Exponent Rule**
> $a^b = e^{b(\ln a)}$, where a and b are real numbers, and $a > 0$.

EXAMPLE 3

Write $(2.3)^{1.5}$ using the exponential function.

Solution $a = 2.3$ and $b = 1.5$ for this problem, so we have $(2.3)^{1.5} = e^{1.5(\ln 2.3)}$. ∎

EXAMPLE 4

Write $(4.32)^{0.663}$ using the exponential function.

Solution $a = 4.32$ and $b = 0.663$, so we have $(4.32)^{0.663} = e^{0.663(\ln 4.32)}$. ∎

EXAMPLE 5

Write $(1 + r/360)^{365t}$ using the exponential function.

Solution $a = (1 + r/360)$ and $b = 365t$, so we have $(1 + r/360)^{365t} = e^{365t\ln(1 + r/360)}$. ∎

Many computer languages allow us to work with both the logarithmic function and the exponential function. Typically, the logarithmic function is specified as LN(X). LN(X) returns the natural logarithm of the number X. The exponential function is usually specified as EXP(X). EXP(X) returns e^x.

EXAMPLE 6

Write each expression below as you would for a computer.

(a) ln 3.14

(b) ln 7

(c) ln (-3)

Solution

(a) LN(3.14)

(b) LN(7)

(c) The function is not defined for -3. ∎

EXAMPLE 7

Write each expression below as you would for a computer.

(a) $e^{5.22}$

(b) e^{360}

(c) $e^{0.66}$

Solution

(a) EXP(5.22)

(b) EXP(360)

(c) EXP(0.66) ∎

EXAMPLE 8

Write each expression below as you would for a computer, using the exponential function.

(a) $(2.3)^{1.5}$

(b) $(14.6)^{-2}$

(c) $(-3)^{4.1}$

Solution

(a) $(2.3)^{1.5} = e^{1.5(\ln 2.3)}$, so we have EXP(1.5*LN(2.3)).

(b) $(14.6)^{-2} = e^{-2(\ln 14.6)}$, so we have EXP($-2$*LN(14.6)).

(c) $(-3)^{4.1} = e^{4.1(\ln -3)}$, but ln$-3$ is not defined. ∎

7.6 EXERCISES

Write each statement below using the natural logarithm notation. See Example 1.

1. $e^{0.5} = 1.649$ **2.** $e^{1.5} = 4.482$

3. $e^{2.5} = 12.182$ **4.** $e^{3.5} = 33.115$

5. $e^{4.5} = 90.017$ **6.** $e^{5.5} = 244.692$

7. $e^{6.5} = 665.142$

Write each expression below using exponent notation. See Example 2.

8. $\ln 0.5 = -0.693$ **9.** $\ln 1.5 = 0.405$

10. $\ln 2.5 = 0.916$ **11.** $\ln 3.5 = 1.253$

12. $\ln 4.5 = 1.504$ **13.** $\ln 5.5 = 1.705$

14. $\ln 6.5 = 1.872$

Write each expression below using the exponential function. See Examples 3–5.

15. 2^3 **16.** 3^4

17. 4^5 **18.** $(2.5)^7$

19. $(8.3)^9$ **20.** $(2.1)^{3.6}$

21. $(7.8)^{4.1}$ **22.** $(19.31)^{-2}$

23. $(7.45)^{-3}$ **24.** $(82.7)^{-16.4}$

25. $(101.3)^{-8.1}$ **26.** $(1 + r)^{360}$

Write each expression below as you would for a computer, using the logarithmic or exponential function. See Examples 6–8.

27. $e^{0.5}$ **28.** $e^{2.5}$ **29.** $e^{-3.5}$ **30.** $e^{-4.5}$

31. $e^{-6.5}$ **32.** $\ln 8.8$ **33.** $\ln 7.3$ **34.** 2^3

35. 3^4 **36.** $(2.1)^{3.6}$ **37.** $\ln 0.5$ **38.** $\ln 2.5$

39. $\ln 5.5$ **40.** $(19.31)^{-8.1}$ **41.** $(1 + r)^{360}$ **42.** $(-5.1)^{3.6}$

Computers do not store tables of values for the exponential and logarithmic functions. Rather, those values are computed when needed by using a finite number of terms from formulas such as the ones that follow:

$e^x = 1 + x + x^2/2! + x^3/3! + \ldots$ (where $n! = n(n - 1)(n - 2) \ldots 1$)

$\ln(1 + x) = x - x^2/2 + x^3/3 - x^4/4 + \ldots$.

✳**43.** **(a)** Compute $e^{2.5}$ using the first four terms of the formula.

     (b) Compute $e^{2.5}$ using the first five terms of the formula.

     (c) Compare your answer to problem 3.

✳**44.** **(a)** Compute $\ln 2.5$ using the first four terms of the formula.

     (b) Compute $\ln 2.5$ using the first five terms of the formula.

     (c) Compare your answer to problem 10.

✳**45.** **(a)** Compute 4^5 using the first five terms of the formulas.

     (b) Compare your answer to the actual value of 4^5.

7.7   Computer Functions

Programming languages like BASIC and Pascal allow us to define and use functions. There are also some built-in functions that we are allowed to use. These differ from one machine to another, but some common ones are presented in this section.

> **Definition**
> The *absolute value* of a real number x, denoted $|x|$, is x if $x \geq 0$ and is $-x$ if $x < 0$.

This definition says that if x is positive or zero ($x \geq 0$), the absolute value is equal to the number x itself. If, however, x is negative ($x < 0$), then we multiply by -1, and the product will be positive. That is, the absolute value of any real number is positive or zero.

EXAMPLE 1

Simplify each expression below.

(a) $|3|$

(b) $|-4|$

(c) $|-3/4|$

(d) $|0|$

Solution

(a) $|3| = 3$

(b) $|-4| = 4$

(c) $|-3/4| = 3/4$

(d) $|0| = 0$ ■

Most computers use ABS(X) as a function that returns the absolute value of a real number X.

EXAMPLE 2

Determine what number is represented by each expression.

(a)   ABS(-1)

(b)   ABS(2)

(c)   ABS(0.5)

Solution

(a)   ABS(-1) = 1

(b)   ABS(2) = 2

(c)   ABS(0.5) = 0.5   ■

The absolute value function could be used to avoid the problem we had with the algorithm in Section 4.3. The problem occurred because the computer had rounded off the binary representation of the decimal value 0.1 The sum came to 1.000000119209 instead of exactly 1. The computed value differed from the expected value of 1 by a very small amount. We could have said IF ABS(X − 1) < 0.0001 THEN and so on. That is, we could check the difference between the computed value of X and the expected value of X. If the difference is sufficiently small, we can ignore it as a round-off error. The absolute value allows us to check for these small errors, regardless of whether X is larger or smaller than we expected it to be.

EXAMPLE 3

Determine whether ABS(X − 5) < 0.0001 for each value of X.

(a) X = 5.000218
(b) X = 4.999995

Solution
(a) ABS(5.000218 − 5) = 0.000218, which is not less than 0.0001.
(b) ABS(4.999995 − 5) = ABS(−0.000005) = 0.000005, which is less than 0.0001. ■

The greatest integer, truncation, and rounding functions are similar to each other, yet subtly different. We present these functions together to emphasize this point.

> **Definition**
> The *greatest integer* in a real number x, denoted $[x]$, is the largest integer less than or equal to x.

One way to think of the greatest integer function is to think of starting at the number and moving to the left on the x-axis of a graph (sometimes called a *number line*) until an integer is encountered.

EXAMPLE 4

Simplify each expression below.

(a) [2.3]
(b) [2]
(c) [−1.3]
(d) [−2.99]

Solution

(a) [2.3] = 2, since 2 is the first integer to the left of 2.3 on the number line.

(b) [2] = 2, since the definition says [*x*] is less than *or equal to x*.

(c) [−1.3] = −2, since −2 is the first integer to the left of −1.3 on the number line.

(d) [−2.99] = −3, since −3 is the first integer to the left of −2.99 on the number line. ∎

Many computer languages (but not all) use INT(X) as a function that returns the greatest integer in the real number X.

EXAMPLE 5

Determine what number is represented by each expression.

(a) INT(23.9)

(b) INT(2.4)

(c) INT(−25.1)

(d) INT(−25.87)

Solution

(a) INT(23.9) = 23

(b) INT(2.4) = 2

(c) INT(−25.1) = −26

(d) INT(−25.87) = −26 ∎

The truncation function is used more with computers than it is in mathematics. It has no mathematical symbol.

Definition
To truncate a real number *x* means to drop any fractional part of the number.

For positive values of *x*, to truncate *x* would give the same answer as finding the greatest integer in *x*. But for negative values of *x*, the two functions behave differently. For example, the greatest integer in −3.1 is −4, but to truncate −3.1 gives −3.

Sometimes TRUNC(X) is used as a computer function to return the truncated value of the real number X.

EXAMPLE 6   Determine what number is represented by each expression. (Compare answers with Example 4.)

(a)   TRUNC(23.9)

(b)   TRUNC(2.4)

(c)   TRUNC(-25.1)

(d)   TRUNC(-25.87)

Solution

(a)   TRUNC(23.9) = 23

(b)   TRUNC(2.4) = 2

(c)   TRUNC(-25.1) = -25

(d)   TRUNC(-25.87) = -25 ■

Notice that truncating a real number is not the same thing as rounding it. To round a real number to the nearest integer, we take the integer that is closest to it (to the left or right) on the number line. We can define the rounding function in terms of the greatest integer function.

> **Definition**
>
> To round off a real number x to the nearest integer, find the greatest integer in $(x + 0.5)$.

This definition is equivalent to the common rule that says if the digit in the tenths position is less than 5, retain the ones digit; if the digit in the tenths position is 5 or more, add 1 to the ones digit.†

Some computers use a function ROUND(X) to return the value of a real number X rounded to the nearest integer.

---

†Sometimes a different rule is used for rounding to the nearest integer: if the tenths digit is less than 5, the ones digit is unchanged. If the tenths digit is greater than 5, 1 is added to the ones digit. If the tenths digit is 5, the ones digit itself determines whether to round up or down. If it is even, it remains unchanged; if it is odd, 1 is added.

EXAMPLE 7

Determine what number is represented by each expression.

(a) ROUND(2.3)

(b) ROUND(2.7)

(c) ROUND(−1.3)

(d) ROUND(−2.99)

Solution

(a) ROUND(2.3) = 2

(b) ROUND(2.7) = 3

(c) ROUND(−1.3) = −1

(d) ROUND(−2.99) = −3 ∎

We don't always want to round off to the nearest integer, however. For example, in financial calculations, we may want to round off to hundredths to indicate dollars and cents. We can define a function to do the rounding. $f(x) = [100x + 0.5]/100$ would round off x to the nearest hundredth. We could write this expression for the computer as INT(100*X + 0.5)/100. Consider X = 2.394. We have

$$INT(100*2.394 + 0.5)/100 =$$

$$INT(239.4 + 0.5)/100 =$$

$$INT(239.9)/100 =$$

$$239/100 =$$

$$2.39.$$

EXAMPLE 8

Verify that the computer function INT(100*X + 0.5)/100 produces the usual result of rounding to the nearest hundredth for each number below.

(a) 34.7994

(b) −6.723

Solution

(a) INT(100*34.7994 + 0.5)/100 = INT(3479.94 + 0.5)/100 =
$$INT(3480.44)/100$$
$$3480/100$$
$$34.80$$

(b) INT(100*(−6.723) + 0.5)/100 = INT(−672.3 + 0.5)/100 =
$$INT(−671.8)/100$$
$$−672/100$$
$$−6.72 ∎$$

We could generalize this process of rounding off by using the rule that follows.

> **To round off a real number x to N decimal places:**
> Use the function $f(x) = [10^N x + 0.5]/10^N$.

Compare the graphs of $Y = INT(X)$, $Y = TRUNC(X)$, and $Y = ROUND(X)$ in Figure 7.8. The open circle indicates that the value is not part of the function.

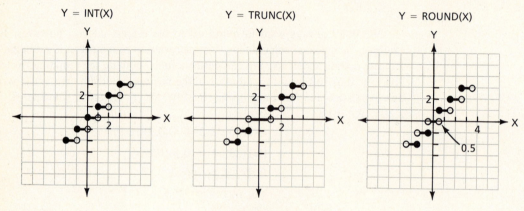

Figure 7.8

7.7 EXERCISES

Simplify each expression. See Example 1.

1. $|2.3|$ **2.** $|-2.3|$ **3.** $|-7|$ **4.** $|7|$

Determine what number is represented by each expression. See Example 2.

5. ABS(5.7) **6.** ABS(-5.7) **7.** ABS(-6.33) **8.** ABS(9.21)

Determine whether ABS(X - 1) < 0.0001 for each problem. See Example 3.

9. X = 1.000001 **10.** X = 0.999999 **11.** X = 1.023

Determine whether ABS(X - 3) < 0.0001 for each problem. See Example 3.

12. X = 3.00001 **13.** X = 2.00999 **14.** X = 2.99991

Simplify each expression. See Example 4.

15. [2.4] **16.** [5.83] **17.** [-5.83] **18.** [-2.4]

Determine what number is represented by each expression. See Example 5.

19. INT(2.4)   **20.** INT(−2.4)   **21.** INT(−19.8)   **22.** INT(19.8)

Determine what number is represented by each expression. See Example 6.

23. TRUNC(2.4)   **24.** TRUNC(−2.4)   **25.** TRUNC(−19.8)   **26.** TRUNC(19.8)

Determine what number is represented by each expression. See Example 7.

27. ROUND(2.4)   **28.** ROUND(−2.4)   **29.** ROUND(−19.8)   **30.** ROUND(19.8)

Verify that the computer function INT(100*X + 0.5)/100 produces the usual result of rounding to the nearest hundredth for each number. See Example 8.

31. 4.2121   **32.** 2.7291   **33.** 3.166

34. −4.2121   **35.** −2.7291   **36.** −3.166

✴**37.** Sketch the graph of **(a)** $y = |x|$.
(b) $y = |x - 1|$.
(c) $y = |x| - 1$.

✴**38.** Sketch the graph of $y = |x| \pm \sqrt{1 - x^2}$.

✴**39.** Sketch the graph of $y = x - [x]$.

CHAPTER REVIEW

A _____ is a correspondence between two sets such that there is associated with each element of the first set exactly one element in the second set. The first set is called the _____, and the second set is called the _____.

In the expression $f(x)$, the letter f denotes the function, and the number in parentheses is called the _____ of the function.

The letter x is called the _____ variable, and the letter y is called the _____ variable.

In a rectangular coordinate system, the horizontal axis is called the _____ and the vertical axis is called the _____. The point of intersection of the two axes is called the _____.

The axes divide the plane into four regions called _____.

If a graph is not the graph of a function, it is possible to place a _____ line somewhere on the plane so that it intersects the graph in more than one point.

The point at which a line crosses the y-axis is called the _____.

The _____ of a line is sometimes described as rise/run.

In the equation $y = mx + b$, m is the _____, and b is the _____.

CHAPTER TEST

1. Determine whether the correspondence is a function:

x	y
-2	4
-1	1
0	0
1	1

If $f(x) = 2x + 3$ and $g(x) = 2x - 3$, evaluate:

2. $f(1)$

3. $f(-2) + g(-2)$

Determine whether the correspondence is a function. If it is a function, specify the domain.

4. $f(x) = 2x + 1$

5. $g(x) = \dfrac{x + 1}{x - 1}$

6. $h(x) = x^2 + 3$

Graph each function on a single plane.

7.

x	y
-1	-1
0	1
1	3

8.

x	y
-1	1
0	1
1	-3

Determine whether each graph represents a function.

9.

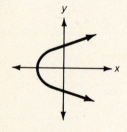

10.

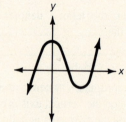

Graph each equation.

11. $y = 3$

12. $y = -3x$

Identify the slope and y-intercept.

13. $y = -x + 4$

14. $x = 2y - 3$

15. Graph: $x + 3y = 6$

For each quadratic function, determine whether the parabola will open upward or downward, find the vertex, and sketch the graph.

16. $y = x^2 - 16$

17. $y = -x^2 + 5x$

18. $y = x^2 + 4x + 4$

19. $y = 2x^2 + 2x + 3$

20. $y = -3x^2 - x + 1$

Sketch the graph of each function.

21. $y = x^3 - 4$ **22.** $y = x^3 - 4x$ **23.** $y = x^3 - 4x^2$

24. $y = x^4 + 3x^2 - 4$ **25.** $y = x^4 + x^3 - 7x^2 - x + 6$

Assume a 280×160 coordinate system on a computer monitor. Also assume that the axes are to be in the center of the screen. Describe how to graph each function using the given scales.

26. $y = x^2 - 4x + 4$ with 20 pixels per unit on each axis.

27. $y = 2x^2 - 4$ with 20 pixels per unit on each axis.

28. $y = x^2$ with 20 pixels per unit on each axis.

29. Write using the natural logarithm notation: $e^{7.5} = 1808.04$.

30. Write using exponential notation: $\ln 7.5 = 2.01$.

31. Write using the exponential function: $(56)^{2.7} = 52{,}493.76$.

Write each expression as you would for a computer using the LN or EXP function.

32. $e^{7.5}$ **33.** $\ln 7.5$

Determine what number is represented by each expression.

34. (a) $|-16.1|$ **35.** (a) $[4.11]$ **36.** (a) INT(2.8)
     (b) ABS(16.1)   (b) $[-4.11]$ (b) TRUNC(2.8)
                                                             (c) ROUND(-2.8)

37. Is ABS(X $-$ 2) < 0.0001 if X $= 2.001$?

38. Verify that the computer function INT(100*X $+$ 0.5)/100 produces the usual result for rounding to the nearest hundredth for X $= 5.731$.

39. The following flowchart shows an algorithm that gives the x-coordinate of the vertex of the parabola $y = ax^2 + bx + c$. Modify it so that the y-coordinate of the vertex and points with an x-coordinate one more or one less than the x-coordinate of the vertex will also be displayed.

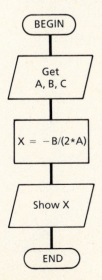

FOR FURTHER READING

Burckhardt, J. J. "Leonhard Euler, 1707–1783," *Mathematics Magazine,* Vol. 56 No. 5 (November 1983), 262–273.

Lifshitz, Maxine Rosman. "Little Known Facts About the Quadratic Function," *The Mathematics Teacher,* Vol. 77 No. 5 (May 1984), 353–356. (Hints for graphing quadratic functions are given.)

Pierce, R. C. "A Brief History of Logarithms," *The Two-Year College Mathematics Journal,* Vol 8 No. 1 (January 1977), 22–26.

Pulfer, Wayne. "The Inverse of a Function (of a Single Variable)," *The Mathematics Teacher,* Vol. 77 No. 1 (January 1984), 34–35. (Inverse functions are related to inverse operations.)

Biography   Leonhard Euler

Leonhard Euler was born April 15, 1707, in Switzerland. His father wanted him to be a minister like himself, but Euler was interested in mathematics.

In 1727, he went to St. Petersburg to assume a position in physiology (which he had studied only a few months). The day that he arrived, however, the Empress Catherine, who had supported the Academy of Sciences, died, and her successors stopped funding it. Euler joined the Russian navy until funds for the academy were restored.

He married Catherine Gsell, daughter of the Swiss artist Georg Gsell. They had 13 children, but only five lived beyond childhood. Euler often wrote his papers with children on his lap and others playing nearby. He is known as the most prolific mathematician ever because he published many books and almost 800 papers. He did much work with functions. He introduced the f notation and the symbol e.

In 1738, he lost the sight in his right eye. According to legend, the loss was caused by intense work on an astronomical problem. In 1741, Euler accepted a position at the Berlin Academy of Sciences. He returned to Russia in 1766 after being told that he could write his own contract.

After he had a cataract removed from his left eye in 1771, an infection left him totally blind. He dictated to his assistants and used a large slate to write formulas for them. There is a story that two of his students disagreed on an answer in the 50th decimal place. Euler did the calculation mentally and obtained the correct answer.

On September 18, 1783, Euler worked on calculations for the orbit of the newly discovered planet Uranus. He played with his grandson, had tea, and died suddenly.

Chapter Eight
Systems of Equations

A system of three equations in three variables represents three planes. The three planes may intersect in a single point, like the floor and two walls of a room.

8.1   Systems of Two Equations

In Chapter 6, we considered equations. Remember that to solve an equation means to find the values that, when substituted for the variables, make the equation true. In this chapter, we consider systems of equations.

> **Definition**
> A system of simultaneous linear equations is a set of two or more linear equations.

Consider $x + y = 5$

$x - y = 1$. For each equation individually there are infinitely many ordered pairs (x, y) that will make the equation true. The equation $x + y = 5$, for example, is true for $(1, 4)$, $(6, -1)$, and $(0, 5)$, to name just three possibilities. The equation $x - y = 1$ is true for $(5, 4)$, $(-1, -2)$, and $(0, -1)$ among other values.

To solve the system means to find the values (in this case, ordered pairs) that will make *both* equations true at the same time. The ordered pair $(3, 2)$ is the solution of this system.

A geometric interpretation for solving a system of equations is possible if we sketch both equations on the same coordinate plane. Each point on the graph of $x + y = 5$ represents a solution of that equation. Likewise, each point on the graph of $x - y = 1$ represents a solution of that equation. The point $(3, 2)$ is the only point which lies on both graphs. See Figure 8.1.

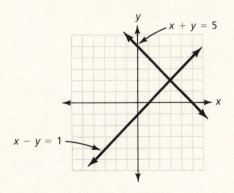

Figure 8.1

> **To solve a system of two simultaneous linear equations in two variables:**
> 1. Graph the first equation on a rectangular coordinate system.
> 2. Graph the second equation on the same coordinate system.
> 3. Identify the point of intersection of the two lines.

EXAMPLE 1

Solve the system $2x - y = 5$
$$x + 2y = -5.$$

Solution

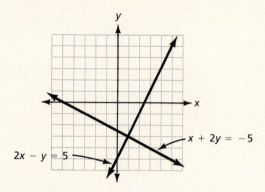

The solution is $(1, -3)$. ■

Graphing as a method of finding the solution of a system of equations has two serious disadvantages.

1. It is difficult to read the graph when the coordinates of the solution are fractions.
2. It is difficult to generalize to three or more variables. The next example illustrates the difficulty of reading fractions on the graph.

EXAMPLE 2

Solve the system $x + 2y = 1$
$$7x + y = 2.$$

Solution

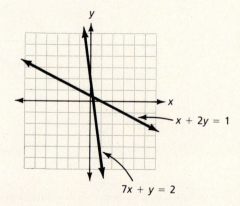

It is impossible to read the solution $(3/13, 5/13)$ from the graph. ■

An algebraic solution allows us to calculate the solution exactly. The method presented here is often called the *elimination method*.

Suppose we reconsider the system $x + y = 5$
$$x - y = 1.$$

If we add the second equation to the first one, we have added equal numbers to both sides of the first equation. The resulting equation is $2x = 6$. It follows that $x = 3$. Since the solution of the system satisfies *both* equations, we can calculate the y value by substituting $x = 3$ into either one. Using the first equation $3 + y = 5$, so $y = 2$.

The fact that we had $+ y$ and $- y$ in this system led to the *elimination* of the variable y from the system. We could solve the resulting equation using the techniques of Chapter 6.

EXAMPLE 3

Solve the system $x + 2y = 1$ **(1)**

$7x + \ y = 2$ by the elimination method.   **(2)**

Solution We would like to eliminate one of the variables. However, we must first replace one of the equations with an equivalent one. It is important to see that we can eliminate *either* variable. If we choose to eliminate x, we must force equation (1) to contain $-7x$. We can make this modification if we multiply both sides of the equation by -7. We have

$$-7x - 14y = -7$$
$$\underline{7x + \ y = \ \ 2.} \text{ Adding the two equations gives}$$
$$-13y = -5, \text{ so } y = 5/13.$$

If we had chosen to eliminate y instead of x, we would have forced equation (2) to contain $-2y$ by multiplying both sides of that equation by -2. That would give us

$$x + 2y = \ \ 1$$
$$\underline{-14x - 2y = -4.} \text{ Adding the two equations, we have}$$
$$-13x \ \ \ \ \ = -3, \text{ so } x = 3/13.$$

Eliminating x produces the y value while eliminating y produces the x value. It does not matter which is done first. ■

EXAMPLE 4

Solve the system $2x - y = \ \ \ 5$ **(1)**

$x + 2y = -5$ by the elimination method.   **(2)**

Solution To eliminate x, we multiply equation (2) by -2.

We have

$$2x - \ y = \ \ 5$$
$$\underline{-2x - 4y = 10.} \text{ Adding the equations gives}$$
$$-5y = 15, \text{ so } y = -3.$$

To eliminate y, we multiply equation (1) by 2. We have

$$4x - 2y = 10$$
$$\underline{x + 2y = -5.}\text{ Adding the two equations gives}$$
$$5x \qquad = \quad 5,\text{ so } x = 1.\text{ The solution, then, is } (1, -3). \blacksquare$$

Sometimes it is necessary to replace both equations in order to eliminate one of the variables.

<table>
<tr><td>**EXAMPLE 5**</td><td>Solve the system $2x + 5y = \quad 1$ **(1)**
$3x - 4y = -10$ by the elimination method.   **(2)**</td></tr>
</table>

Solution   To eliminate x, force equation (1) to contain $+6x$ and force equation (2) to contain $-6x$. That is, multiply equation (1) by 3 and equation (2) by -2. We have

$$6x + 15y = \quad 3$$
$$\underline{-6x + \quad 8y = 20.}\text{ Adding the two equations, we have}$$
$$23y = 23,\text{ so } y = 1.$$

To eliminate y, multiply equation (1) by 4 and equation (2) by 5 to get $20y$ and $-20y$. Notice that the signs are already opposite:

$$8x + 20y = \quad 4$$
$$\underline{15x - 20y = -50.}\text{ Adding the two equations, we have}$$
$$23x \qquad = -46,\text{ so } x = -2.$$

The solution, then, is $(-2, 1)$. $\blacksquare$

<table>
<tr><td>**EXAMPLE 6**</td><td>Solve the system $5x + 3y = 7$ **(1)**
$4x - 2y = 5$ by the elimination method.   **(2)**</td></tr>
</table>

Solution   To eliminate x, multiply equation (1) by -4 and equation (2) by 5. We have

$$-20x - 12y = -28$$
$$\underline{20x - 10y = \quad 25.}\text{ Adding the two equations gives}$$
$$-22y = -3,\text{ so } y = 3/22.$$

To eliminate y, multiply equation (1) by 2 and equation (2) by 3. We have

$$10x + 6y = 14$$
$$\underline{12x - 6y = 15.}\text{ Adding the two equations gives}$$
$$22x \qquad = 29,\text{ so } x = 29/22.\text{ The solution is } (29/22, 3/22). \blacksquare$$

When the first value is a fraction rather than an integer (as in Example 6), we solve the system by going through the elimination process for each variable. When the first value is an integer, as it was in Examples 4 and 5, it is probably easier to find the second value by substituting the first into *either* equation. The two equations will produce the same answer, so choose the one that looks like it will be easier to solve. The procedure can be generalized.

To solve a system of two simultaneous linear equations in two variables by the elimination method:
1. If the equations are not in the form $ax + by = c$, write equivalent equations that are in that form.
2. Choose one of the variables as the one to eliminate (called the *pivotal variable*).
3. If necessary, multiply one or both equations in order to make the coefficients of the pivotal variable have the same absolute value, but opposite signs.
4. Add the two equations.
5. Solve for the remaining variable.
6. (a) If the value of this variable is an integer, substitute it into either equation to find the other variable.
   (b) If the value of the first variable is a fraction, return to the original problem, and repeat steps 2–5 using the other variable as the pivotal variable.

EXAMPLE 7

Solve $3x + 4y = 10$ (1)

$\quad\quad 5x - 7 = 3y$ by the elimination method. (2)

Solution

1. Replace equation (2) by the equivalent equation
   $$5x - 3y = 7. \quad\quad (3)$$
2. Choose y as the pivotal variable (we could also work the problem by choosing x).
3. Multiply equation (1) by 3 and equation (3) by 4.
   $$9x + 12y = 30$$
   $$20x - 12y = 28.$$
4. $29x \quad\quad = 58$ is the result of adding the two equations.
5. So, $x = 2$.
6. Since 2 is an integer, we substitute $x = 2$ into the first equation.
   $3(2) + 4y = 10$ or $6 + 4y = 10$ or $4y = 4$, so $y = 1$.
   The solution is $(2, 1)$. ∎

EXAMPLE 8

Solve $3x + 2y - 5 = 0$ (1)

$6x + 4y = 10.$ (2)

Solution

1. Replace equation (1) by the equivalent equation

$3x + 2y = 5.$ (3)

2. We choose x as the pivotal variable.
3. Multiply equation (3) by -2.   We have $-6x - 4y = -10$

$6x + 4y = 10.$

4. Adding the two equations gives

$0 = 0.$

5. Eliminating x also eliminated y.

Every solution of one equation is a solution of the other. There are thus an infinite number of solutions of the system. We say that the system is *dependent*. Graphically, one equation would produce the same line as the other. This happens because one equation is a multiple of the other (i.e., they are equivalent equations). ■

EXAMPLE 9

Solve $-2x + 4y = 7$ (1)

$x - 2y = 3.$ (2)

Solution

1. The equations are in the correct form.
2. We choose to eliminate x.
3. Multiply equation (2) by 2.   We have $-2x + 4y = 7$

$2x - 4y = 6.$

4. Add the two equations to get $0 = 13.$

5. Eliminating x simultaneously eliminated y again, but this time, the resulting statement ($0 = 13$) is false. There are no values of x and y for which the equation will be true. We say that the system is *inconsistent*. Graphically, the two equations will produce lines that are parallel to each other and, therefore, never intersect. ■

8.1 EXERCISES

Graph each pair of equations and determine the solution of the system from the graph. See Examples 1 and 2.

1. $x + y = 8$
$2x - y = 1$

2. $2x + y = 8$
$x + 2y = 10$

3. $3x - 2y = 1$
$4x - y = 1$

4. $2x + 3y = 1$
$3x - 2y = -5$

5. $x + 2y = 1$
$2x - y = 0$

6. $6x + 6y = 7$
$6x - 6y = 1$

7. $x + 3y = 4$
$\quad 2x + 6y = 7$

8. $2x - 5y = 10$
$\quad 4x - 10y = 20$

9. $3x - 4y = 12$
$\quad 9x - 12y = 36$

10. $x + 7y = 7$
$\quad 2x + 14y = 15$

Solve each system of equations by the elimination method. See Examples 3–9.

11. $x + y = 8$
$\quad 2x - y = 1$

12. $2x + y = 8$
$\quad x + 2y = 10$

13. $3x - 2y = 1$
$\quad 4x - y = 1$

14. $2x + 3y = 1$
$\quad 3x - 2y = -5$

15. $x + 2y = 1$
$\quad\quad 2x = y$

16. $6x + 6y = 7$
$\quad\quad 6x = 6y + 1$

17. $x + 3y = 4$
$\quad 2x + 6y = 7$

18. $2x = 5y + 10$
$\quad 4x = 10y + 20$

19. $3x - 4y = 12$
$\quad 9x - 12y = 36$

20. $x + 7y - 7 = 0$
$\quad 2x + 14y - 15 = 0$

21. $x + 3y = 7$
$\quad 2x + y + 1 = 0$

22. $2x = 3y + 8$
$\quad 5x + 2y = 1$

23. $3x + 2y = 0$
$\quad\quad 3x = 2y$

24. $3x - 4y = 1$
$\quad\quad 4x = 3y - 1$

✸25. If $2^{(2x + y)} = 256$ and $3^{(y - x)} = 243$, find x and y.

✸26. A two-digit number is such that the sum of the digits is nine. When the order of the digits is reversed, the new number is nine less than three times the original number. What was the original number?

✸27. If the three-digit number $1a2$ is added to 345, the sum is the three-digit number $4b7$, which is divisible by nine. Find the digits a and b.

8.2   Systems of Three Equations

Our strategy in solving a system of three equations in three variables is to consider the equations two at a time. By eliminating one variable from a pair, we reduce the pair of equations to one equation with two variables. Once we have eliminated the same variable from two pairs, we can find the solution to the system formed by the resulting equations.

Consider the following system of three equations with three variables:

$$x - 3y + 2z = 5 \tag{1}$$

$$x + 2y + 3z = 7 \tag{2}$$

$$2x - y + z = 4. \tag{3}$$

We want to eliminate one of the variables from all three equations. Since x is listed first in each equation, y is listed second in each, and z is listed third in each, this system is ready for the elimination step. We choose x as the variable to eliminate. Working with the equations two at a time, we pair the first two to get the system

$$x - 3y + 2z = 5 \tag{1}$$
$$x + 2y + 3z = 7. \tag{2}$$

Multiply equation (1) by -1 to get $-x + 3y - 2z = -5.$

Equation (2) is $\underline{x + 2y + 3z = 7.}$

Adding, we have $5y + z = 2. \tag{4}$

We have not yet used equation (3) of the original system. We must eliminate x from it also. We can pair it with either of the other two equations. We choose the first, so that we have the system:

$$x - 3y + 2z = 5 \tag{1}$$
$$2x - y + z = 4. \tag{3}$$

Multiply equation (1) by -2 to get $-2x + 6y - 4z = -10.$

Equation (3) is $\underline{2x - y + z = 4.}$

Adding, we have $5y - 3z = -6. \tag{5}$

The results of these two elimination procedures give us a system of two equations in two variables. That is, we have

$$5y + z = 2 \tag{4}$$
$$5y - 3z = -6. \tag{5}$$

If we solve this system for y and z, we will have two of the three values in the original system.

We eliminate y by multiplying equation (5) by -1 and adding the two equations:

$$5y + z = 2$$
$$\underline{-5y + 3z = 6}$$
$$4z = 8, \text{ so } z = 2.$$

We solve for y by substituting the value $z = 2$ into equation (4). Thus, $5y + 2 = 2$, so $5y = 0$ and $y = 0$.

Now that we have the values of y and z, we return to the original system of three equations to solve for x. We can use any one of the three equations, because all three will produce the same solution. We choose the first. $x - 3y + 2z = 5$ becomes

$$x - 3(0) + 2(2) = 5, \text{ or}$$
$$x \qquad\quad + 4 = 5$$
$$\text{so } x \qquad\qquad = 1.$$

We now have the complete solution $(1, 0, 2)$. To check the solution, we substitute the values into the original system:

$$1 - 3(0) + 2(2) = 5 \quad \text{or} \quad 1 + 4 = 5$$
$$1 + 2(0) + 3(2) = 7 \quad \text{or} \quad 1 + 6 = 7$$
$$2(1) - 0 + 2 \quad\; = 4 \quad \text{or} \quad 2 + 2 = 4.$$

This procedure is summarized as follows:

To solve a system of three simultaneous linear equations in three variables:

1. Arrange the terms of each equation so that the variables appear in the same order for each equation.
2. Choose a pivotal variable for the elimination step.
3. Use two of the three equations to eliminate the pivotal variable. (Multiply one or both equations to make the coefficients have the same absolute value but different signs, and add.)
4. Pair the remaining equation of the system with one of the other two and eliminate the pivotal variable.
5. Pair the resulting equations from steps 3 and 4. Solve as a system of two equations in two variables.
6. Return to the original system of three equations. Choose one of the equations and substitute the values found in step 5. Solve the equation for the remaining variable.

EXAMPLE 1

Solve:
$$\begin{align} 2x + y - 3z &= 2 \tag{1}\\ 3x - y + 4z &= 9 \tag{2}\\ 4x + 2y - 3z &= 7 \tag{3} \end{align}$$

Solution

1. The variables appear in the same order for each equation, so the system is ready for the elimination step.
2. We choose to eliminate y.

3. Pair the first and second equations and eliminate y:

$$2x + y - 3z = 2$$
$$\underline{3x - y + 4z = 9.} \quad \text{Adding, we have}$$
$$5x \quad + z = 11. \tag{4}$$

4. Pair equation (3) with equation (1) and eliminate y:

Multiply equation (1) by -2. We have $\quad -4x - 2y + 6z = -4.$

Equation (3) is $\quad\quad\quad \underline{4x + 2y - 3z = 7.}$

Adding the two equations gives $\quad\quad\quad\quad\quad 3z = 3$ or
$$z = 1. \tag{5}$$

5. We pair the results of steps 3 and 4:

$$5x + z = 11 \tag{4}$$
$$z = 1. \tag{5}$$

Solving this system, we have $z = 1$ and $5x + 1 = 11$, or $5x = 10$, so $x = 2$.

6. We return to the original system and choose the first equation. Substituting $x = 2$ and $z = 1$ (from step 5), we have

$$2(2) + y - 3(1) = 2$$
$$4 + y - 3 \quad = 2$$
$$1 + y \quad\quad = 2$$
$$y \quad\quad\quad = 1.$$

The complete solution is $(2, 1, 1)$. ∎

EXAMPLE 2

Solve:
$$2x - 3y - 2z = 5 \tag{1}$$
$$x + 2y + 2z = 0 \tag{2}$$
$$y - z = 3. \tag{3}$$

Solution We choose to eliminate x.

From (1) and (2):
$$2x - 3y - 2z = 5$$
$$\underline{-2x - 4y - 4z = 0}$$
$$-7y - 6z = 5. \tag{4}$$

From (3): It is not necessary to pair equation (3) with another, since x does not appear. We have

$$y - z = 3. \tag{5}$$

Pairing equations (4) and (5), we have

$$-7y - 6z = 5 \tag{4}$$
$$y - z = 3 \tag{5}$$

We choose to eliminate y:

$$-7y - 6z = 5$$
$$\underline{7y - 7z = 21}$$
$$-13z = 26$$
$$z = -2.$$

From equation (5), we have $y - (-2) = 3$, so $y = 1$.

Finally, equation (2) gives us $x + 2(1) + 2(-2) = 0$ or

$$x + 2 - 4 = 0$$
$$x = 2.$$

The complete solution is $(2, 1, -2)$. ▧

8.2 EXERCISES

Solve each system by the elimination method. See Examples 1 and 2.

1. $x - y + 2z = -1$
$2x + y - z = 9$
$x - 2y - z = 0$

2. $2x + y + 3z = 0$
$x - 3y + 2z = 7$
$3x - 2y - z = 7$

3. $2x - 3y + z = 0$
$3x + 2y - z = 4$
$2x - 2y + z = 1$

4. $3x + y + 2z = 3$
$2x - y - 3z = 15$
$2x + 2y + 3z = -3$

5. $3x + 2z = 3$
$x - y - 3z = 15$
$x + y = 0$

6. $x + y - z = 2$
$x - y + z = 4$
$-x + y + z = 6$

7. $2x + 2y + z = 0$
$x - y - 1 = 0$
$3x - 3y + 2z = 3$

8. $2x + 3y = z + 2$
$4x - 3y = 1 - z$
$2x - 3y = z - 2$

9. $x + z = y + 3$
$2x + z = 3 - y$
$2x - z = 2y$

10. $3x + 4y + z - 6 = 0$
$6x - 4y - 2z - 7 = 0$
$3x - 4y - 3z - 4 = 0$

11. $3x - z + 2y = 3$
$2x - y + 3z = 8$
$3y + 2z + x = 1$

12. $x + 2y + 2z = 2$
$2x + y - 7 = 2z$
$2x + 2y = z + 8$

✸**13.** Of 100 programmers, 2/3 of those who know Pascal know BASIC, and 3/4 of those who know BASIC know Pascal. If there are 12 who know neither language, how many know both?

✸**14.** A rectangular box (see the following page) has dimensions such that the sum of the height and width is 7 cm, the sum of the width and length is 15 cm, and the sum of the length and height is 16 cm.

(a) Find the length, width, and height of the box.

(b) How far is it from the upper back left corner to the lower front right corner?

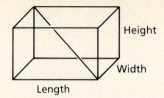

Height

Width

Length

✳15. We represent three variables using ordered triples in space with three axes. In a system of three equations with three variables, each equation represents a plane. You might visualize two walls and the floor as an example of three planes that intersect in a point.

  (a) What type of figure is formed by the intersection of two planes?

  (b) In a system of three equations, how many pairs of equations are there?

  (c) Why do we consider only two pairs in the solution of a system of three equations?

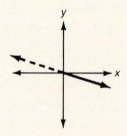

8.3   Gauss Elimination

When we used the elimination method to solve a system of three equations, we were able to choose from several options. We could decide which of three variables to eliminate, and we could decide which equations to pair in order to complete the elimination. We chose the options that made the arithmetic the easiest. Computer considerations are different.

A computer solution need not take into account the difficulty of the arithmetic. It should, however, be orderly. The most common computer method for solving a system of linear equations is a variation of the elimination procedure called "Gauss elimination with back substitution." It is named after a German mathematician, Carl Friedrich Gauss, although it was known to others before him. The method is presented in this section for a system of three equations in three variables, although it can be generalized to n equations in m variables.

The basic idea is to write the system as an equivalent system that has the pattern:

$$x + b_1y + c_1z = d_1$$

$$y + c_2z = d_2$$

$$z = d_3.$$

That is, the x terms are to be eliminated from the equations after the first, and y terms are to be eliminated from equations after the second. Back substitution means that once we have the equations in this form, we can take the value for z in the third equation and substitute it back into the second equation to solve for y. Knowing y and z, we can substitute these two values back into the first equation to solve for x.

EXAMPLE 1

Solve the following system by Gauss elimination with back substitution.

$$x - 3y + 2z = 5 \qquad \textbf{(1)}$$

$$x + 2y + 3z = 7 \qquad \textbf{(2)}$$

$$2x - y + z = 4 \qquad \textbf{(3)}$$

Solution   We will consider the problem in five phases.

1. For phase one, the variables are arranged in the same order for each equation, as they are in this problem.

2. For phase two, we work with the variable x.
    (a) Equation (1) has 1 as the coefficient of x. (If it had not been 1, we would have divided both sides of the equation by this coefficient.) We now use equation (1) or $x - 3y + 2z = 5$ to eliminate x from the other two equations.
    (b) We eliminate x from equation (2):

$$
\begin{array}{rcr}
x + 2y + 3z = & 7 & \\
-x + 3y - 2z = & -5 & \quad (-1 \text{ times equation (1))} \\
\hline
5y + z = & 2. & \qquad \textbf{(4)}
\end{array}
$$

We eliminate x from equation (3):

$$
\begin{array}{rcr}
2x - y + z = & 4 & \\
-2x + 6y - 4z = & -10 & \quad (-2 \text{ times equation (1))} \\
\hline
5y - 3z = & -6. & \qquad \textbf{(5)}
\end{array}
$$

3. For phase three of the solution, we work with the variable y. We use the system as it is at the end of phase two. That is, we now have the system

$$x - 3y + 2z = 5 \qquad \textbf{(1)}$$

$$5y + z = 2 \qquad \textbf{(4)}$$

$$5y - 3z = -6. \qquad \textbf{(5)}$$

(a) We write an equation that is equivalent to equation (4), but has 1 as the coefficient of y. That is, we divide both sides of the equation by 5. We have

$$y + (1/5)z = 2/5. \tag{6}$$

(b) We use this equation to eliminate y from equation (5).

$$\begin{array}{rl} 5y - 3z = & -6 \\ \underline{-5y - z = -2} & \quad (-5 \text{ times equation (6)}) \\ -4z = & -8 \end{array} \tag{7}$$

4. For phase four, we work with the variable z. We use the system as it appears at the end of phase three. That is, we have

$$x - 3y + 2z = 5 \tag{1}$$
$$y + (1/5)z = 2/5 \tag{6}$$
$$- 4z = -8. \tag{7}$$

(a) We write an equation that is equivalent to equation (7), but has 1 as the coefficient of z. That is, we divide both sides of the equation by -4. We have

$$z = 2. \tag{8}$$

5. For phase five, we do the back substitution steps. We use the system as it is at the end of phase four. That is, the system now appears as

$$x - 3y + 2z = 5 \tag{1}$$
$$y + (1/5)z = 2/5 \tag{6}$$
$$z = 2. \tag{8}$$

(a) We substitute the value $z = 2$ from equation (8) into equation (6):

$y + (1/5)(2) = 2/5$ or $y + 2/5 = 2/5$, so $y = 0$.

(b) We substitute the values $z = 2$ and $y = 0$ into equation (1):

$x - 3(0) + 2(2) = 5$ or $x + 4 = 5$, so $x = 1$.

The final solution to the problem is $(1, 0, 2)$.

Check: $1 - 3(0) + 2(2) = 5$ or $1 + 4 = 5$
$1 + 2(0) + 3(2) = 7$ or $1 + 6 = 7$
$2(1) - 0 + 2 = 4$ or $2 + 2 = 4$. ∎

Because the same steps can be followed for any system of three equations in three variables, the algorithm is easily adaptable for a computer.

To solve a system of three simultaneous linear equations in three variables (x, y, z) using the method of Gauss elimination with back substitution:

1. If the variables do not appear in the same order for each equation, rearrange the terms so that they do.
2. Work with x.
   a. Divide both sides of the first equation by the coefficient of x. If this step requires division by zero, interchange the equation with one of those below it.
   b. Use the resulting equation to eliminate x from the second and third equations.
3. Using the system that results from step 2, work with y.
   a. Divide both sides of the second equation by the coefficient of y. If this step requires division by zero, interchange the equation with the one below it.
   b. Use the resulting equation to eliminate y from the third equation.
4. Using the system that results from step 3, work with z.
   a. If this step requires division by zero, the system is either dependent or inconsistent.
   b. Otherwise, divide both sides of the third equation by the coefficient of z.
5. Using the system that results from step 4, do back substitution.
   a. Substitute the value of z from the third equation into the second, and solve for y.
   b. Substitute the values of y and z into the first equation, and solve for x.

EXAMPLE 2

Solve the following system by the method of Gauss elimination with back substitution.

$$2x + y - 3z = 2 \tag{1}$$
$$3x - y + 4z = 9 \tag{2}$$
$$4x + 2y - 3z = 7 \tag{3}$$

Solution

1. The variables appear in the same order for each equation, so we are ready to begin the elimination.
2. (a) Divide equation (1) by 2:

$$x + (1/2)y - (3/2)z = 1 \tag{4}$$

(b) Eliminate x from the second equation:

$$3x - y + 4z = 9$$
$$\underline{-3x - (3/2)y + (9/2)z = -3} \quad (-3 \text{ times equation (4))}$$
$$-(5/2)y + (17/2)z = 6. \tag{5}$$

Eliminate x from the third equation.

$$4x + 2y - 3z = 7$$
$$\underline{-4x - 2y + 6z = -4} \quad (-4 \text{ times equation (4))}$$
$$3z = 3. \tag{6}$$

3. Consider the resulting system

$$x + (1/2)y - (3/2)z = 1 \tag{4}$$
$$- (5/2)y + (17/2)z = 6 \tag{5}$$
$$3z = 3 \tag{6}$$

(a) Divide equation (5) by $-5/2$ (i.e., multiply by $-2/5$). We have

$$y - (17/5)z = -12/5 \tag{7}$$

(b) The variable y has been eliminated from equation (6).

4. Consider the resulting system

$$x + (1/2)y - (3/2)z = 1 \tag{4}$$
$$y - (17/5)z = -12/5 \tag{7}$$
$$3z = 3 \tag{6}$$

5. Divide both sides of equation (6) by 3:

$$z = 1. \tag{8}$$

6. Consider the resulting system

$$x + (1/2)y - (3/2)z = 1 \tag{4}$$
$$y - (17/5)z = -12/5 \tag{7}$$
$$z = 1 \tag{8}$$

Do back substitution: $y - (17/5)(1) = -12/5$, or $y - 17/5 = -12/5$, so $y = 5/5$ or 1.

$$x + 1/2(1) - 3/2(1) = 1, \text{ or}$$
$$x + 1/2 - 3/2 \qquad = 1$$
$$x - 2/2 \qquad = 1$$
$$x \qquad = 2$$

The final solution is (2, 1, 1).

$$\text{Check:} \quad 2(2) + 1 - 3(1) = 2 \quad \text{or} \quad 4 + 1 - 3 = 2$$
$$3(2) - 1 + 4(1) = 9 \quad \text{or} \quad 6 - 1 + 4 = 9$$
$$4(2) + 2(1) - 3(1) = 7 \quad \text{or} \quad 8 + 2 - 3 = 7 \quad ▪$$

Actually, we have oversimplified the problem in this presentation. In actual use by a computer, even the order of the equations may affect the results. Consider, for example, a machine capable of a six-digit calculation. When Gauss elimination is used on the system

$$-10^{-8}x + y = 1$$
$$x + y = 2,$$

a solution of (0, 1) is obtained after rounding off. However, when the order of the equations is changed, the solution (1, 1) is obtained. This example points out that we have only scratched the surface of the mathematics actually used in computer programming. Numerical analysis is the branch of mathematics that considers these problems in detail.

8.3 EXERCISES

Solve each system by the method of Gauss elimination with back substitution. See Examples 1 and 2.

1.
$$x - y + 2z = -1$$
$$2x + y - z = 9$$
$$x - 2y - z = 0$$

2.
$$2x + y + 3z = 0$$
$$x - 3y + 2z = 7$$
$$3x - 2y - z = 7$$

3.
$$2x - 3y + z = 0$$
$$3x + 2y - z = 4$$
$$2x - 2y + z = 1$$

4.
$$3x + y + 2z = 3$$
$$2x - y - 3z = 15$$
$$2x + 2y + 3z = -3$$

5.
$$3x \quad + 2z = 3$$
$$x - y - 3z = 15$$
$$x + y \quad = 0$$

6.
$$x + y - z = 2$$
$$x - y + z = 4$$
$$-x + y + z = 6$$

7.
$$2x + 2y + z = 0$$
$$x - y \quad = 1$$
$$3x - 3y + 2z = 3$$

8.
$$2x + 3y = z + 2$$
$$4x - 3y = 1 - z$$
$$2x - 3y = z - 2$$

9.
$$x + z = y + 3$$
$$2x + z = 3 - y$$
$$2x - z = 2y$$

10.
$$3x + 4y + z - 6 = 0$$
$$6x - 4y - 2z - 7 = 0$$
$$3x - 4y - 3z - 4 = 0$$

11.
$$3x - z + 2y = 3$$
$$2x - y + 3z = 8$$
$$3y + 2z + x = 1$$

12.
$$x + 2y + 2z = 2$$
$$2x + y - 7 = 2z$$
$$2x + 2y \quad = z + 8$$

13. Use pseudocode or a flowchart to show an algorithm for solving a system of three equations in three variables using Gauss elimination.

✱**14.** Use Gauss elimination to solve the system $-10^{-8}x + y = 1$

$$x + y = 2.$$

Round off each value to six decimal places.

✱**15.** Use Gauss elimination to solve the system $x + y = 2$

$$-10^{-8}x + y = 1.$$

Round off each value to six decimal places.

✱**16.** Gauss elimination is often done using only the coefficients of the variables. For example, $2x + 3y = 7$ is written 2   3   7. Making the

$$3x - 2y = 4,\qquad\qquad 3\quad -2\quad 4$$

coefficient of x equal to one in the first equation, and eliminating x from the second equation gives 1   3/2   7/2. Making the

$$0\quad -13/2\quad -13/2$$

coefficient of y equal to one in the second equation gives 1   3/2   7/2

$$0\quad 1\quad 1.$$

It follows that $y = 1$ and back substitution gives $x = 2$. Do problem 1 using only the coefficients.

8.4   The Gauss-Seidel Method

At one time it was thought that because the effects of round-off error are less serious for iteration techniques than for the Gauss elimination method, iteration would be a better method. Now that the problem has been studied more thoroughly, numerical analysts agree that when implemented properly, Gauss elimination is generally the preferred procedure.

Iteration is still an effective technique for certain types of problems, and the Gauss-Seidel method for a system of two equations with two variables is presented in this section as an example of this type of solution. The method can be generalized for larger systems. The Gauss-Seidel method is particularly effective for a large system when a number of the coefficients are equal to zero.

For an iteration technique, it is important to know under what conditions the procedure will produce a solution. For the Gauss-Seidel method, we start with the variables in the same order in both equations. That is, the equations should be in the form

$$a_1x + b_1y = c_1$$
$$a_2x + b_2y = c_2.$$

If $|a_1b_2| > |b_1a_2|$, we say the Gauss-Seidel method *converges* to a solution. That is, after a finite number of iterations, we will have the answer to the desired degree of precision. If this condition is not met, it may be possible to reorder the equations so that it will be met. If reordering does not cause the condition to be met, the solution, if it exists, is sought by another method.

The basic idea behind the Gauss-Seidel method is to solve the first equation for x and the second for y. We begin with an estimate for y and use that estimate to solve for x. That value for x can be used to improve the estimate for y and so forth until the desired degree of precision is reached. The algorithm follows.

To solve a system of two simultaneous linear equations in two variables (x, y) using the Gauss-Seidel method:

1. Write the equations in the form
$$a_1x + b_1y = c_1$$
$$a_2x + b_2y = c_2.$$
2. Check for convergence.
   a. If $|a_1b_2| > |b_1a_2|$, the method converges to a solution.
   b. If $|a_1b_2| \leq |b_1a_2|$, reorder the equations and check again.
3. If the method converges, solve the first equation for x and the second one for y.
4. Choose the degree of precision desired (tenths, hundredths, thousandths, and so on).
5. Start with an estimated value for y. It need not be accurate. Zero is often used.
6. Use the estimate of step 5 to solve for x in the first equation. Round off the answer to the desired precision.
7. Use the (unrounded) value of x from the previous step to solve for y in the second equation. Round off the answer to the desired degree of precision.
8. Use the (unrounded) value of y from step 7 to solve for x in the first equation. Round off the answer to the desired degree of precision.
9. Repeat steps 7 and 8 until the same rounded x value is obtained twice in succession *and* the same rounded y value is obtained twice in succession (or the difference between successive values is less than some predetermined amount).

EXAMPLE 1

Solve by the Gauss-Seidel method if possible:
$$2x - y = 5$$
$$x + 2y = -5.$$

Solution

1. The equations are in the correct form.
2. We check for convergence: Since $|2(2)| > |1(-1)|$, the method converges.
3. Solve the first equation for x: $2x = y + 5$, so $x = (y + 5)/2$.
   Solve the second equation for y: $2y = -x - 5$, so $y = (-x - 5)/2$.
4. We want the answer correct to the nearest hundredth.
5. We begin with $y = 0$ as an estimate.
6. When $y = 0$, the first equation gives $x = 5/2 = 2.5$.
7. When $x = 5/2$, the second equation gives $y = (-2.5 - 5)/2$ or $-7.5/2 = -3.75$.
8. When $y = -3.75$, the first equation gives $x = (-3.75 + 5)/2$ or $(1.25)/2 = 0.625$, which is rounded to 0.63.

9. (a) When $x = 0.625$, $y = (-0.625 - 5)/2 = (-5.625)/2 = -2.8125$,
        which is rounded to -2.81.

    (b) When $y = -2.8125$, $x = (-2.8125 + 5)/2 = 1.09375$, which is
        rounded to 1.09.

The table that follows shows these results as well as the values that are obtained
as the process is continued. The iteration number is given by n.

n	x	y	
0		0	
1	2.5	-3.75	
2	0.63	-2.81	
3	1.09	-3.05	
4	0.98	-2.99	
5	1.01	-3.00	
6	1.00	-3.00	(the y value is repeated, but x is not)
7	1.00	-3.00	(both x and y are repeated)

We conclude that to the nearest hundredth, $x = 1.00$ and $y = -3.00$.
Check: $2(1.00) - (-3.00) = 5$ or $2 + 3 = 5$
         $1.00 + 2(-3.00) = -5$ or $1 - 6 = -5$ ■

EXAMPLE 2

Solve by the Gauss-Seidel method, if possible: $x + 2y = 1$
                                                   $7x + y = 2$.

Solution

1. The equations are in the proper form.

2. $|1(1)| \le |2(7)|$, so we reorder the equations as $7x + y = 2$
                                                            $x + 2y = 1$.

    $|2(7)| > |1(1)|$, so the method will converge to a solution.

3. $x = (-y + 2)/7$
    $y = (-x + 1)/2$

4. We want the solution correct to the nearest thousandth.

5. Begin with an initial estimate of $y = 0$.

6–9. The table that follows shows the results of the iteration procedure.

n	x	y	
0		0	
1	0.286	0.357	
2	0.235	0.383	
3	0.231	0.384	
4	0.231	0.385	(the x value has repeated, but the y has not)
5	0.231	0.385	(both the x and y values have repeated)

We conclude that $x = 0.231$ and $y = 0.385$ to the nearest thousandth. Since these values are *approximate* solutions, a check will probably lead to some discrepancy in the original equations.

Check: $0.231 + 2(0.385) = 1$ or $0.231 + 0.77 = 1.001$

$7(0.231) + 0.385 = 2$ or $1.617 + 0.385 = 2.002$ ■

EXAMPLE 3

Solve by the Gauss-Seidel method, if possible: $x + y = 5$

$x - y = 1.$

Solution

1. The equations are in the correct form.
2. Since $|1(-1)| \leq |1(1)|$, we reorder the equations $x - y = 1$

$x + y = 5.$

Since $|1(1)| \leq |1(-1)|$, we conclude that the Gauss-Seidel method will not converge to a solution for this system of equations. ■

It is important to realize that we have not said that the system in Example 3 cannot be solved. The solution is $(3, 2)$, but this solution must be obtained by some other method, such as the elimination method.

EXAMPLE 4

Solve by the Gauss-Seidel method, if possible: $2x - 4y = 6$

$-2x + 4y = 7.$

Solution

1. The equations are in the correct form.
2. Since $|2(4)| \leq |-2(-4)|$, we reorder the equations $-2x + 4y = 7$

$2x - 4y = 6.$

Since $|-2(-4)| \leq |2(4)|$, the Gauss-Seidel method will not converge to a solution. ■

Notice that the system in Example 4 has no solution. The system is inconsistent.

8.4 EXERCISES

Solve the following systems of equations to the nearest hundredth using the Gauss-Seidel method where possible. See Examples 1–4.

1. $2x - y = 5$
   $x + 2y = 3$

2. $x + y = 7$
   $3x - y = 7$

3. $3x - y = 12$
   $3x + 5y = 10$

4. $2x - y = 3$
   $3x + y = 4$

5. $10x + 6y = 8$
   $5x + 3y = 4$

6. $3x + 2y = 5$
   $6x - 4y = -1$

7. $2x + y = 6$
$x + 4y = 9$

8. $3x = y + 2$
$x + 2y = 1$

9. $x = y - 1$
$x + y = 2$

10. $x + 3y - 5 = 0$
$3x + y - 5 = 0$

11. $2y + x = 6$
$3x + y = 7$

12. $3x - y = 8$
$4x + 2y = 11$

13. $x + 3y = 3$
$4x + 2y = 11$

14. $2x + 3y = 4$
$2x - 3y = 0$

15. $4x + 6y = 7$
$6x - 9y = 5$

➡**16.** Use pseudocode or a flowchart to show the algorithm for solving a system of two equations in two variables using the Gauss-Seidel method.

✳**17.** Try to use the Gauss-Seidel method on the following system without changing the order of the equations: $2x + 3y = 5$
$x + y = 2.$

✳**18.** The Gauss-Seidel method can be used on a system of n equations in n variables for $n > 2$. The method converges if the system is "diagonally row dominant." This means that when the equations are written with the variables in the same order for each one, the coefficient of the ith variable in the ith equation (for each i) is larger in absolute value than the sum of the absolute values of the coefficients of the other variables in that row. The method is to solve the ith equation for the ith variable for each i. Begin by assigning values (often 0) to each variable other than the first in order to solve the first equation for the first variable. (If the system is not diagonally row dominant, the method may or may not converge.)

(a) Solve by the Gauss-Seidel method, if possible:
$3x + y - z = 5$
$x + 3y + z = 3$
$x - y - 3z = 3$

(b) $x \qquad - 2z = 6$
$2y + z = 2$
$2x + y \qquad = 6$

8.5 Cramer's Rule

If a system of equations is written so that the variables occur in the same order in each equation, we can obtain a solution by manipulating the coefficients. In the late seventeenth century, Gottfried Wilhelm von Leibniz in Germany and Seki Kowa in Japan independently discovered one such method. Their idea was not widely used, however, until 1750 when Gabriel Cramer from Switzerland rediscovered the method and popularized it. The method is now known as "Cramer's rule."

Cramer's rule is based on the concept of a determinant, which is defined on page 319.

Definition

An $n \times n$ determinant is a real number represented by a square array (or arrangement) of n^2 numbers.

A pair of bars is used with the array to indicate that the numbers are to be considered as an array and not as individual numbers.

For example, $\begin{vmatrix} 2 & 3 \\ 1 & 2 \end{vmatrix}$ is a 2×2 determinant.

$\begin{vmatrix} 3 & 2 & 0 \\ 0 & 1 & 3 \\ 2 & 0 & 1 \end{vmatrix}$ is a 3×3 determinant.

We say that the entries along a slanted line from upper left to lower right are on the *main diagonal*. This main diagonal is used in evaluating a determinant.

$$\begin{vmatrix} \text{Main diagonal} & \\ & \end{vmatrix}$$

Evaluating a 2×2 determinant is easier than evaluating a 3×3 determinant, so we consider that case first.

To evaluate a 2×2 determinant:
1. Form the product of the entries along the main diagonal.
2. Form the product of the entries along the other diagonal (from upper right to lower left).
3. Subtract the product found in step 2 from the product found in step 1.

EXAMPLE 1

Evaluate $\begin{vmatrix} 5 & -1 \\ -5 & 2 \end{vmatrix}$.

Solution $\begin{vmatrix} 5 & -1 \\ -5 & 2 \end{vmatrix}$

$\underset{\uparrow}{5(2)} \qquad \underset{\uparrow}{-} \qquad \underset{\uparrow}{(-1)(-5)}$

{from main diagonal}   subtract   {from other diagonal}

$10 - 5 = 5$, so the value of the determinant is 5. ∎

EXAMPLE 2

Evaluate $\begin{vmatrix} 2 & -1 \\ -5 & 2 \end{vmatrix}$.

Solution $2(2) - (-1)(-5) = 4 - 5 = -1$, so the value of the determinant is -1. ∎

To evaluate a 3×3 determinant, we copy the first two columns next to the original determinant. There are now two complete diagonals (i.e., with three entries) parallel to the main diagonal. We find the product of the entries along *each* of these diagonals, as well as the product of the entries along *each* of the other complete diagonals.

> **To evaluate a 3×3 determinant:**
> 1. Copy the first two columns to the right of the original determinant.
> 2. Find the product of the entries along the main diagonal and along each complete diagonal parallel to it. Add these products.
> 3. Find the product of the entries along each of the other complete diagonals. Add these products.
> 4. Subtract the sum found in step 3 from the sum found in step 2.

This procedure is known as Sarrus' rule.

EXAMPLE 3

Evaluate $\begin{vmatrix} 1 & -3 & 2 \\ 1 & 2 & 3 \\ 2 & -1 & 1 \end{vmatrix}$.

Solution   We write $\begin{vmatrix} 1 & -3 & 2 \\ 1 & 2 & 3 \\ 2 & -1 & 1 \end{vmatrix} \begin{matrix} 1 & -3 \\ 1 & 2 \\ 2 & -1 \end{matrix}$

Using the main diagonal and diagonals parallel to it, we have:

$$
\begin{aligned}
2(1)(-1) &= -2 \\
-3(3)(2) &= -18 \\
1(2)(1) &= \underline{2} \\
& -18
\end{aligned}
$$

Using the other diagonals, we have:

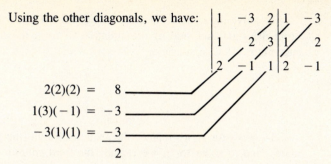

$$2(2)(2) = 8$$
$$1(3)(-1) = -3$$
$$-3(1)(1) = -3$$
$$\underline{}$$
$$2$$

The value of the determinant is $-18 - 2 = -20$. ∎

EXAMPLE 4

Evaluate $\begin{vmatrix} 5 & -3 & 2 \\ 7 & 2 & 3 \\ 4 & -1 & 1 \end{vmatrix}$.

Solution   We have $\begin{vmatrix} 5 & -3 & 2 \\ 7 & 2 & 3 \\ 4 & -1 & 1 \end{vmatrix} \begin{matrix} 5 & -3 \\ 7 & 2 \\ 4 & -1 \end{matrix}$

$$10 - 36 - 14 = -40$$
$$16 - 15 - 21 = -20$$

The value of the determinant is $-40 - (-20) = -40 + 20 = -20$. ∎

Cramer's rule says that the value of each variable of a system of linear equations can be written as a ratio of determinants. Each variable will have the same denominator, but the numerator will be different for each one. First we write the equations so that the variables occur in the same order for each equation, and the constant term is on the other side of the equation.

For example, consider the system

$$x - 3y + 2z = 5$$
$$x + 2y + 3z = 7$$
$$2x - y + z = 4.$$

All three variables will have the same denominator. It is the determinant formed by listing the coefficients of x in the first column, the coefficients of y in the second column, and the coefficients of z in the third column. The denominator, then, is the determinant

$$\begin{vmatrix} 1 & -3 & 2 \\ 1 & 2 & 3 \\ 2 & -1 & 1 \end{vmatrix}.$$

To solve for x, we replace the first column (the x column) by the constants from the righthand side of the equation to form the determinant that is the numerator. That is, we have

$$\begin{vmatrix} 5 & -3 & 2 \\ 7 & 2 & 3 \\ 4 & -1 & 1 \end{vmatrix}.$$

To solve for y, we replace the second column (the y column) by the constants; and to solve for z, we replace the third column (the z column) by the constants. We have

constants replace coefficients of x

constants replace coefficients of y

constants replace coefficients of z

$$x = \frac{\begin{vmatrix} 5 & -3 & 2 \\ 7 & 2 & 3 \\ 4 & -1 & 1 \end{vmatrix}}{\begin{vmatrix} 1 & -3 & 2 \\ 1 & 2 & 3 \\ 2 & -1 & 1 \end{vmatrix}} \quad y = \frac{\begin{vmatrix} 1 & 5 & 2 \\ 1 & 7 & 3 \\ 2 & 4 & 1 \end{vmatrix}}{\begin{vmatrix} 1 & -3 & 2 \\ 1 & 2 & 3 \\ 2 & -1 & 1 \end{vmatrix}} \quad z = \frac{\begin{vmatrix} 1 & -3 & 5 \\ 1 & 2 & 7 \\ 2 & -1 & 4 \end{vmatrix}}{\begin{vmatrix} 1 & -3 & 2 \\ 1 & 2 & 3 \\ 2 & -1 & 1 \end{vmatrix}}.$$

To solve a system of simultaneous linear equations by Cramer's rule:

1. If the variables do not appear in the same order in each equation, rearrange the terms so that they do.

2. Each variable is given as the ratio of two determinants.
   a. The determinant in the denominator has the coefficients of the first variable (in order) as the first column, the coefficients of the second variable (in order) as the second column, and so on.
   b. The determinant in the numerator is like the one in the denominator except for the entries in the column that represents the variable sought. That column is replaced by the constants.

3. Evaluate the determinants. If the solution produces a denominator of zero, the system is inconsistent if any numerator has a nonzero value. If the system is dependent, all of the numerators, as well as the denominator, will be equal to zero.

EXAMPLE 5

Use Cramer's rule to solve: $\quad 2x - y = 5$
$$x + 2y = -5.$$

Solution   We have

$$x = \frac{\begin{vmatrix} 5 & -1 \\ -5 & 2 \end{vmatrix}}{\begin{vmatrix} 2 & -1 \\ 1 & 2 \end{vmatrix}} \qquad y = \frac{\begin{vmatrix} 2 & 5 \\ 1 & -5 \end{vmatrix}}{\begin{vmatrix} 2 & -1 \\ 1 & 2 \end{vmatrix}}.$$

Evaluating the determinants gives $x = \dfrac{10 - (5)}{4 - (-1)} = \dfrac{10 - 5}{4 + 1} = \dfrac{5}{5} = 1,$

and $y = \dfrac{-10 - 5}{4 - (-1)} = \dfrac{-15}{5} = -3.$

Check: $2(1) - (-3) = \quad 5 \quad$ or $\quad 2 + 3 \quad\quad = \quad 5$
$\qquad\quad\; 1 + 2(-3) = -5 \quad$ or $\quad 1 + (-6) = -5$ ■

EXAMPLE 6

Use Cramer's rule to solve:
$$\begin{aligned} x + \; y + 2z &= 1 \\ 2x - \; y + \; z &= 1 \\ 2x + 2y - \; z &= 2. \end{aligned}$$

Solution

$$x = \frac{\begin{vmatrix} 1 & 1 & 2 \\ 1 & -1 & 1 \\ 2 & 2 & -1 \end{vmatrix}}{\begin{vmatrix} 1 & 1 & 2 \\ 2 & -1 & 1 \\ 2 & 2 & -1 \end{vmatrix}} \qquad y = \frac{\begin{vmatrix} 1 & 1 & 2 \\ 2 & 1 & 1 \\ 2 & 2 & -1 \end{vmatrix}}{\begin{vmatrix} 1 & 1 & 2 \\ 2 & -1 & 1 \\ 2 & 2 & -1 \end{vmatrix}} \qquad z = \frac{\begin{vmatrix} 1 & 1 & 1 \\ 2 & -1 & 1 \\ 2 & 2 & 2 \end{vmatrix}}{\begin{vmatrix} 1 & 1 & 2 \\ 2 & -1 & 1 \\ 2 & 2 & -1 \end{vmatrix}}$$

We evaluate the determinant in the denominator:

$$\begin{vmatrix} 1 & 1 & 2 \\ 2 & -1 & 1 \\ 2 & 2 & -1 \end{vmatrix}\begin{matrix} 1 & 1 \\ 2 & -1 \\ 2 & 2 \end{matrix} = (1 + 2 + 8) - (-4 + 2 - 2)$$
$$= 11 - (-4) = 15.$$

The numerator of x is:

$$\begin{vmatrix} 1 & 1 & 2 \\ 1 & -1 & 1 \\ 2 & 2 & -1 \end{vmatrix}\begin{matrix} 1 & 1 \\ 1 & -1 \\ 2 & 2 \end{matrix} = (1 + 2 + 4) - (-4 + 2 - 1)$$
$$= 7 - (-3) = 10.$$

The numerator of y is:

$$\begin{vmatrix} 1 & 1 & 2 \\ 2 & 1 & 1 \\ 2 & 2 & -1 \end{vmatrix}\begin{matrix} 1 & 1 \\ 2 & 1 \\ 2 & 2 \end{matrix} = (-1 + 2 + 8) - (4 + 2 - 2) = 9 - 4 = 5.$$

The numerator of z is:

$$\begin{vmatrix} 1 & 1 & 1 \\ 2 & -1 & 1 \\ 2 & 2 & 2 \end{vmatrix} \begin{matrix} 1 & 1 \\ 2 & -1 \\ 2 & 2 \end{matrix} = (-2 + 2 + 4) - (-2 + 2 + 4) = 4 - 4 = 0.$$

Therefore we have $x = 10/15 = 2/3$, $y = 5/15 = 1/3$, and $z = 0/15 = 0$.

Check: $2/3 + 1/3 + 2(0)$ $= 1$ or $2/3 + 1/3 = 1$

$\quad\quad\quad 2(2/3) - 1/3 + 0$ $= 1$ or $4/3 - 1/3 = 1$

$\quad\quad\quad 2(2/3) + 2(1/3) - 0 = 2$ or $4/3 + 2/3 = 2$ ∎

The method for evaluating determinants introduced in this section cannot be generalized to determinants with dimensions larger than 3×3. There are more general methods (see problem 27 in the following exercises), but Cramer's rule is not an efficient method to use on a computer. Consider a system of 12 equations with 12 variables (not uncommon in practice). Cramer's rule would require evaluating 13 determinants. There would be over 6×10^9 terms, each obtained by 11 multiplications. That is, there would be almost 7×10^{10} multiplications. A computer capable of performing 100,000 multiplications per second, operating 24 hours a day, would take over a week to solve the system. By contrast, Gauss elimination would require 794 multiplications and would take about 0.008 seconds. Furthermore, the effects of round-off error are more serious with Cramer's rule.

8.5 EXERCISES

Evaluate each determinant below. See Examples 1 and 2.

1. $\begin{vmatrix} 2 & 3 \\ 1 & -1 \end{vmatrix}$

2. $\begin{vmatrix} 3 & -1 \\ 2 & 3 \end{vmatrix}$

3. $\begin{vmatrix} 1 & -2 \\ -3 & 4 \end{vmatrix}$

4. $\begin{vmatrix} 1 & 0 \\ 0 & 1 \end{vmatrix}$

5. $\begin{vmatrix} -1 & 2 \\ 3 & -4 \end{vmatrix}$

6. $\begin{vmatrix} -7 & -5 \\ 6 & 8 \end{vmatrix}$

Evaluate each determinant below. See Examples 3 and 4.

7. $\begin{vmatrix} 2 & 1 & -1 \\ -1 & 2 & 1 \\ 1 & -1 & 2 \end{vmatrix}$

8. $\begin{vmatrix} 2 & 3 & 0 \\ 1 & -2 & 3 \\ 3 & 0 & -1 \end{vmatrix}$

9. $\begin{vmatrix} 1 & 0 & 0 \\ 0 & 1 & 0 \\ 0 & 0 & 1 \end{vmatrix}$

10. $\begin{vmatrix} 0 & -1 & 3 \\ 1 & -3 & 0 \\ 0 & 3 & -1 \end{vmatrix}$

Use Cramer's rule to solve each system of equations. See Example 5.

11. $x + y = 8$
$\quad\quad 2x - y = 1$

12. $3x - 2y = 1$
$\quad\quad 4x - y = 1$

13. $x + 2y = 1$
$2x - y = 0$

14. $x + 3y = 4$
$2x + 6y = 7$

15. $2x - 5y = 10$
$4x - 10y = 20$

16. $2x + 3y = 7$
$3x - y = 5$

17. $3x + 2y = 0$
$x - 5y = 9$

Use Cramer's rule to solve each system of equations. See Example 6.

18. $2x - 3y + z = 0$
$3x + 2y - z = 4$
$2x - 2y + z = 1$

19. $2x + 2y + z = 0$
$x - y = 1$
$3x - 3y + 2z = 3$

20. $x + y - z = 2$
$x - y + z = 4$
$-x + y + z = 6$

21. $x + 3y - 2z = 6$
$2x - 2y + z = -1$
$2x + 6y - 4z = 3$

22. $x - y + z = 3$
$2x + y + z = 3$
$2x - 2y + z = 0$

23. $2x - y + z = 5$
$x + 3y - z = -1$
$x + y + 4z = 1$

24. $2x + y - z = 4$
$x - 3y + z = -3$
$4x + 2y - 2z = 8$

✳25. When a triangle is drawn on a Cartesian plane so that its vertices are at (x_1, y_1), (x_2, y_2), and (x_3, y_3), its area is given by $\begin{vmatrix} x_1 & y_1 & 1 \\ x_2 & y_2 & 1 \\ x_3 & y_3 & 1 \end{vmatrix}$.

(a) Find the area of a triangle whose vertices are $(0, 0)$, $(4, 5)$, and $(2, 7)$.

(b) Find the area of a triangle whose vertices are $(2, -5)$, $(6, 2)$, and $(4, 1)$.

✳26. (a) Evaluate $\begin{vmatrix} 1 & 4 & -2 \\ 1 & -1 & 3 \\ 0 & 2 & 1 \end{vmatrix}$.

(b) Interchange two rows and evaluate. How is the determinant affected?

(c) Evaluate $\begin{vmatrix} 1 & 1 & 0 \\ 4 & -1 & 2 \\ -2 & 3 & 1 \end{vmatrix}$ where the entries in row i are the entries that were in column i of the determinant in part (a).

How is the determinant affected by this change?

✳27. A double subscript is often used to indicate row and column of a particular entry of a determinant. For example, a_{11} indicates the entry in the first row and first column while a_{23} indicates the entry in the second row and third column. One general method for evaluating determinants larger than 2×2 is known as Chio's rule. If A is an $n \times n$ determinant with $a_{11} \neq 0$, then $A = 1/(a_{11})^{n-2} B$, where B is an $(n - 1) \times (n - 1)$ determinant with entries

$$b_{ij} = \begin{vmatrix} a_{11} & a_{1(j+1)} \\ a_{(i+1)1} & a_{(i+1)(j+1)} \end{vmatrix}.^\dagger$$

---

†Miller, Eldon. "Evaluating an Nth Order Determinant in N Easy Steps," *The MATYC Journal*, Vol. 12 No. 2 (Spring 1978), 123–128.

With a little practice, you will see the pattern used to locate the entries of each b_{ij}, and since 2×2 determinants are easy to evaluate, you can quickly write down B using Chio's rule. Then Chio's rule can be applied to B to obtain a $(n - 2) \times (n - 2)$ determinant, and so on.

Consider the example:

$$\begin{vmatrix} 2 & -1 & 0 & 3 \\ 1 & 0 & -3 & 2 \\ 0 & 3 & -2 & 1 \\ -3 & 2 & 1 & 0 \end{vmatrix} = 1/4 \begin{vmatrix} 1 & -6 & 1 \\ 6 & -4 & 2 \\ 1 & 2 & 9 \end{vmatrix} = 1/4 \begin{vmatrix} 32 & -4 \\ 8 & 8 \end{vmatrix} =$$

$(1/4)(256 + 32) = (1/4)(288) = 72.$

Use Chio's rule to evaluate:

(a) $\begin{vmatrix} -1 & 2 & 0 & -2 \\ 2 & -3 & 1 & 3 \\ 4 & -2 & 3 & -1 \\ 3 & -1 & 2 & 0 \end{vmatrix}$

(b) $\begin{vmatrix} 3 & 2 & -1 & 0 \\ 1 & -1 & 2 & 3 \\ 0 & 1 & -3 & 2 \\ -1 & 3 & 0 & 1 \end{vmatrix}$

✳**28.** When Chio's rule (See problem 27) is used to reduce an $n \times n$ determinant to an $(n - 1) \times (n - 1)$ determinant, how many 2×2 determinants must be evaluated?

8.6 Matrices

Another method for solving a system of linear equations involves using matrix equations. Because of the importance of matrix algebra to computer programming applications in economics, atomic physics, and statistics, this method is presented as a means of introducing the branch of mathematics known as *linear algebra*.

The word *matrix* was introduced in 1850 by the mathematician J. J. Sylvester. He wanted to refer to an array of numbers as a single object, but he could not use the word *determinant*, because the array did not have a value. Matrices (the plural of matrix) eventually became more important than determinants in the study of mathematics. To distinguish a matrix from a determinant, brackets are used instead of the bars.

Definition

An $m \times n$ matrix is a rectangular array (or arrangement) of numbers with m (horizontal) rows and n (vertical) columns.

For example, $\begin{bmatrix} 1 & -2 & 3 \\ -4 & 5 & 6 \end{bmatrix}$ is a 2 × 3 matrix and $\begin{bmatrix} 1 & 0 \\ 2 & -1 \\ -3 & 6 \end{bmatrix}$ is a 3 × 2 matrix.

The entries of a matrix are often written using double subscripts. The notation a_{ij} means the entry in row i and column j. In the matrix $\begin{bmatrix} 1 & -2 & 3 \\ -4 & 5 & 6 \end{bmatrix}$

$a_{12} = -2$ and $a_{23} = 6$.

If we are to add two matrices, the two matrices must have the same dimensions (i.e., the same number of rows in each and the same number of columns in each). We say that "corresponding entries" are added. Corresponding entries have the same row number and the same column number.

To add two matrices (A and B) with the same dimensions:

Add each entry a_{ij} of matrix A to the corresponding entry b_{ij} of matrix B to get the entry c_{ij} of the sum.

Matrices can be subtracted by subtracting corresponding entries.

EXAMPLE 1

Add, if possible: $\begin{bmatrix} 1 & 2 \\ 0 & -3 \end{bmatrix} + \begin{bmatrix} -4 & 5 \\ -1 & 0 \end{bmatrix}$.

Solution

$$\begin{bmatrix} 1 & 2 \\ 0 & -3 \end{bmatrix} + \begin{bmatrix} -4 & 5 \\ -1 & 0 \end{bmatrix} = \begin{bmatrix} -3 & 7 \\ -1 & -3 \end{bmatrix}$$

$(1 + -4)$ $(2 + 5)$

$(0 + -1)$ $(-3 + 0)$ ∎

EXAMPLE 2

Subtract, if possible: $\begin{bmatrix} 4 & -2 & 3 \\ -1 & 0 & 1 \end{bmatrix} - \begin{bmatrix} 2 & 0 & 5 \\ 3 & -1 & 4 \end{bmatrix}$.

Solution

$$\begin{bmatrix} 4 & -2 & 3 \\ -1 & 0 & 1 \end{bmatrix} - \begin{bmatrix} 2 & 0 & 5 \\ 3 & -1 & 4 \end{bmatrix} = \begin{bmatrix} 2 & -2 & -2 \\ -4 & 1 & -3 \end{bmatrix}$$ ∎

EXAMPLE 3 Add, if possible: $\begin{bmatrix} 1 & 0 \\ 3 & -2 \\ 0 & 4 \end{bmatrix} + \begin{bmatrix} 1 & 0 & 2 \\ 0 & -3 & 4 \end{bmatrix}$.

Solution The addition is not defined. The first matrix is a 3×2 matrix, while the second one is a 2×3 matrix. ■

In order to multiply two matrices, the first matrix must have exactly as many columns as the second has rows. The product of an $m \times n$ matrix and an $n \times p$ matrix is an $m \times p$ matrix. That is, when

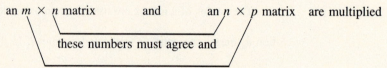

an $m \times n$ matrix   and   an $n \times p$ matrix   are multiplied

these numbers must agree and

these numbers give the dimensions of the product.

Thus, we can multiply a 2×3 matrix by a 3×2 matrix, but we cannot multiply a 2×3 matrix by another 2×3 matrix.

The entry that goes in row i, column j of the product is obtained by multiplying entries in row i of the first matrix by entries in column j of the second matrix, and adding these products. Figure 8.2 may help to explain the process.

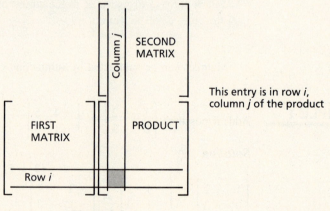

Figure 8.2

Consider the product of $\begin{bmatrix} 1 & 0 \\ 3 & -2 \\ 0 & 4 \end{bmatrix}$ and $\begin{bmatrix} 1 & 0 & 2 \\ 0 & -3 & 4 \end{bmatrix}$.

For the purpose of illustration, we arrange the matrices as in Figure 8.2.

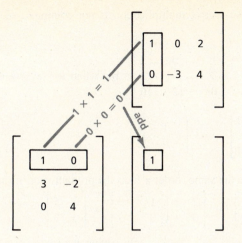

Figure 8.3

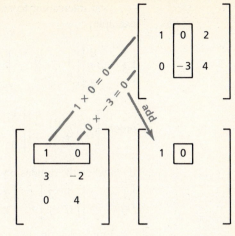

Figure 8.4

By partitioning the first matrix into rows and the second matrix into columns, we can find each entry of the 3×3 matrix that is the product (see Figures 8.3 and 8.4). The final result is the matrix that follows:

$$\begin{bmatrix} 1(1) + 0(0) & 1(0) + 0(-3) & 1(2) + 0(4) \\ 3(1) + (-2)(0) & 3(0) + (-2)(-3) & 3(2) + (-2)(4) \\ 0(1) + 4(0) & 0(0) + 4(-3) & 0(2) + 4(4) \end{bmatrix} \quad \text{or}$$

$$\begin{bmatrix} 1 & 0 & 2 \\ 3 & 6 & -2 \\ 0 & -12 & 16 \end{bmatrix}.$$

EXAMPLE 4

Multiply $\begin{bmatrix} 1 & -2 \\ 0 & 3 \end{bmatrix}\begin{bmatrix} 2 & 3 \\ 0 & 1 \end{bmatrix}.$

Solution   Partition the first matrix into rows and the second into columns.

$$\begin{bmatrix} 1 & -2 \\ 0 & 3 \end{bmatrix} \begin{bmatrix} 2 & 3 \\ 0 & 1 \end{bmatrix} = \begin{bmatrix} 1(2) + (-2)(0) & 1(3) + (-2)(1) \\ 0(2) + 3(0) & 0(3) + 3(1) \end{bmatrix}$$

so the product is $\begin{bmatrix} 2 & 1 \\ 0 & 3 \end{bmatrix}.$ ∎

EXAMPLE 5

Multiply $\begin{bmatrix} 2 & 3 \\ 4 & 5 \end{bmatrix}\begin{bmatrix} 1 & 0 \\ 0 & 1 \end{bmatrix}.$

Solution   We have $\begin{bmatrix} 2 & 3 \\ 4 & 5 \end{bmatrix}\begin{bmatrix} 1 & 0 \\ 0 & 1 \end{bmatrix} = \begin{bmatrix} 2+0 & 0+3 \\ 4+0 & 0+5 \end{bmatrix} = \begin{bmatrix} 2 & 3 \\ 4 & 5 \end{bmatrix}.$ ∎

It is interesting to see that matrix multiplication is not commutative. For example,

$$\begin{bmatrix} 1 & 0 \\ 3 & -2 \\ 0 & 4 \end{bmatrix} \begin{bmatrix} 1 & 0 & 2 \\ 0 & -3 & 4 \end{bmatrix} = \begin{bmatrix} 1 & 0 & 2 \\ 3 & 6 & -2 \\ 0 & -12 & 16 \end{bmatrix}$$ but when we reverse the

order, $$\begin{bmatrix} 1 & 0 & 2 \\ 0 & -3 & 4 \end{bmatrix} \begin{bmatrix} 1 & 0 \\ 3 & -2 \\ 0 & 4 \end{bmatrix} = \begin{bmatrix} 1 & 8 \\ -9 & 22 \end{bmatrix}.$$

Even when the dimensions are the same, order affects the product. For instance,

$$\begin{bmatrix} 1 & -2 \\ 0 & 3 \end{bmatrix} \begin{bmatrix} 2 & 3 \\ 0 & 1 \end{bmatrix} = \begin{bmatrix} 2 & 1 \\ 0 & 3 \end{bmatrix}$$ but $$\begin{bmatrix} 2 & 3 \\ 0 & 1 \end{bmatrix} \begin{bmatrix} 1 & -2 \\ 0 & 3 \end{bmatrix} = \begin{bmatrix} 2 & 5 \\ 0 & 3 \end{bmatrix}.$$

From now on, the examples will involve only square matrices. A square matrix is one that has the same number of rows as columns. An $n \times n$ matrix with 1's along the main diagonal and 0's for all of the other entries is called the *Identity* matrix for $n \times n$ matrices. When the product of two matrices is the identity, each matrix is the *inverse* of the other.
For example,

$$\begin{bmatrix} 7 & -1 \\ -6 & 1 \end{bmatrix} \begin{bmatrix} 1 & 1 \\ 6 & 7 \end{bmatrix} = \begin{bmatrix} 1 & 0 \\ 0 & 1 \end{bmatrix},$$

so we say that the matrices $$\begin{bmatrix} 7 & -1 \\ -6 & 1 \end{bmatrix}$$ and $$\begin{bmatrix} 1 & 1 \\ 6 & 7 \end{bmatrix}$$ are inverses.

$$\begin{bmatrix} 1 & 0 & 3 \\ 1 & 1 & 3 \\ 0 & 2 & 2 \end{bmatrix} \begin{bmatrix} -2 & 3 & -3/2 \\ -1 & 1 & 0 \\ 1 & -1 & 1/2 \end{bmatrix} = \begin{bmatrix} 1 & 0 & 0 \\ 0 & 1 & 0 \\ 0 & 0 & 1 \end{bmatrix},$$ so the two matrices that

were multiplied are inverses.

When the two factors of a matrix multiplication are inverses, or when one of the factors is the identity matrix, the multiplication is commutative, but remember that matrix multiplication, in general, is not.

In computer programming, a subscripted variable is called an array. A 12×31 array, for example, might store the income of a business for each day of the year. The subscripts could indicate the month and day. An entry of 340 in position (3, 15) would indicate that on the fifteenth day of March the business had an income of $340.

8.6 EXERCISES

Perform the following matrix additions and subtractions, if possible. See Examples 1–3.

1. $\begin{bmatrix} 2 & -1 \\ 3 & 0 \end{bmatrix} + \begin{bmatrix} 0 & 3 \\ 7 & -5 \end{bmatrix}$

2. $\begin{bmatrix} 12 & 5 \\ -3 & 4 \end{bmatrix} - \begin{bmatrix} 6 & -2 \\ 1 & 7 \end{bmatrix}$

3. $\begin{bmatrix} 3 & 0 & -5 \\ 2 & -4 & 7 \\ 6 & 3 & -8 \end{bmatrix} + \begin{bmatrix} 7 & -8 & 5 \\ 4 & 2 & -9 \\ -6 & 9 & 1 \end{bmatrix}$

4. $\begin{bmatrix} -2 & 3 & 5 \\ -6 & 8 & 0 \\ 3 & -4 & 4 \end{bmatrix} - \begin{bmatrix} 7 & 2 & -4 \\ 7 & -7 & 8 \\ -5 & 5 & 9 \end{bmatrix}$

5. $\begin{bmatrix} 2 & -9 \\ 0 & 6 \\ 3 & -4 \end{bmatrix} + \begin{bmatrix} 7 & 8 \\ 0 & -3 \\ 1 & 5 \end{bmatrix}$

6. $\begin{bmatrix} 9 & -7 & 5 \\ 0 & 3 & -1 \end{bmatrix} - \begin{bmatrix} -2 & 0 & 1 \\ 1 & -6 & 3 \end{bmatrix}$

7. $\begin{bmatrix} 1 & 3 & -2 \\ 0 & -5 & 6 \end{bmatrix} + \begin{bmatrix} 1 & 3 \\ -2 & 0 \\ -5 & 6 \end{bmatrix}$

8. $\begin{bmatrix} 1 & 4 \\ 9 & 0 \end{bmatrix} + \begin{bmatrix} 2 & 3 \\ 5 & 6 \\ 8 & 7 \end{bmatrix}$

Perform the matrix multiplications, if possible. See Examples 4 and 5.

9. $\begin{bmatrix} 9 & -1 \\ 1 & -9 \end{bmatrix} \begin{bmatrix} 4 & 6 \\ -2 & 1 \end{bmatrix}$

10. $\begin{bmatrix} 4 & 6 \\ -2 & 1 \end{bmatrix} \begin{bmatrix} 9 & -7 \\ 1 & -9 \end{bmatrix}$

11. $\begin{bmatrix} 3 & 5 \\ 5 & 8 \end{bmatrix} \begin{bmatrix} -8 & 5 \\ 5 & -3 \end{bmatrix}$

12. $\begin{bmatrix} 2 & -1 & 5 \\ -3 & 0 & 4 \\ 1 & 1 & 7 \end{bmatrix} \begin{bmatrix} -4 \\ 3 \\ 6 \end{bmatrix}$

13. $\begin{bmatrix} 1 & 0 & 2 \\ -1 & 1 & 3 \\ 3 & 2 & 4 \end{bmatrix} \begin{bmatrix} 2 \\ 1 \\ 5 \end{bmatrix}$

14. $\begin{bmatrix} 5 & 9 \\ 4 & 7 \end{bmatrix} \begin{bmatrix} -7 & 9 \\ 4 & -5 \end{bmatrix}$

15. $\begin{bmatrix} 1 & 0 \\ 2 & -3 \end{bmatrix} \begin{bmatrix} 1 & 2 \\ 0 & -3 \end{bmatrix}$

16. $\begin{bmatrix} 2 & 3 \\ 4 & 0 \\ 1 & -1 \end{bmatrix} \begin{bmatrix} 2 & 4 & 1 \\ 3 & 0 & -1 \end{bmatrix}$

17. $\begin{bmatrix} 2 & 4 & 1 \\ 3 & 0 & -1 \end{bmatrix} \begin{bmatrix} 2 & 3 \\ 4 & 0 \\ 1 & -1 \end{bmatrix}$

18. $\begin{bmatrix} 1 & 0 & 2 \\ 3 & 1 & -1 \end{bmatrix} \begin{bmatrix} 4 \\ -7 \\ 5 \end{bmatrix}$

19. $\begin{bmatrix} 1 & 0 & 2 \\ 3 & 2 & -1 \\ -2 & 0 & 3 \end{bmatrix} \begin{bmatrix} 0 & 1 & -1 \\ -1 & 0 & 1 \\ 1 & -1 & 0 \end{bmatrix}$

20. $\begin{bmatrix} 1 & 0 & 1 \\ 0 & -1 & 1 \\ -1 & 1 & 0 \end{bmatrix} \begin{bmatrix} 1 & 0 & -1 \\ 0 & -1 & 1 \\ 1 & 1 & 0 \end{bmatrix}$

21. $\begin{bmatrix} 0 & 1 & -1 \\ -1 & 0 & 1 \\ 1 & -1 & 0 \end{bmatrix} \begin{bmatrix} 1 & 0 & 2 \\ 3 & 2 & -1 \\ -2 & 0 & 3 \end{bmatrix}$

22. $\begin{bmatrix} 1 & 0 & 0 \\ 0 & 1 & 0 \\ 0 & 0 & 1 \end{bmatrix} \begin{bmatrix} 1 & 2 & 3 \\ 4 & 5 & 6 \\ 7 & 8 & 9 \end{bmatrix}$

23. $\begin{bmatrix} 2 & 2 & 0 \\ 3 & 0 & 1 \\ 3 & 1 & 1 \end{bmatrix} \begin{bmatrix} 1/2 & 1 & -1 \\ 0 & -1 & 1 \\ -3/2 & -2 & 3 \end{bmatrix}$

24. $\begin{bmatrix} 1/2 & 1 & -1 \\ 0 & -1 & 1 \\ -3/2 & -2 & 3 \end{bmatrix} \begin{bmatrix} 2 & 2 & 0 \\ 3 & 0 & 1 \\ 3 & 1 & 1 \end{bmatrix}$

▶25. Use pseudocode or a flowchart to show an algorithm for adding two $n \times n$ matrices.

➠**26.** Use pseudocode or a flowchart to show an algorithm for multiplying two $n \times n$ matrices.

✳**27.** A square matrix is said to be a *magic square* matrix if the sum of the elements in any row, column, or diagonal is the same as the sum in any other row, column,

or diagonal. For example, $\begin{bmatrix} 6 & 1 & 5 \\ 3 & 4 & 5 \\ 3 & 7 & 2 \end{bmatrix}$ is a magic square matrix.

(a) Can you form a magic square matrix so that the sum of each row, column, or diagonal is 18?

(b) Can you form a 3×3 magic square matrix using the digits 1–9 so that each digit is used only once?

✳**28.** A 64K memory chip is a square array of data storage cells with 256 rows and 256 columns. Each cell can store one bit.

(a) How many cells are there in a 64K chip?

(b) K is short for *kilo*. For computer applications, kilo means 1024 (or 2^{10}) rather than 1000. Use powers of 2 to verify that $256 \times 256 = 64(1024)$.

(c) How many rows and columns would you expect to find on a 256K chip?

✳**29.** Computer memory size is often specified in *bytes*. A byte usually consists of eight bits—one from each of eight different chips.

(a) How many 64K chips would be required to store 64K bytes (about the number required for a word-processing program)?

(b) How many 64K chips would be required to store 256K bytes?

(c) How many 256K chips would be required to store 256K bytes?

✳**30.** Each byte can be accessed individually by specifying an ''address.'' Addresses are usually given in hexadecimal notation. One popular microcomputer has its memory divided into three categories:

RAM (Random Access Memory) occupies 0–BFFF.
I/O (Input-Output locations) occupies C000–CFFF.
ROM (Read Only Memory) occupies D000–FFFF.

(a) How many bytes are allocated for each category?

(b) What are the addresses for each category in decimal notation?

8.7   Inverse Matrices and Matrix Equations

In the previous section, we saw several examples of matrices that are inverses of each other. In this section, we will see how to find the inverse of a matrix.

To find the inverse of a matrix, we write the matrix and the identity matrix side by side. To make the computation easier, we can include both matrices within a single pair of brackets. For example, to find the inverse of $\begin{bmatrix} 2 & 5 \\ 1 & 3 \end{bmatrix}$ we write $\left[\begin{array}{cc|cc} 2 & 5 & 1 & 0 \\ 1 & 3 & 0 & 1 \end{array}\right]$.

As we change the matrix on the left into the identity, the identity matrix will be transformed into the inverse of the matrix on the left. The transformation is done in two phases. The *first phase* will be to change the *first column*

to $\begin{matrix} 1 \\ 0 \end{matrix}$ and the *second phase* will be to change the *second column* to $\begin{matrix} 0 \\ 1 \end{matrix}$. The entry replaced by 1 for each phase is called the *pivot*. In each phase, it is important to change the pivot to a 1 as the first step, so that it can be used to replace the other entries of the column by 0's. Consider the following example.

Phase 1: The 2 in the first column is the pivot. We want to replace it by 1, so we divide each entry of the row by two. We have $\left[\begin{array}{cc|cc} 1 & 5/2 & 1/2 & 0 \end{array} \right]$.

We can now use this row to replace the bottom row with a row that has 0 as the first entry. We need a -1 to add to the first $+1$ in that row, so we multiply the pivot row by -1.

We have

$$
\begin{array}{cccc}
1 & 3 & 0 & 1 \quad \text{(the row we want to change)} \\
-1 & -5/2 & -1/2 & -0 \quad (-1 \times \text{ pivot row}) \\
\hline
0 & 1/2 & -1/2 & 1 \quad \text{(sum).}
\end{array}
$$

The sum of these two rows replaces the second row. We have

$$
\left[\begin{array}{cc|cc}
1 & 5/2 & 1/2 & 0 \\
0 & 1/2 & -1/2 & 1
\end{array} \right].
$$

Since the first column now has the desired form, phase 1 is complete. We use this matrix for phase 2.

Phase 2: We begin by finding a new pivot. The entry 1/2 in the second column is the pivot for phase 2. We want to change it to a 1. We divide each entry of the row by 1/2 (or multiply by 2) to obtain $\left[\begin{array}{cc|cc} 0 & 1 & -1 & 2 \end{array} \right]$.

We can now use this row to replace the 5/2 in the second column by 0. We have

$$
\begin{array}{cccc}
1 & 5/2 & 1/2 & 0 \quad \text{(the row we want to change)} \\
-0 & -5/2 & +5/2 & -5 \quad (-5/2 \times \text{ new pivot row}) \\
\hline
1 & 0 & 3 & -5 \quad \text{(sum).}
\end{array}
$$

The sum of these two rows replaces the first row. We have $\left[\begin{array}{cc|cc} 1 & 0 & 3 & -5 \\ 0 & 1 & -1 & 2 \end{array} \right]$. Since the second column is in the desired form, phase 2 is complete. In fact, the solution is complete, for we have the identity matrix   on   the   left.   That   means   the   matrix   on   the   right,

$\begin{bmatrix} 3 & -5 \\ -1 & 2 \end{bmatrix}$, is the inverse of $\begin{bmatrix} 2 & 5 \\ 1 & 3 \end{bmatrix}$. We can verify that these two matrices

are inverses of each other by checking to see that their product is the identity matrix.

$$\begin{bmatrix} 2 & 5 \\ 1 & 3 \end{bmatrix}\begin{bmatrix} 3 & -5 \\ -1 & 2 \end{bmatrix} = \begin{bmatrix} 6-5 & -10+10 \\ 3-3 & -5+6 \end{bmatrix} = \begin{bmatrix} 1 & 0 \\ 0 & 1 \end{bmatrix}.$$

We generalize this example to state an algorithm for finding the inverse of any $n \times n$ matrix for which the inverse exists.

> **To find the inverse of a square matrix:**
> 1. Write the matrix and the identity matrix side by side.
> 2. Transform the matrix into the identity matrix by working with one column at a time. Repeat steps a and b for each column.
>   a. Change the pivot to 1. That is, divide each entry of the row with the pivot by the pivot.
>   b. Change each entry above or below the pivot to 0. That is, multiply each entry of the pivot row by the additive inverse of the entry to be changed. Add the result to the row with the entry to be changed. Replace that row with the sum.
> 3. When the original matrix on the left has been replaced by the identity matrix, the matrix on the right will be the inverse of the original matrix.

EXAMPLE 1

Find the inverse of $\begin{bmatrix} 2 & 1 & -3 \\ 3 & -1 & 4 \\ 4 & 2 & -3 \end{bmatrix}$.

Solution

$$\left[\begin{array}{ccc|ccc} 2 & 1 & -3 & 1 & 0 & 0 \\ 3 & -1 & 4 & 0 & 1 & 0 \\ 4 & 2 & -3 & 0 & 0 & 1 \end{array}\right]$$

               (1)
               (2)
               (3)

Phase 1: Divide the first row by two to obtain the **first row.** We have

1	1/2	−3/2	1/2 0 0	(This is the pivot row)	(4)
0	−5/2	17/2	−3/2 1 0	(Row (2) + (−3) × pivot row)	(5)

$$3 \quad -1 \quad 4 \qquad 0\ 1\ 0$$
$$\underline{-3 \quad -3/2 \quad 9/2 \quad -3/2\ 0\ 0}$$
$$0 \quad -5/2 \quad 17/2 \quad -3/2\ 1\ 0$$

0	0	3	−2 0 1	(Row (3) + (−4) × pivot row)	(6)

$$
\begin{array}{rrrrrr}
4 & 2 & -3 & 0 & 0 & 1 \\
\hline
-4 & -2 & 6 & -2 & 0 & 0 \\
\hline
0 & 0 & 3 & -2 & 0 & 1
\end{array}
$$

Phase 2: Divide the second row from phase 1 by $-5/2$ (or multiply by $-2/5$) to get the **second row** for phase 2. This row $(0 \quad 1 \quad -17/5 \quad 3/5 \quad -2/5 \quad 0)$ is the new pivot row. We have

$$
\begin{bmatrix}
1 & 0 & 1/5 & 1/5 & 1/5 & 0 \\
\\
\\
\\
0 & 1 & -17/5 & 3/5 & -2/5 & 0 \\
0 & 0 & 3 & -2 & 0 & 1
\end{bmatrix}
$$

(Row (4) + $(-1/2) \times$ pivot row)   **(8)**

$$
\begin{array}{rrrrrr}
1 & 1/2 & -3/2 & 1/2 & 0 & 0 \\
-0 & -1/2 & 17/10 & -3/10 & 1/5 & 0 \\
\hline
1 & 0 & 1/5 & 1/5 & 1/5 & 0
\end{array}
$$

(The pivot row)   **(7)**

(Row (6) does not change)   **(9)**

We use this matrix for phase 3.

Phase 3: Divide the third row from phase 2 by 3 to get the **third row** for phase 3. This row $(0 \quad 0 \quad 1 \quad -2/3 \quad 0 \quad 1/3)$ is the pivot row. We have

$$
\begin{bmatrix}
1 & 0 & 0 & 1/3 & 1/5 & -1/15 \\
\\
\\
\\
0 & 1 & 0 & -5/3 & -2/5 & 17/15 \\
\\
\\
\\
0 & 0 & 1 & -2/3 & 0 & 1/3
\end{bmatrix}
$$

(Row (8) + $(-1/5) \times$ pivot row)   **(11)**

$$
\begin{array}{rrrrrr}
1 & 0 & 1/5 & 1/5 & 1/5 & 0 \\
0 & 0 & -1/5 & 2/15 & 0 & -1/15 \\
\hline
1 & 0 & 0 & 1/3 & 1/5 & -1/15
\end{array}
$$

(Row (7) + $17/5 \times$ pivot row)   **(12)**

$$
\begin{array}{rrrrrr}
0 & 1 & -17/5 & 3/5 & -2/5 & 0 \\
0 & 0 & 17/5 & -34/15 & 0 & 17/15 \\
\hline
0 & 1 & 0 & -5/3 & -2/5 & 17/15
\end{array}
$$

(This is the pivot row)   **(10)**

Phase 3 is complete, so we have the inverse of

$$
\begin{bmatrix} 2 & 1 & -3 \\ 3 & -1 & 4 \\ 4 & 2 & -3 \end{bmatrix} \text{ is } \begin{bmatrix} 1/3 & 1/5 & -1/15 \\ -5/3 & -2/5 & 17/15 \\ -2/3 & 0 & 1/3 \end{bmatrix}.
$$

That these two matrices are inverses of each other can be verified by checking that their product is the identity matrix. ∎

If the number in the pivot position happens to be zero, it is impossible to divide by 0 to change it to a 1. It is permissible, however, to change the order of the rows. So that you do not undo what you have already done, you should interchange the offending row with one below it.

EXAMPLE 2

Find the inverse of $\begin{bmatrix} 0 & 2 \\ 3 & 1 \end{bmatrix}$.

Solution $\left[\begin{array}{cc|cc} 0 & 2 & 1 & 0 \\ 3 & 1 & 0 & 1 \end{array}\right]$

Phase 1: Since 0 is in the pivot position, and it is impossible to divide by 0, we interchange the rows.

$\left[\begin{array}{cc|cc} 3 & 1 & 0 & 1 \\ 0 & 2 & 1 & 0 \end{array}\right]$ (1)
(2)

Phase 1: $\left[\begin{array}{cc|cc} 1 & 1/3 & 0 & 1/3 \\ 0 & 2 & 1 & 0 \end{array}\right]$ (1/3 × (Row (1))) (3)
(Row (2) does not change) (4)

Phase 2: $\left[\begin{array}{cc|cc} 1 & 0 & -1/6 & 1/3 \\ 0 & 1 & 1/2 & 0 \end{array}\right]$ (Row (3) + (−1/3) × Row (5)) (6)
(1/2 × Row (4)) (5)

So $\begin{bmatrix} -1/6 & 1/3 \\ 1/2 & 0 \end{bmatrix}$ is the inverse of $\begin{bmatrix} 0 & 2 \\ 3 & 1 \end{bmatrix}$. ∎

Some matrices do not have inverses. Even interchanging rows in such a matrix will not produce an inverse.

EXAMPLE 3

Find the inverse, if it exists, of $\begin{bmatrix} 1 & -1 & 0 \\ 0 & 1 & 1 \\ -1 & 1 & 0 \end{bmatrix}$.

Solution $\left[\begin{array}{ccc|ccc} 1 & -1 & 0 & 1 & 0 & 0 \\ 0 & 1 & 1 & 0 & 1 & 0 \\ -1 & 1 & 0 & 0 & 0 & 1 \end{array}\right]$ (1)
(2)
(3)

Phase 1: $\left[\begin{array}{ccc|ccc} 1 & -1 & 0 & 1 & 0 & 0 \\ 0 & 1 & 1 & 0 & 1 & 0 \\ 0 & 0 & 0 & 1 & 0 & 1 \end{array}\right]$ (Pivot row) (4)
(Row (2) does not change) (5)
(Row (3) + pivot row) (6)

Phase 2: $\left[\begin{array}{ccc|ccc} 1 & 0 & 1 & 1 & 1 & 0 \\ 0 & 1 & 1 & 0 & 1 & 0 \\ 0 & 0 & 0 & 1 & 0 & 1 \end{array}\right]$ (Row (4) + pivot row) (8)
(Pivot row) (7)
(Row (6) does not change) (9)

Phase 3: Zero appears in the pivot position, and there is no row below row (9) to interchange with it. You may have foreseen this problem at the end of phase 1. If so, it was not necessary to go through phase 2. ■

Matrices can be used to solve a system of equations, although it is a method rarely used on computers. Consider the equation $AX = B$ where A, B, and X are matrices. If we multiply both sides by the inverse of A (denoted A^{-1}),

we have $$A^{-1}AX = A^{-1}B.$$

Since $A^{-1}A =$ the identity matrix, I, we have $$IX = A^{-1}B.$$

But $IX = X$, so $$X = A^{-1}B.$$

A system of n equations in n variables can be written in matrix form. Matrix A is a square matrix whose entries are the coefficients of the variables in the system, matrix X is a single column whose entries are the variables of the system, and matrix B is a single column whose entries are the constants of the system. Remember that multiplication of matrices is not, in general, commutative. Thus, if we use the factor A^{-1} as the first factor on the lefthand side of the equation, we use A^{-1} as the first factor on the righthand side also.

To solve a system of n equations in n variables using matrix equations:

1. If the variables are not in the same order for each equation, write an equivalent system in which they are.
2. Write the matrix equation $AX = B$ which represents the system.
   (a) A is the matrix of coefficients of the variables, in order.
   (b) X is a column matrix containing the variables, in order.
   (c) B is a column matrix containing the constants of the equations, in order.
3. Find the inverse (A^{-1}) of matrix A.
4. Multiply both sides of the matrix equation by A^{-1}, using A^{-1} as the first factor.
5. Each variable corresponds to an entry in the product matrix $A^{-1}B$.

EXAMPLE 4

Solve using matrix equations: $$4x + 7y = 1$$ $$3x + 5y = 1.$$

Solution   We have $\begin{bmatrix} 4 & 7 \\ 3 & 5 \end{bmatrix} \begin{bmatrix} x \\ y \end{bmatrix} = \begin{bmatrix} 1 \\ 1 \end{bmatrix}$.

We find the inverse of $\begin{bmatrix} 4 & 7 \\ 3 & 5 \end{bmatrix}$: $\left[\begin{array}{cc|cc} 4 & 7 & 1 & 0 \\ 3 & 5 & 0 & 1 \end{array}\right]$.

Phase 1: $\left[\begin{array}{cc|cc} 1 & 7/4 & 1/4 & 0 \\ 0 & -1/4 & -3/4 & 1 \end{array}\right]$ *Phase 2:* $\left[\begin{array}{cc|cc} 1 & 0 & -5 & 7 \\ 0 & 1 & 3 & -4 \end{array}\right]$

Thus, the inverse of $\begin{bmatrix} 4 & 7 \\ 3 & 5 \end{bmatrix}$ is $\begin{bmatrix} -5 & 7 \\ 3 & -4 \end{bmatrix}$.

We multiply both sides of the matrix equation by this inverse:

$$\begin{bmatrix} -5 & 7 \\ 3 & -4 \end{bmatrix} \begin{bmatrix} 4 & 7 \\ 3 & 5 \end{bmatrix} \begin{bmatrix} x \\ y \end{bmatrix} = \begin{bmatrix} -5 & 7 \\ 3 & -4 \end{bmatrix} \begin{bmatrix} 1 \\ 1 \end{bmatrix}.$$ We have

$$\begin{bmatrix} x \\ y \end{bmatrix} = \begin{bmatrix} 2 \\ -1 \end{bmatrix}.$$

The solution of the system is $(2, -1)$, which is easily verified:

$4(2) + 7(-1) = 1$ or $8 - 7 = 1$

$3(2) + 5(-1) = 1$ or $6 - 5 = 1$. ■

8.7 EXERCISES

Find the inverse of each matrix. See Examples 1–3.

1. $\begin{bmatrix} 5 & 4 \\ 6 & 5 \end{bmatrix}$

2. $\begin{bmatrix} 3 & -2 \\ -2 & 1 \end{bmatrix}$

3. $\begin{bmatrix} 2 & 1 \\ -5 & -3 \end{bmatrix}$

4. $\begin{bmatrix} 2 & 3 \\ 3 & 4 \end{bmatrix}$

5. $\begin{bmatrix} 1 & 3 \\ 2 & 3 \end{bmatrix}$

6. $\begin{bmatrix} 1 & 2 \\ -3 & 0 \end{bmatrix}$

7. $\begin{bmatrix} 2 & 4 \\ 0 & -1 \end{bmatrix}$

8. $\begin{bmatrix} 1/2 & 1 \\ -1 & 1/2 \end{bmatrix}$

9. $\begin{bmatrix} 1 & 1 & 1 \\ 1 & 3 & 3 \\ 2 & 4 & 3 \end{bmatrix}$

10. $\begin{bmatrix} 1 & 0 & 1 \\ 0 & 2 & 1 \\ 3 & 2 & 3 \end{bmatrix}$

11. $\begin{bmatrix} 1 & 2 & 1 \\ 2 & 2 & 2 \\ 2 & 3 & 3 \end{bmatrix}$

12. $\begin{bmatrix} 5 & 0 & -2 \\ 0 & -2 & 1 \\ -7 & 0 & 3 \end{bmatrix}$

13. $\begin{bmatrix} 2 & 3 & 2 \\ 3 & 5 & 2 \\ 4 & 6 & 5 \end{bmatrix}$

14. $\begin{bmatrix} 6 & -10 & -7 \\ -1 & 2 & 1 \\ -5 & 11 & 5 \end{bmatrix}$

15. $\begin{bmatrix} 1 & 4 & 2 \\ 2 & 2 & 1 \\ 8 & 1 & 1 \end{bmatrix}$

Use matrix equations to solve the following systems of equations. See Example 4.

16. $2x + 3y = 1$
$3x - 2y = -5$

17. $6x + 6y = 7$
$6x - 6y = 1$

18. $2x - 3y = 1$
$5x - 7y = 3$

19. $x - 5y = 6$
$2x - 9y = 11$

20. $x + y = 1$
$2x - y = 1$

21. $2x + 3y = 2$
$2x - 3y = 0$

22. $2x + 2y = 1$
$4x + y = 1$

23. $x + 3y + 3z = 2$
$2x + 4y + 3z = 5$
$x + 2y + 3z = 1$

24. $3x + 3y + 3z = 6$
$4x + 3y + 3z = 7$
$3x + 3y + 2z = 7$

25. $x + 2y + z = 3$
$3x + 4y + 2z = 8$
$x + y + z = 2$

＊**26.** The inverse of a 2×2 matrix with integral entries often contains fractions. The denominator of these fractions can be thought of as the determinant formed by the entries of the original matrix. Can you discover a quick way to invert 2×2 matrices using this fact?

＊**27.** Consider the magic square matrix $\begin{bmatrix} 3 & 1 & 2 \\ 1 & 2 & 3 \\ 2 & 3 & 1 \end{bmatrix}$.

Is its inverse a magic square matrix?

＊**28.** Is the inverse of the matrix $\begin{bmatrix} 4 & 1 & 4 \\ 3 & 3 & 3 \\ 2 & 5 & 2 \end{bmatrix}$ a magic square matrix?

＊**29.** To determine whether a matrix has an inverse, we can use the following rule: If the determinant of a square matrix is not zero, then the matrix has an inverse. If the determinant is zero, then the matrix does not have an inverse. For each matrix, determine whether the inverse exists.

a. $\begin{bmatrix} 1 & 0 \\ 0 & -1 \end{bmatrix}$

b. $\begin{bmatrix} -1/4 & 2 \\ 1 & -8 \end{bmatrix}$

c. $\begin{bmatrix} 1/2 & -1/3 & 1/6 \\ 0 & 6 & 0 \\ -3 & 2 & -1 \end{bmatrix}$

d. $\begin{bmatrix} 1 & -2 & 3 \\ -3 & 1 & -2 \\ 2 & -3 & 1 \end{bmatrix}$

CHAPTER REVIEW

A set of two or more linear equations is called a ＿＿＿＿＿＿＿ of simultaneous linear equations.

A system of equations is said to be ＿＿＿＿＿＿＿ if there are an infinite number of solutions.

A system of equations is said to be ＿＿＿＿＿＿＿ if there is no solution.

The method of ＿＿＿＿＿＿＿ elimination with ＿＿＿＿＿＿＿ ＿＿＿＿＿＿＿ is often used to find solutions to systems of simultaneous linear equations on a computer.

An $n \times n$ _____ is a real number represented by a square array of n^2 numbers.

_____ rule says that the value of each variable of a system of linear equations can be written as a ratio of determinants.

An $m \times n$ _____ is a rectangular array of numbers with m rows and n columns.

When the product of two square matrices is the identity matrix, the two matrices are called _____ of each other.

CHAPTER TEST

Graph each pair of equations and determine the solution of the system from the graph.

1. $x - 2y = 0$
  $2x + y = 10$

2. $3x - 2y = 1$
  $2x - 3y = 9$

Solve each system of equations by the elimination method.

3. $3x - 2y = 1$
  $2x - 3y = 9$

4. $y = x + 1$
  $x = y + 1$

5. $y = 4$
  $x + y = 6$

Solve each system of equations by the elimination method.

6. $x - 2y - 3z = 2$
  $2x + 3y - z = 6$
  $3x - y - 2z = 4$

7. $x - y - z = 4$
  $-x + y - z = 4$
  $-x - y + z = 0$

8. $x + y = 3$
  $y + z = 5$
  $x + z = 4$

9. $x = y + 5$
  $y = z - 1$
  $x = z + 4$

10. $2x + y - 2z = 9$
  $x + 2y + 2z = 0$
  $2x - y + 2z = -1$

Solve each system by the method of Gauss elimination with back substitution.

11. $x - 2y - 3z = 2$
  $2x + 3y - z = 6$
  $3x - y - 2z = 4$

12. $x - y - z = 4$
  $-x + y - z = 4$
  $-x - y + z = 0$

13. $x + y = 3$
  $y + z = 5$
  $x + z = 4$

14. $x = y + 5$
  $y = z - 1$
  $x = z + 4$

15. $2x + y - 2z = 9$
  $x + 2y + 2z = 0$
  $2x - y + 2z = -1$

Solve each system of equations to the nearest hundredth using the Gauss-Seidel method where possible.

16. $3x + y = 17$
  $x - 3y = 2$

17. $x + y = 4$
  $7x - 3y = 11$

18. $x + y = 5$
  $x - 3y = 0$

19. $2x + 2y = 7$
  $x + 3y = 8$

20. $x - y = 1$
  $x - 3y = 14$

Evaluate each determinant.

21. $\begin{vmatrix} -2 & 1 \\ 3 & 1 \end{vmatrix}$

22. $\begin{vmatrix} 1 & 0 & 2 \\ 3 & -1 & 1 \\ 0 & 2 & -1 \end{vmatrix}$

23. $\begin{vmatrix} 1 & -1 & 2 \\ -3 & 2 & 2 \\ 3 & 4 & -3 \end{vmatrix}$

Use Cramer's rule to solve each system of equations.

24. $2x + 2y = 7$
  $x + 3y = 8$

25. $x + y = 3$
  $y + z = 5$
  $x + z = 4$

Perform the matrix addition and subtraction, if possible.

26. $\begin{bmatrix} 1 & 0 \\ 2 & -2 \end{bmatrix} + \begin{bmatrix} 2 & -3 \\ 0 & 3 \end{bmatrix}$

27. $\begin{bmatrix} 1 & 1 \\ 0 & -2 \\ 4 & 7 \end{bmatrix} - \begin{bmatrix} 1 & 0 \\ 2 & 5 \\ 0 & -3 \end{bmatrix}$

28. $\begin{bmatrix} 1 & 2 & 0 \\ 3 & -1 & 4 \end{bmatrix} + \begin{bmatrix} 1 & 2 \\ 2 & 5 \\ -1 & 4 \end{bmatrix}$

Perform the matrix multiplication, if possible.

29. $\begin{bmatrix} 1 & 0 \\ 3 & 2 \end{bmatrix} \begin{bmatrix} 2 & -5 \\ 6 & 0 \end{bmatrix}$

30. $\begin{bmatrix} 2 & 3 & 0 \\ 1 & -2 & 4 \\ 3 & 0 & -1 \end{bmatrix} \begin{bmatrix} 3 & -2 & 0 \\ 1 & 0 & -1 \\ 2 & 3 & 4 \end{bmatrix}$

Find the inverse of each matrix.

31. $\begin{bmatrix} 2 & 5 \\ 1 & 3 \end{bmatrix}$

32. $\begin{bmatrix} 1 & 5 & 2 \\ -1 & 3 & 1 \\ -3 & 4 & 1 \end{bmatrix}$

Use matrix equations to solve the following systems of equations.

33. $x + 2y = 3$
  $2x + 3y = 4$

34. $x + y = 2$
  $2x + 3y = 4$

35. $x + y + z = 2$
  $x + 3y + 3z = 4$
  $2x + 4y + 3z = 4$

36. The flowchart below shows an algorithm for solving a system of equations $a_1x + b_1y = c_1$ using
$$a_2x + b_2y = c_2$$
Cramer's rule when the solution is unique. Modify it so that inconsistent or dependent systems are identified.

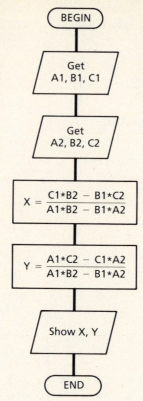

BEGIN

Get
A1, B1, C1

Get
A2, B2, C2

$$X = \frac{C1*B2 - B1*C2}{A1*B2 - B1*A2}$$

$$Y = \frac{A1*C2 - C1*A2}{A1*B2 - B1*A2}$$

Show X, Y

END

FOR FURTHER READING

Conklin, Kenneth R. "Using Determinants and Computers to Recognize Dependent and Inconsistent Linear Systems," *The Mathematics Teacher,* Vol. 74 No. 8 (November 1981), 641–646.

Hanson, Robert. "Integer Matrices Whose Inverses Contain Only Integers," *The Two-Year College Mathematics Journal,* Vol. 13 No. 1 (January 1982), 18–21. (A method for constructing such matrices is given.)

Lancaster, Ronald J. "Magic Square Matrices," *The Mathematics Teacher,* Vol. 72 No. 1 (January 1979), 30–32.

Rogues, Alban J. "Determinants: A Short Program," *The Two-Year College Mathematics Journal,* Vol. 10 No. 5 (November 1979), 340–342. (A FORTRAN program for evaluating determinants by Chio's rule is given.)

Biography   **Carl Friedrich Gauss**

Carl Friedrich Gauss was born April 30, 1777 in Germany. His father was a bricklayer responsible for the men who worked under him. He was astounded when the three-year-old Carl, watching him work on the payroll, noticed an error and gave the correct sum. When Gauss was about ten, the schoolmaster assigned his students the problem of adding all of the natural numbers from 1 to 100. The exercise was designed to keep the class busy. Gauss, however, quickly noticed that $1 + 100 = 2 + 99 \ldots = 101$, and that there are 50 pairs or a sum of $50(101) = 5050$. The teacher had hardly finished making the assignment when Gauss turned in his slate with the correct answer.

When he proved, in 1796, that a polygon with 17 sides of equal length could be constructed using only compass and straight edge, he decided to study mathematics instead of languages. Ferdinand, Duke of Braunschweig, financed Gauss' education. When he was awarded the doctoral degree in 1801, the Duke granted him a regular stipend, so that he could continue his mathematical research without having to teach.

His reputation was enhanced by his astronomical work. Only a few observations of the position of Ceres, the first asteroid discovered, were possible. Gauss, however, was able to compute the orbit. Furthermore, with his method, problems that had previously taken several days were solvable in several hours.

In 1805, Gauss married Johanna Osthoff. They had three children. She died in 1809, and for the sake of his small children, he married Minna Waldek, a friend of his first wife. They also had three children.

Gauss lived until February 23, 1855. He requested that his tombstone bear a 17-sided polygon. A 17-pointed star was used instead, so that it would not be mistaken for a circle.

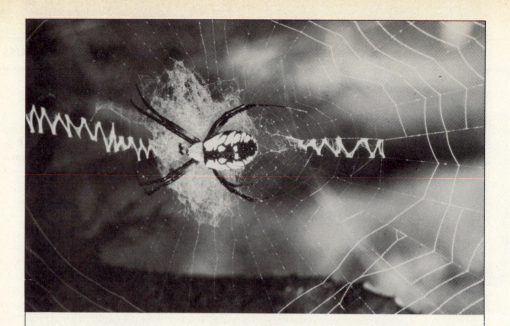

Chapter Nine
Linear Programming

The simplex algorithm for solving linear programming problems is sometimes described as a method of looking for an optimum solution to a problem by moving from one corner of a polyhedron to another, just as an insect might crawl along its surface.

9.1   Linear Programming

Linear programming is a branch of mathematics that has developed since the 1940s. The word *programming* has nothing to do with computer programming, although computers are often used to solve problems that would be overwhelming otherwise. A linear programming problem is basically one of finding the maximum or minimum value of a linear function (i.e., one in which no variable has an exponent other than one). There are restrictions on the values the variables may assume. These *constraints* are written as linear equations or inequalities.

The problems we consider in this chapter are simpler than those solved in actual practice, but they illustrate the general approach. Linear programming often deals with production problems in which the goal is to produce maximum output with minimum resources. We must assume the *proportionality property*. This means that if the resources are increased by a certain factor, the output is increased by the same factor. But discounts are often given on quantity purchases, so if it is possible to double the money spent on supplies, it may be possible to more than double the output. In practice, then, the proportionality property may not apply.

In this section, our goal will be to read linear programming problems and state them mathematically. Consider the problem that follows.

"A bakery sells chocolate cakes and carrot cakes. The profit on a chocolate cake is $3, while the profit on a carrot cake is $2. A chocolate cake requires 10 minutes in the mixer and 40 minutes in the oven. A carrot cake requires 5 minutes in the mixer and 36 minutes in the oven. The mixer can run at most 240 minutes a day without overheating, and the oven can operate no more than 1280 minutes a day.

The manager expects to sell at least as many carrot cakes as chocolate cakes. How many of each type should the baker make in a day in order to produce the most profit, assuming all cakes can be sold?"

Our objective or goal in this problem is to maximize profit. There are three variables.

Let x = number of chocolate cakes.

Let y = number of carrot cakes.

Let P = amount of profit.

Since the profit on each chocolate cake is $3 and the bakery will sell x of those, the profit on chocolate cakes is $3x$. Since the profit on each carrot cake is $2 and the bakery will sell y of those, the profit on carrot cakes is $2y$. Total profit is given by the equation $P = 3x + 2y$. This equation is called the *objective function*.

The profit is to be maximized subject to certain restrictions. The restrictions are time restrictions for the operation of the equipment. The mixer uses 10 minutes for each of the x chocolate cakes or $10x$ minutes, while it uses 5 minutes for each of the y carrot cakes or $5y$ minutes. Since the mixer can run "at most" 240 minutes, we can say $10x + 5y \leq 240$. For the oven, we have $40x + 36y \leq 1280$. Notice that the phrase "no more than" is translated as $\leq$. Finally, the statement that there are to be "at least as many carrot cakes as chocolate cakes" can be written $y \geq x$. We also know that $x \geq 0$ and $y \geq 0$ for this problem, because a negative number of cakes will not be produced.

Phrases like "at least," "at most," and "no more than" are common in linear programming problems. A is *at least* B means $A \geq B$. A is *at most* B means $A \leq B$. A is no more than B means $A \leq B$.

Linear programming problems often appear to be very long. There are a number of conditions to be stated. As we translate these problems into mathematical form, it helps to consider each problem in four parts. First, ask what the objective or goal is. Often it will be to maximize profit or to minimize cost. Second, identify the variables. Look at the question and at the objective in step 1 to help you see what is unknown in the problem. The third step is to write the objective function. The objective function is a formula for the quantity that is to be maximized or minimized. Finally, we are ready to write the constraints as the fourth step. The constraints are inequalities that put restrictions on the values that the variables may assume. These steps are summarized in the following table.

> **To write the mathematical formulation for a linear programming problem:**
> 1. State the objective or goal.
> 2. Identify the variables.
> 3. Write the objective function.
> 4. Write the constraints.

EXAMPLE 1

Give the mathematical formulation of the following problem: A dietitian wishes to provide at least 20 mg of fat and at least 17 mg of carbohydrates by combining crabmeat and fish. A spoonful of crabmeat contains 2 mg of fat and 1 mg of carbohydrates, while a spoonful of fish contains 5 mg of fat and 6 mg of carbohydrates. Crabmeat costs 70¢ a spoonful and fish costs 30¢ a spoonful. How much crabmeat and how much fish should be used to produce the least expensive meal that satisfies the dietitian's requirements?

Solution

1. The objective or goal is to minimize cost.

2. We will need a variable for cost. And since the question is how much crabmeat and how much fish, we need variables for those quantities.

Let C = cost of the mixture.
Let x = number of spoonfuls of crabmeat.
Let y = number of spoonfuls of fish.

3. The objective function will be a formula for cost. Since each spoonful of crabmeat costs 70¢, x spoonfuls will cost $70x$. Likewise, y spoonfuls of fish will cost $30y$. Therefore, $C = 70x + 30y$.

4. The constraints are restrictions on the amounts of fat and carbohydrates that must be included:

Fat comes from both crabmeat and fish. There are 2 mg of fat in each of the x spoonfuls of crabmeat and 5 mg in each of the y spoonfuls of fish. Also there must be at least 20 mg of fat. We have $2x + 5y \geq 20$.

Carbohydrates come from both crabmeat and fish. There is 1 mg of carbohydrates in each of the x spoonfuls of crabmeat and 6 mg of carbohydrates in each of the y spoonfuls of fish. There must be at least 17 mg of carbohydrates, so we have $x + 6y \geq 17$.

Furthermore, $x \geq 0$ and $y \geq 0$ for this problem, since we cannot mix in a negative amount of either crabmeat or fish.

The statement of the problem is now complete. It will be solved in Section 9.4. ■

9.1 EXERCISES

Give the mathematical formulation for each problem. See Example 1.

1. The owner of a candy store has 110 chocolate-covered cherries and 104 orange creams. She can sell a box containing 10 cherries and 24 creams and a box containing 20 cherries and 8 creams. If the profit is $3 on the first-size box and $2 on the second-size box, find the number of boxes of each type that she should sell in order to maximize profit.

2. The owner of a candy store has 90 chocolate-covered cherries and 176 orange creams. She can sell a large box containing 10 cherries and 24 creams and a small box containing 20 cherries and 8 creams. The profit on a large box is $2.50, and the profit on a small box is $2. Find the number of boxes of each type she should sell in order to maximize profit.

3. A company makes granola by mixing cereal and fruit. They have 6000 pounds of cereal and 4000 pounds of fruit in stock. A regular box of granola contains 3/4 lb. of cereal and 1/2 lb. of fruit, while a premium box contains 1/2 lb. of cereal and 1 lb. of fruit. The profit on a regular box is $.20 and on a premium box it is $.50. Find the number of boxes of each type that should be produced to maximize profit.

4. A company makes granola by mixing cereal and fruit. They have 6000 pounds of cereal and 4000 pounds of fruit in stock. A regular box of granola contains 3/4 lb. of cereal and 1/2 lb. of fruit, while a premium box contains 1/2 lb. of cereal and 1 lb. of fruit. The profit on a regular box is $.50 and on a premium box it is $.30. If only 4000 lbs. of granola can be shipped, find the number of boxes of each type that should be produced to maximize profit.

5. A company produces metal picture frames. A 5″ x 7″ frame requires 1/2 minute of shaping and 1/3 minute of plating. An 8″ x 10″ frame requires 3/4 minute of shaping and 1 minute of plating. The machines used for shaping and plating can be used for up to 8 hours (480 minutes) a day. If the profit is $1 on each 5″ x 7″ frame and $2 on each 8″ x 10″ frame, find the number of each size that should be produced to maximize profit.

6. A company produces metal picture frames. A 5″ x 7″ frame requires 1/2 minute of shaping and 1/3 minute of plating. An 8″ x 10″ frame requires 3/4 minute of shaping and 1 minute of plating. The machines used for shaping and plating can each be used for up to 8 hours (480 minutes) a day. The total number of frames produced cannot exceed 720. If the profit is $1 on each 5″ x 7″ frame and $2 on each 8″ x 10″ frame, find the number of each size that should be produced to maximize profit.

7. A dietitian wants to use a mixture of processed cheddar and Swiss cheeses. One ounce of cheddar has 210 mg of calcium and 300 mg of vitamin A, while one ounce of Swiss has 270 mg of calcium and 330 mg of vitamin A. The dietitian wants the mixture to contain at least 1170 mg of calcium and at least 1560 mg of vitamin A. The cost of cheddar is $.25 per ounce and the cost of Swiss is $.20 per ounce. How many ounces of each cheese should he use in order to minimize cost?

8. A dietitian wants to use a mixture of processed cheddar and Swiss cheeses. One ounce of cheddar has 210 mg of calcium and 300 mg of vitamin A, while one ounce of Swiss has 270 mg of calcium and 330 mg of vitamin A. Furthermore, each ounce of cheddar contains 105 calories, and each ounce of Swiss contains 105 calories. The dietitian wants the mixture to contain at least 1170 mg of calcium and at least 1560 mg of vitamin A. How many ounces of each cheese should he use in order to minimize calories?

9. A hiker wants to make a trail mix of chocolate candy and peanuts. One serving of chocolate candy supplies 104 mg of calcium and 1 mg of iron, while one serving of peanuts supplies 26 mg of calcium and 2 mg of iron. The hiker wants the mixture to contain at least 234 mg of calcium and at least 4 mg of iron. Candy costs $.30 per serving, and peanuts cost $.20 per serving. Find the number of servings of each food that should be mixed in order to minimize cost.

10. A hiker wants to make a trail mix of chocolate candy and peanuts. One serving of chocolate candy supplies 104 mg of calcium and 1 mg of iron, while one serving of peanuts supplies 26 mg of calcium and 2 mg of iron. The hiker wants the mixture to contain at least 234 mg of calcium and at least 4 mg of iron. Furthermore, there must be at least four servings of the mixture. Candy costs $.30 per serving, and peanuts cost $.20 per serving. Find the number of servings of each food that should be mixed in order to minimize cost.

11. A baker wishes to enrich his breads by adding wheat germ and nonfat dry milk to the flour. One cup of wheat germ contains 1 1/2 mg of vitamin B_1 and 1/2 mg of vitamin B_2. One cup of nonfat dry milk contains 1/2 mg of vitamin B_1 and 1 1/2 mg of vitamin B_2. The baker wants his recipe to be enriched by at least 1 mg of each vitamin. Wheat germ costs $.41 per cup and nonfat dry milk costs $.34 per cup. How many cups of wheat germ and how many cups of nonfat dry milk should be added in order to minimize cost?

12. A baker wishes to enrich his breads by adding wheat germ and nonfat dry milk to the flour. One cup of wheat germ contains 1 1/2 mg of vitamin B_1 and 1/2 mg of vitamin B_2. One cup of nonfat dry milk contains 1/2 mg of vitamin B_1 and 1 1/2

mg of vitamin B_2. The baker wants his recipe to be enriched by at least 1 mg of each vitamin. Furthermore, he must add at least 1 1/2 cups of the mixture to the recipe. Wheat germ costs \$.41 per cup and nonfat dry milk costs \$.34 per cup. How many cups of wheat germ and how many cups of nonfat dry milk should be added in order to minimize cost?

9.2 Linear Inequalities

In Section 9.4 we will solve the problems of Section 9.1 by graphing. In this section and the next, we examine some graphing techniques that will be useful for these problems.

Since a line represents an equation of the form $y = mx + b$, we can say that the line represents ordered pairs (x, y) such that y and $mx + b$ are equal. For any given value of x, the points *above* the line have y values greater than the y value of the point *on* the line (i.e., $y > mx + b$). Points *below* the line have y values smaller than the y value of the point *on* the line ($y < mx + b$). This observation makes it easy to graph linear inequalities.

For instance, to graph $y \geq 2x - 3$, we first graph the equation $y = 2x - 3$. The solution of $y \geq 2x - 3$ includes the points *above* the line, as well as the points *on* the line, so we shade that area of the plane to represent the solution. See Figure 9.1.

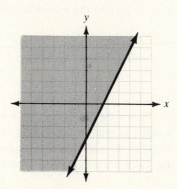

Figure 9.1

EXAMPLE 1   Sketch the graph of $y < \dfrac{1}{2}x - 2$.

Solution   Sketch the graph of $y = \dfrac{1}{2}x - 2$ using a dotted line, since the values *on* the line are not included in the solution. Then we shade the area *below* the line in order to get the values of y such that for each value of x, the y values are *less than* the y value on the line. See the following figure.

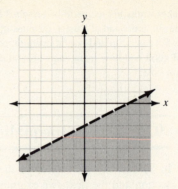

Linear inequalities are not always expressed with y on one side and $mx + b$ on the other. Consider $2y - 3x \geq 6$. The constraints for a linear programming problem are often written like this one, with both variables on the same side of the inequality. Rather than rewrite the inequality as $y \geq \frac{3}{2}x + 3$ to use the slope and intercept, we might keep the form $2y - 3x \geq 6$ to use both the x- and y-intercepts. Two points determine a line, and the two intercepts are particularly easy to work with. When $x = 0$, the line crosses the y-axis, giving the y-intercept. When $y = 0$, the line crosses the x-axis, giving the x-intercept.

To find the y-intercept for the line $2y - 3x = 6$, we let $x = 0$

$$2y \quad = 6, \text{ so } y = 3.$$

To find the x-intercept for the line $2y - 3x = 6$, we let $y = 0$

$$- 3x = 6, \text{ so } x = -2.$$

Thus, the graph has $(0, 3)$ and $(-2, 0)$ as two of its points.

Since the coefficient of y is positive, we know that if the equation were written so that the coefficient of y were 1, we would still have a $\geq$ inequality. That is, for any value of x, values of y above the line will still be greater than the value of y on the line. See Figure 9.2.

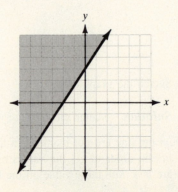

Figure 9.2

The steps used in graphing an inequality are summarized as follows.

> **To graph a linear inequality:**
> 1. If necessary, write an equivalent inequality that has y on the lefthand side with a positive coefficient.
> 2. Replace the inequality symbol with an equals sign and find the two intercepts, or find the slope and y-intercept.
> 3. Graph the equation. It will be the boundary line.
>   a. Use a solid line for a $\geq$ or a $\leq$ inequality.
>   b. Use a dotted line for a $>$ or a $<$ inequality.
> 4. Shade the area above the boundary line for a $\geq$ or a $>$ inequality. Shade the area below the boundary line for a $\leq$ or a $<$ inequality.

EXAMPLE 2

Sketch the graph of $4x - 3y < 12$.

Solution   We multiply both sides of the inequality by -1 so the coefficient of y will be positive. We have $-4x + 3y > -12$. (Remember that multiplication by a negative reverses the direction of the inequality symbol.) Consider $-4x + 3y = -12$. When $x = 0$, $y = -4$. When $y = 0$, $x = 3$. Thus $(0, -4)$ and $(3, 0)$ are two points on the boundary line. We use a dotted line and shade the area above the line. See the following figure.

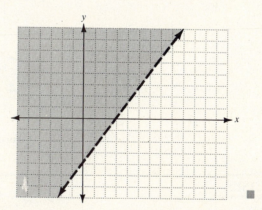

EXAMPLE 3

Sketch the graph of $2x - 5y \geq 10$.

Solution   The inequality is equivalent to $-2x + 5y \leq -10$. Consider $-2x + 5y = -10$. When $x = 0$, $y = -2$. When $y = 0$, $x = 5$. The intercepts are $(0, -2)$ and $(5, 0)$. We use a solid line for the boundary line and shade the area below the line. See the following figure.

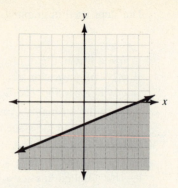

Two or more inequalities may be graphed on the same plane. The solution to the system of inequalities is represented by that region of the plane in which the shading for the individual solutions overlaps. For example, if $4x - 3y > 12$ and $2x - 5y \geq 10$ are shown on the same plane, we have the graph shown in Figure 9.3. The darkest region indicates the values that satisfy *both* inequalities.

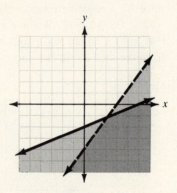

Figure 9.3

9.2 EXERCISES

Sketch the graph of each inequality. See Examples 1–3.

1. $y \geq 2x + 1$

2. $y < 3x - 2$

3. $y > \dfrac{1}{2}x - 1$

4. $y \leq \dfrac{2}{3}x + 3$

5. $y < x + 2$

6. $y \leq -x + 4$

7. $y \geq -2x - 3$

8. $x + y \geq 4$

9. $2x + 3y \leq 6$

10. $3x - 5y > 10$

11. $x - 2y < 12$

12. $2x + y \leq 4$

13. $x - y > 7$

14. $y \geq x$

15. $x \leq y$

16. $x > -3$ **17.** $y \leq 1$ **18.** $x < 2$

19. $y + 1 > x$ **20.** $x + 1 \leq y$ **21.** $3x - y \geq 2$

22. $3y - x < 2$ **23.** $y - 3 \leq x$ **24.** $x - 2 > y$

Sketch each pair of inequalities on the same plane, and shade the region that satisfies both inequalities.

✳**25.** $2x - y \leq 5$
$\quad\quad x + 2y \leq 5$

✳**26.** $x - y \leq 1$
$\quad\quad x + y \leq 5$

✳**27.** $3x - y \geq 4$
$\quad\quad x + 3y \geq 8$

✳**28.** $2x + 5y \leq 9$
$\quad\quad 3x - 4y \leq 10$

9.3   Graphing

A linear programming problem usually involves more than one inequality or constraint. To solve such a problem, we show all of the constraints on the same plane. The region of the plane in which the points satisfy all of the constraints is called the *feasible region*. The graph quickly becomes cluttered if we try to shade the appropriate area for each inequality. Therefore, we will shade just a narrow band near the boundary for each one, and we will label each equation. Consider the inequalities

$$x \geq 0, y \geq 0$$
$$x \leq 4, y \leq 4$$
$$x + y \leq 5.$$

In Figure 9.4 we show the graph of $x \leq 4$:

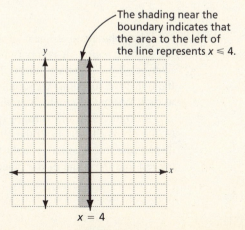

The shading near the boundary indicates that the area to the left of the line represents $x \leq 4$.

$x = 4$

Figure 9.4

The other inequalities are then shown on this same graph. See Figure 9.5.

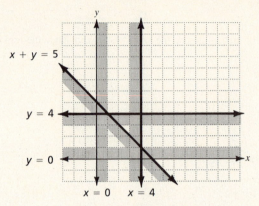

Figure 9.5

Finally, we shade the area in which the points satisfy all of the inequalities at the same time. It is a region bounded on all sides by the boundary shading we have done, and none of the boundary lines cross through it. See Figure 9.6.

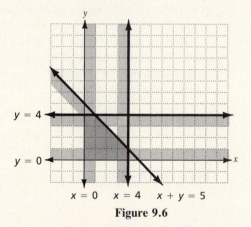

Figure 9.6

The solution to a linear programming problem always lies at a vertex of the feasible region. A *vertex* is a point at which two sides of the feasible region intersect (i.e., a corner point). To find the maximum (or minimum) value of the objective function, we simply examine the value of the objective function at each vertex to see which produces the maximum (or minimum) value.

In Figure 9.6, the vertices (plural of vertex) are located at $(0, 0)$, $(4, 0)$, $(4, 1)$, $(1, 4)$ and $(0, 4)$. If the objective function were $P = 3x + 2y$, we could display the information in table form. The x- and y-coordinates are shown on the left and the value of the objective function on the right.

x	y	$P = 3x + 2y$
0	0	$0 + 0 = 0$
4	0	$12 + 0 = 12$
4	1	$12 + 2 = 14$
1	4	$3 + 8 = 11$
0	4	$0 + 8 = 8$

The maximum value of P occurs when $x = 4$ and $y = 1$.

The steps that we used in this example are summarized in the table that follows.

To find the maximum (or minimum) value of the objective function:

1. Sketch all of the constraints on the same graph.
2. Shade the feasible region.
3. Identify the vertices.
4. Calculate the value of the objective function at each vertex.
5. Identify the maximum (or minimum) value of the objective function.

EXAMPLE 1

Find the values of x and y that maximize $P = 2x + y$ subject to the constraints

$x \geq 0,\ y \geq 0,$

$x + 2y \leq 10,$ and

$2y + 2 \geq x.$

Solution   The constraints $x \geq 0$ and $y \geq 0$ will restrict the feasible region to the first quadrant. $x + 2y \leq 10$ has $(0, 5)$ and $(10, 0)$ as the intercepts of the boundary line. The solution is the area below the line. $2y + 2 \geq x$ can be written $2y - x \geq -2$, and the intercepts of the boundary line are $(0, -1)$ and $(2, 0)$. The solution is the area above the line. The graph is shown in the following figure.

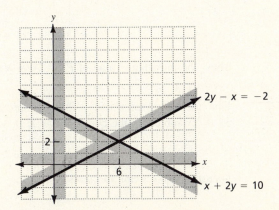

The feasible region is that area in which the points satisfy all of the inequalities. It is shown in gray in the figure below.

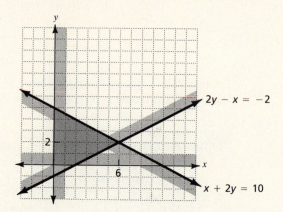

We know three of the four vertices from having used the x- and y-intercepts to graph the constraints. They are $(0, 0)$, $(0, 5)$, and $(2, 0)$. The fourth vertex is located at the intersection of $x + 2y = 10$ and $2y - x = -2$. We find that point by elimination.

$$x + 2y = 10$$
$$\underline{-x + 2y = -2}$$
$$4y = 8 \text{ so } y = 2$$

If $y = 2$, we have $x + 2(2) = 10$, so $x = 6$. The vertex, then, is $(6, 2)$. We list these vertices and the value of $P = 2x + y$ at each one.

x	y	$P = 2x + y$
0	0	$0 + 0 = 0$
0	5	$0 + 5 = 5$
2	0	$4 + 0 = 4$
6	2	$12 + 2 = 14$

Thus, the maximum value of $P = 2x + y$ occurs when $x = 6$ and $y = 2$. ∎

EXAMPLE 2   Find the values of x and y that will produce the minimum value for $C = 3x + 2y$ subject to the constraints $x \geq 0$, $y \geq 0$,
$$x \leq 7, y \leq 6, \text{ and}$$
$$x + y \geq 1.$$

Solution   We graph the feasible region.

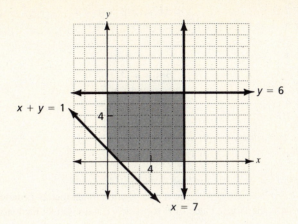

The vertices and objective function are shown in the table below.

x	y	$C = 3x + 2y$
0	1	$0 + 2 = 2$
1	0	$3 + 0 = 3$
7	0	$21 + 0 = 21$
7	6	$21 + 12 = 33$
0	6	$0 + 12 = 12$

The minimum value of C occurs when $x = 0$ and $y = 1$. ▪

9.3 EXERCISES

Find the values of x and y that will maximize $P = 2x + 3y$ subject to the given constraints. See Example 1.

1. $x \geq 0, y \geq 0$
    $x + y \leq 6$
    $y - x \leq 2$

2. $x \geq 0, y \geq 0$
    $x + 2y \leq 8$
    $2y \leq 3x$

3. $x \geq 0, y \geq 0$
    $x \leq 5, y \leq 6$
    $x + y \leq 8$

4. $x \geq 0, y \geq 0$
    $y - x \leq 3$
    $y + x \leq 6$
    $y - x \geq -2$

5. $x \geq 0, y \geq 0$
    $3y - x \leq 6$
    $3y + x \leq 18$

6. $x \geq 0, y \geq 0$
    $x \leq 8, y \leq 8$
    $2y - x \leq 12$

Find the values of x and y that will minimize $C = 2x + 3y$ subject to the constraints given. See Example 2.

7. $x \geq 0, y \geq 0$
    $x + y \geq 4$

8. $x \geq 0, y \geq 0$
    $3y + x \geq 9$
    $y + 2x \geq 8$

9. $x \geq 0, y \geq 0$
    $3y - 2x \geq 2$
    $y + 2x \geq 6$

10. $x \geq 0, y \geq 0$
$x + y \geq 5$
$y + 3x \geq 9$
$y + 3 \geq x$

11. $x \geq 0, y \geq 0$
$x + y \geq 4$
$y + 3x \geq 6$
$4y + x \geq 7$

12. $x \geq 0, y \geq 0$
$x + y \geq 3$
$x \leq 8, y \leq 6$

13. $x \geq 0, y \geq 0$
$y + 3x \geq 9$
$y - x \leq 4$
$x + y \leq 10$

14. $x \geq 3, y \geq 0$
$3y + x \leq 27$
$3y + 7x \leq 63$

15. $x \geq 1, y \geq 3$
$x + y \geq 6$
$x + y \leq 10$

✵**16.** Graph the inequalities $x \geq 0, y \geq 0$,
$x \leq 4, y \leq 4,$ and
$x + y \leq 5.$
Then graph $P = 3x + 2y$ for $P = 2, 6, 10, 14,$ and 18.
Imagine $P = 3x + 2y$ moving from one position to another as P increases from 0 to 18. What is the largest value of P for which x and y satisfy the constraints?

✵**17.** Graph the inequalities $x \geq 0, y \geq 0$,
$x + 2y \leq 10,$ and
$2y + 2 \geq x.$
Then graph $P = 2x + y$ for $P = 0, 4, 7, 11,$ and 14. What is the largest value of P for which x and y satisfy the constraints?

✵**18.** Based on problems 16 and 17, explain how you know that maximum profit will always occur at a vertex of the feasible region.

9.4   Word Problems

Now that we know how to solve a linear programming problem, we return to the two examples in Section 9.1. The first example concerned a bakery.

"A bakery sells chocolate cakes and carrot cakes. The profit on a chocolate cake is $3, while the profit on a carrot cake is $2. A chocolate cake requires 10 minutes in the mixer and 40 minutes in the oven. A carrot cake requires 5 minutes in the mixer and 36 minutes in the oven. The mixer can run at most 240 minutes a day without overheating, and the oven can operate no more than 1280 minutes a day. The manager expects to sell at least as many carrot cakes as chocolate cakes. How many of each type should the baker make in a day in order to produce the most profit, assuming all cakes can be sold?"

The problem led to the mathematical formulation shown in Example 1 below.

EXAMPLE 1

Maximize $P = 3x + 2y$ subject to the constraints:

$$x \geq 0, y \geq 0$$
$$10x + 5y \leq 240$$
$$40x + 36y \leq 1280$$
$$y \geq x.$$

Solution We show these inequalities on a graph. $x \geq 0$, $y \geq 0$ restricts the graph to the first quadrant. The inequality $10x + 5y \leq 240$ has $(0, 48)$ and $(24, 0)$ as the intercepts of the boundary. The inequality $40x + 36y \leq 1280$ has $(0, 35\ 5/9)$ and $(32, 0)$ as the intercepts of the boundary. The boundary of $y \geq x$ has a y-intercept of $(0, 0)$ and a slope of 1. Because the numbers we have to graph are fairly large, we will "scale" the axes. That is, instead of using one block per unit, let one block represent, in this case, four units. See the following figure.

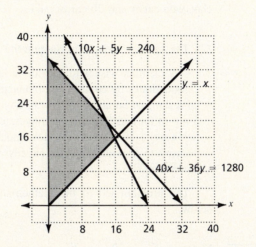

From the graph we can read the values of three of the four vertices. They are $(0, 0)$, $(16, 16)$, and $(0, 35\ 5/9)$. The fourth vertex is located at the intersection of the two lines $10x + 5y = 240$ and $40x + 36y = 1280$. Before we solve this system, we replace the two equations with equivalent ones that have smaller coefficients. That is, we divide both sides of the first equation by 5 to get $2x + y = 48$, and we divide both sides of the other one by 4 to get $10x + 9y = 320$. We now eliminate x from the system. That is, we have

$$-10x - 5y = -240$$
$$\underline{10x + 9y = \quad 320}$$
$$4y = \quad 80, \text{ so } y = 20.$$

Since $2x + y = 48$, we have $2x + 20 = 48$, or $2x = 28$, so $x = 14$. Now that we know the values of all four vertices, we can compute the value of the objective function.

x	y	$P = 3x + 2y$
0	0	$0 + 0 = 0$
16	16	$48 + 32 = 80$
0	$35\frac{5}{9}$	$0 + 71\frac{1}{9} = 71\frac{1}{9}$
14	20	$42 + 40 = 82$

(For our purposes, we retain the fraction 5/9, even though the baker would not make 5/9 cake!)

The maximum value for profit is 82. Therefore the bakery should produce 14 chocolate cakes and 20 carrot cakes. ∎

The second example in Section 9.1 concerned a dietitian.

"A dietitian wishes to provide at least 20 mg of fat and at least 17 mg of carbohydrates by combining crabmeat and fish. A spoonful of crabmeat contains 2 mg of fat and 1 mg of carbohydrates, while a spoonful of fish contains 5 mg of fat and 6 mg of carbohydrates. Crabmeat costs $.70 a spoonful and fish costs $.30 a spoonful. How much crabmeat and how much fish should be used to provide the least expensive meal that satisfies the dietitian's requirements?"

The problem led to the mathematical formulation shown in Example 2 below.

EXAMPLE 2

Minimize $C = .70x + .30y$ subject to the constraints:

$$x \geq 0, \ y \geq 0$$
$$2x + 5y \geq 20$$
$$x + 6y \geq 17.$$

Solution   The constraints $x \geq 0$ and $y \geq 0$ restrict the feasible region to the first quadrant. The boundary of $2x + 5y \geq 20$ has intercepts at $(0, 4)$ and $(10, 0)$. The boundary of $x + 6y \geq 17$ has intercepts at $(0, 17/6)$ and $(17, 0)$. Since the numbers are fairly small, no change of scale is necessary. See the following figure.

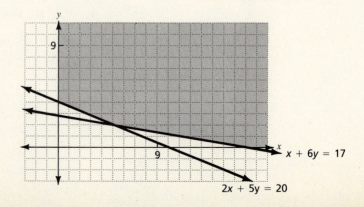

Notice that the feasible region is unbounded. We know two of the three vertices. They are $(0, 4)$ and $(17, 0)$. The third vertex is located at the intersection of $x + 6y = 17$ and $2x + 5y = 20$. We solve this system of equations by eliminating x. We have $-2x - 12y = -34$

$$\frac{2x + 5y = 20}{- 7y = -14}, \text{ so } y = 2.$$

Since $x + 6y = 17$, when $y = 2$, we have $x + 12 = 17$, so $x = 5$. Thus, the third vertex is located at $(5, 2)$.

We show this information in the table below.

x	y	$C = .7x + .3y$
0	4	$0 \quad + 1.2 = \quad 1.20$
17	0	$11.90 + 0 \quad = 11.90$
5	2	$3.50 + .60 = \quad 4.10$

The minimum cost occurs when $x = 0$ and $y = 4$. That is, the dietitian should use four spoonfuls of fish and no crabmeat. ∎

9.4 EXERCISES

Solve the problems of Section 9.1 Exercises. See Examples 1 and 2.

9.5   The Simplex Method

To solve a problem involving three variables by graphing would be very difficult. The feasible region would be a three-dimensional figure. For a problem involving four variables, graphing would be impossible. In 1947, George Dantzig developed a procedure called the *simplex method* for solving linear programming problems. This method will work if the following four conditions are met:

1. The problem is to maximize the objective function.
2. All of the constraints are of the form $a_1x_1 + a_2x_2 + \ldots + a_nx_n \leq b$.
3. The x_i are variables such that $x_i \geq 0$.
4. $b \geq 0$.

The third stipulation requires that the variables be positive or zero. This condition has been met for all examples thus far, and we will assume it for the remaining problems of this chapter. We need no longer state $x \geq 0$ and $y \geq 0$ as two of the constraints.

Because the simplex method is a rather long procedure, the steps are discussed in this section, and the rationale behind it is introduced in the following section.

To use the simplex method, we first rewrite each constraint as an equation by adding a new variable ($x_i \geq 0$) to the smaller side of the inequality. These new variables are called *slack variables*. We construct a table (often called a *tableau*) by listing the variables, including the slack variables, across the top of the table. Beneath each variable, we list the coefficients of that variable in the equations obtained from the constraints. We also list the constants in a column. If a variable does not appear in one of the equations, we use a 0 as the coefficient. We then draw a rectangle around all of these coefficients. Just below the rectangle, we list the coefficients of the objective function. Finally, we list the slack variables along the lefthand side of the rectangle. Each slack variable is listed beside the coefficients of the equation in which it is used. We also list the objective variable (usually P or C) last. Once we have completed these steps, we have the *initial tableau*. The initial tableau can be used to find the solution to the linear programming problem. These steps are summarized in the table that follows.

To construct the initial simplex tableau:
1. Introduce slack variables in order to write each constraint as an equation.
2. Begin the tableau by listing the variables at the top.
3. List the coefficients from each equation under the appropriate variable.
4. Draw a rectangle around the coefficients.
5. Beneath the rectangle, list the coefficients of the objective function in the appropriate columns.
6. List the slack variables followed by the objective variable on the lefthand side of the tableau.

EXAMPLE 1

A microcomputer manufacturer produces three models of a computer: a 16K model, a 64K model, and a 128K model. The profit on a 16K model is $50, on a 64K model it is $70, and on a 128K model it is $90. The machines are manufactured in three divisions of the factory. Because of the scarcity of labor, only 200 work hours per week are available in division I. In division II, 150 work hours per week are available, and 300 work hours per week are available in division III. A 16K model requires 1 hour of production in each division. A 64K model requires 1 hour in division I, 2 hours in division II, and 1 hour in division III. A 128K model requires 1 hour in division I, 1 hour in division II, and 3 hours in division III. Find the number of each model that should be produced in order to maximize profit, assuming all computers can be sold.

Solution   The mathematical formulation of this problem is as follows.

We are to maximize profit.

Let x = number of 16K models.

Let y = number of 64K models.

Let z = number of 128K models.

Let P = profit.

The objective function is $P = 50x + 70y + 90z$.

The constraints are on time in each division:

Division I: $x + y + z \leq 200$

Division II: $x + 2y + z \leq 150$

Division III: $x + y + 3z \leq 300$.

The simplex method can be used on this problem because it meets all four conditions required. That is, the objective is to maximize profit, and the constraints are all the $\leq$ variety with nonnegative constants on the righthand side of each inequality.

The steps of the simplex method will be numbered in the example to make it easier to follow.

1.   Rewrite the constraints as equations.

$$\begin{aligned}
x + y + z + x_1 &= 200 \\
x + 2y + z \quad\;\; + x_2 \quad\;\; &= 150 \\
x + y + 3z \quad\quad\;\; + x_3 &= 300
\end{aligned}$$

where x_1, x_2, and x_3 are slack variables that are greater than or equal to zero.

2. and **3.**   Construct the tableau:

x	y	z	x_1	x_2	x_3		
1	1	1	1	0	0	200	coefficients from 1st equation
1	2	1	0	1	0	150	coefficients from 2nd equation
1	1	3	0	0	1	300	coefficients from 3rd equation

variables, including slack variables

4.   ↖ A rectangle is drawn around the coefficients.

Steps **5** and **6** are illustrated in the following tableau.

	x	y	z	x_1	x_2	x_3	
x_1	1	1	1	1	0	0	200
x_2	1	2	1	0	1	0	150
x_3	1	1	3	0	0	1	300
P	50	70	90	0	0	0	0

← coefficients of objective function slack variables and objective variable

We now have the completed initial tableau. ■

This tableau is used to construct the *first tableau*. For each tableau, the variables on the left are called "basic" variables or "nonzero" variables. The value of each basic variable is given by the value in the same row on the righthand side of the tableau. The algorithm for using one tableau to get another follows.

To use one simplex tableau to find another:

1. Find the largest positive entry on the bottom row. This entry identifies the "pivotal column." Divide each *positive* entry of this column into the rightmost entry of the same row for the rows in the rectangle. The smallest quotient identifies the "pivotal row." Circle the entry in the pivotal row and pivotal column. It is called the *pivot*.

2. Divide each entry in the pivotal row by the pivot. This step produces a 1 in the pivot position for the next tableau.

3. Change all the other entries in the pivotal column to 0 by multiplying the new pivotal row by the additive inverse of the entry to be changed and adding to the row to be changed.

4. Replace the variable on the lefthand side of the pivotal row by the variable at the top of the pivotal column. All other variables retain their positions in the tableau.

5. Repeat steps 1–4 until all the entries on the bottom row are negative.

To illustrate these steps, we return to Example 1. The initial tableau was:

	x	y	z	x_1	x_2	x_3		
x_1	1	1	1	1	0	0	200	(1)
x_2	1	2	1	0	1	0	150	(2)
x_3	1	1	③	0	0	1	300	(3)
P	50	70	90	0	0	0	0	(4)

Step 1:   ↖— 90 is the largest entry so the z column is pivotal.

Compare $200 \div 1 = 200$
$150 \div 1 = 150$
$300 \div 3 = 100$. Since 100 is the smallest quotient, the third row is the pivotal row, and 3 is the pivot. We now begin to construct a new tableau.

Step 2: Divide the third row by 3:

$$1/3 \quad 1/3 \quad 1 \quad 0 \quad 0 \quad 1/3 \quad 100 \tag{5}$$

This is the third row for the new tableau.

Step 3: Replace the first row:

$$
\begin{array}{rrrrrrrl}
\text{i.e.,} \quad 1 & 1 & 1 & 1 & 0 & 0 & 200 & \text{Row (1)} \\
-1/3 & -1/3 & -1 & 0 & 0 & -1/3 & -100 & -1 \times \text{Row (5)} \\
\hline
2/3 & 2/3 & 0 & 1 & 0 & -1/3 & 100 & \qquad \textbf{(6)}
\end{array}
$$

Replace the second row:

$$
\begin{array}{rrrrrrrl}
1 & 2 & 1 & 0 & 1 & 0 & 150 & \text{Row (2)} \\
-1/3 & -1/3 & -1 & 0 & 0 & -1/3 & -100 & -1 \times \text{Row (5)} \\
\hline
2/3 & 5/3 & 0 & 0 & 1 & -1/3 & 50 & \qquad \textbf{(7)}
\end{array}
$$

Replace the fourth row:

$$
\begin{array}{rrrrrrrl}
50 & 70 & 90 & 0 & 0 & 0 & 0 & \text{Row (4)} \\
-30 & -30 & -90 & 0 & 0 & -30 & -9000 & -90 \times \text{Row (5)} \\
\hline
20 & 40 & 0 & 0 & 0 & -30 & -9000 & \qquad \textbf{(8)}
\end{array}
$$

Step 4: Replace the variable in the pivotal row on the left by the variable in the pivotal column. We have as the first tableau:

	x	y	z	x_1	x_2	x_3		
x_1	2/3	2/3	0	1	0	−1/3	100	**(6)**
x_2	2/3	5/3	0	0	1	−1/3	50	**(7)**
z	1/3	1/3	1	0	0	1/3	100	**(5)**
P	20	40	0	0	0	−30	−9000	**(8)**

Step 5: We must repeat steps 1–4, since there are still positive entries on the bottom row.

The new solution is based on the first tableau above.

Step 1: The y column is the pivotal column, because the entry in that column is the largest on the bottom row. Compare

$$100 \div 2/3 = 150$$
$$50 \div 5/3 = 30$$
$$100 \div 1/3 = 300.$$

The second row is the pivotal row, and the entry 5/3 is the pivot.

Step 2: Divide the second row by 5/3 (or multiply by 3/5). The new second row is:

$$2/5 \quad 1 \quad 0 \quad 0 \quad 3/5 \quad -1/5 \quad 30. \tag{9}$$

Step 3: Replace the first row:

2/3	2/3	0	1	0	−1/3	100	Row (6)
−4/15	−2/3	0	0	−2/5	2/15	−20	(−2/3) × Row (9)
6/15	0	0	1	−2/5	−3/15	80	(10)

Replace the third row:

1/3	1/3	1	0	0	1/3	100	Row (5)
−2/15	−1/3	0	0	−1/5	1/15	−10	(−1/3) × Row (9)
3/15	0	1	0	−1/5	6/15	90	(11)

Replace the fourth row:

20	40	0	0	0	−30	−9000	Row (8)
−16	−40	0	0	−24	8	−1200	−40 × Row (9)
4	0	0	0	−24	−22	−10200	(12)

Step 4: Replace the x_2 on the left by y. The second tableau is:

	x	y	z	x_1	x_2	x_3		
x_1	2/5	0	0	1	−2/5	−1/5	80	(10)
y	2/5	1	0	0	3/5	−1/5	30	(9)
z	1/5	0	1	0	−1/5	2/5	90	(11)
P	4	0	0	0	−24	−22	−10200	(12)

Step 5: Steps 1–4 must be repeated, since there is still a positive entry on the last row. The x column is the pivotal column. Compare $80 \div 2/5 = 200$, $30 \div 2/5 = 75$, and $90 \div 1/5 = 450$. The second row is the pivotal row with 2/5 as the pivot. The second row is divided by 2/5 (or multiplied by 5/2) to obtain the new second row:

$$1 \quad 5/2 \quad 0 \quad 0 \quad 3/2 \quad -1/2 \quad 75. \tag{13}$$

The new first row is Row (10) + (−2/5) × Row (13).

The new third row is Row (11) + (−1/5) × Row (13).

The new fourth row is Row (12) + (−4) × Row (13).

The variable x replaces the y on the lefthand side.

The third tableau is

	x	y	z	x_1	x_2	x_3		
x_1	0	-1	0	1	-1	0	50	**(14)**
x	1	5/2	0	0	3/2	$-1/2$	75	**(13)**
z	0	$-1/2$	1	0	$-1/2$	1/2	75	**(15)**
P	0	-10	0	0	-30	-20	-10500	**(16)**

Now that all of the entries in the bottom row are negative, we can read the solution to the problem from the tableau. Each variable on the left has as its value the last entry on the same row. That is, $x_1 = 50$, $x = 75$, $z = 75$, and $P = 10,500$ (notice the change of sign in the entry for the objective variable). The remaining variables are all zero. Therefore $y = 0$, $x_2 = 0$, and $x_3 = 0$.

9.5 EXERCISES

Solve the following problems by the simplex method. You may assume that $x \geq 0$ and $y \geq 0$. See Example 1.

1. Find the values of x and y that maximize $P = 3x + 4y$ subject to $x + y \leq 6$
$$y - x \leq 2.$$

2. Find the values of x and y that maximize $P = 4x + 3y$ subject to $x + y \leq 6$
$$y - x \leq 2.$$

3. Find the values of x and y that maximize $P = 2x + 3y$ subject to $x + 2y \leq 8$
$$-3x + 2y \leq 0.$$

4. Find the values of x and y that maximize $P = 4x + 5y$ subject to $x \leq 5$, $y \leq 6$
$$x + y \leq 8.$$

5. Find the values of x and y that maximize $P = x + 2y$ subject to $x \leq 8$, $y \leq 8$
$$2y - x \leq 12.$$

6. A company that manufactures t.v. sets finds that by testing their merchandise before shipping, they can increase profits. The profit is increased on a 13″ model by $1 and on a 19″ model by $2. It takes 30 seconds to perform test A on a 13″ model and 40 seconds to perform test A on a 19″ model. It takes 15 seconds to perform test B on a 13″ model and 10 seconds to perform test B on a 19″ model. If at most 36,000 seconds a day can be devoted to test A and at most 15,000 seconds a day can be devoted to test B, how many of each model should be tested in order to maximize the increase in profits?

7. An athlete wants to maximize his carbohydrate intake. A serving of spaghetti has 35 g of carbohydrates and a serving of pizza has 23 g of carbohydrates. A serving of spaghetti costs $.50 and a serving of pizza costs $.75. If he has only $5.25 to spend and can eat no more than eight servings, how many servings of each food should he buy?

8. A woman wants to lose weight by exercising. Golfing burns 250 calories per hour while swimming burns 350 calories per hour. If she has at most six hours a day to exercise and wants to golf at least as long as she swims, how many hours should she spend on each sport to maximize weight loss?

9. Three people build end tables in three styles. Worker A does the cutting and lathe work and can work 40 hours per week. Worker B does the building and can work 44 hours per week. Worker C does the finishing and can work 40 hours per week.

> For style 1, cutting requires 2 hours
> building requires 1 hour
> finishing requires 2 hours, and profit is $20.
>
> For style 2, cutting requires 1 hour
> building requires 2 hours
> finishing requires 2 hours, and profit is $20.
>
> For style 3, cutting requires 3 hours
> building requires 4 hours
> finishing requires 2 hours, and profit is $30.

How many of each style should they produce to maximize profits, assuming they can sell all they make?

10. A business advertises in three newspapers. An ad costs $5 per week in paper A, $7 per week in paper B, and $9 per week in paper C. It is estimated that an ad in A will bring in 7 new customers, an ad in B will bring in 10, and an ad in C will bring in 13. In a certain period, no more than 12 ads can be placed. There is a maximum of $72 available for ads, and there must be at least as many ads in A as in B. How many ads should run in each paper in order to bring in the most new customers?

9.6   Justification of the Simplex Method

Consider the problem: Maximize $P = 3x + 2y$ subject to the constraints

$$2x + y \leq 7$$
$$3x + y \leq 9.$$

To solve it using the simplex method, we introduce slack variables and write:

$$2x + y + x_1 \qquad = 7$$
$$3x + y \qquad + x_2 = 9.$$

The graph is shown in Figure 9.7.

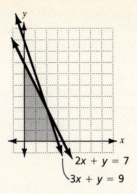

$2x + y = 7$

$3x + y = 9$

Figure 9.7

Notice that on the boundary line $2x + y = 7$, the slack variable $x_1 = 0$ (since $2x + y = 7$, and $2x + y + x_1 = 7$, we must have $x_1 = 0$). Likewise, on the boundary line $3x + y = 9$, the variable $x_2 = 0$. For a graphical solution we would consider vertices, and at each vertex exactly two of the variables have values of zero. The following table lists each vertex and the variables that are zero there.

x	y	zero variables
0	0	x, y
3	0	y, x_2
2	3	x_1, x_2
0	7	x, x_1

The simplex method was devised to consider solutions in which two of the variables are zero, so it, too, involves the vertices of the graph. For the simplex method, we begin with $x = 0$ and $y = 0$ (the least desirable profit) and proceed to change the zero variables, looking for a better profit. If x and y are zero, then the other variables (x_1 and x_2) must be nonzero values as determined from the equations

$$2x + y + x_1 \qquad\quad = 7 \qquad\qquad \textbf{(1)}$$

$$3x + y \qquad + x_2 = 9 \qquad\qquad \textbf{(2)}$$

$$3x + 2y \qquad\qquad = P. \qquad\qquad \textbf{(3)}$$

This is the information we display in the initial simplex tableau by listing the nonzero variables x_1 and x_2 on the left, so that they correspond to their values $x_1 = 7$ and $x_2 = 9$ on the right. And since $P = 0$, we replace that variable with 0.

For the first tableau, we want to increase one of the zero variables. The logical one to change would be x, because an increase in x is multiplied by 3 in the profit equation, but an increase in y is multiplied by only 2. Hence, we choose to increase the variable with the largest coefficient in the profit equation. That is why we locate the largest number in the bottom row of the simplex tableau. Since this variable will *not* be zero in the next tableau, it will replace one of the variables on the left.

We need to determine which variable it will replace, and that variable will become zero in the next tableau. The next solution then, will be $y = 0$ and either $x_1 = 0$ or $x_2 = 0$, with the remaining variables as the nonzeroes. We try both possibilities.

Suppose $x_1 = 0$. Then from equations (1) and (2), we have

$$2x + 0 + 0 = 7, \text{ so } x = 3.5, \text{ and}$$

$$3(3.5) + 0 + x_2 = 9, \text{ so } x_2 = -1.5.$$

But x_2 was introduced as a *positive* number added to the inequality $3x + y \leq 9$ to *increase* the lefthand side. Thus, we cannot use $x_1 = 0$.

The other possibility was to try $x_2 = 0$. We have

$$2x + 0 + x_1 = 7.$$

$$3x + 0 + 0 = 9, \text{ so } x = 3, \text{ and the other equation gives } x_1 = 1.$$

This solution is reasonable.

What we have done is to choose the smaller of the values 7/2 and 9/3 as the value for x, since the larger led to a contradiction. In the simplex method, this step corresponds to choosing the lowest quotient when the constants are divided by the entries in the pivotal column. Once we have decided that $x_2 = 0$, we can replace the x_2 on the left by x.

Changing the pivot to 1 and replacing entries above and below it by zeroes merely corresponds to eliminating a variable, as we did in the Gauss elimination procedure. The tableau is then ready to begin the process again.

In the final tableau, the profit appears in the lower righthand corner of the tableau, but the sign is negative. This row began as the equation $3x + 2y = P$. If we left P in the equation as a variable, as the equations are combined, we would actually get $P - n$, where n is a number. But the other variables in the profit equation are zero, so $P - n = 0$, and $P = n$. Since we did not write the variable P in the tableau, all that shows up is the $- n$.

Rather than reason along these lines every time we solve a linear programming problem, we recognize that every linear programming problem that satisfies the conditions listed at the beginning of Section 9.5 follows the same pattern and the work is shortened considerably by mechanically applying the steps of the simplex method.

In 1984, Narendra K. Karmarkar of Bell Laboratories described a new algorithm for linear programming problems. He said, "The simplex method is fine for problems with a few thousand variables, but after 16 or 20 thousand, it runs out of steam."[†] The new algorithm is more complicated, but requires fewer steps. The larger the problem, the more difference there is between the number of steps for the two methods. News of Karmarkar's algorithm stirred interest in the business world as plans for testing and computer implementation of it were made.

9.6 EXERCISES

Use the simplex method to solve problems 1–6 of Section 9.1 Exercises.

7. Find values of x, y, and z to maximize $P = 3x + 2y + z$ subject to

$$x + 2y + z \leq 3$$
$$x + y + 2z \leq 5$$
$$x + 3y + z \leq 7.$$

8. Find values of x, y, and z to maximize $P = x + 2y + 3z$ subject to

$$x + 2y + z \leq 3$$
$$x + y + 2z \leq 5$$
$$x + 3y + z \leq 7.$$

9. Find values of x, y, and z to maximize $P = x + 2y + z$ subject to

$$x + z \leq 13$$
$$x + y \leq 11$$
$$y + z \leq 7.$$

10. Find values of x, y, and z to maximize $P = x + y + 2z$ subject to

$$x +   z \leq 13$$
$$x + y \leq 11$$
$$y + z \leq 7.$$

11. Find values of x, y, and z to maximize $P = 2x + y + z$ subject to

$$x + y + z \leq 3$$
$$4x + 2y + z \leq 8$$
$$y + z \leq 5.$$

[†]"New Algorithm Stirs Interest," *Siam News*, Vol 18 # 1 (January 1985), 1.

9.7   Minimization Problems

Consider the problem: Find the values of x and y to minimize $C = 7x + 3y$ subject to

$$2x + 5y \geq 20$$
$$x + 6y \geq 17.$$

The simplex method works for a problem that satisfies four conditions.

1. It must be a maximization problem.
2. All of the constraints are of the form $a_1x_1 + a_2x_2 + \ldots a_nx_n \leq b$.
3. The x_i are all variables such that $x_i \geq 0$.
4. $b \geq 0$.

The simplex method will not work for a minimization problem without some modification of the problem. Notice, however, that as a quantity, A, becomes large, its opposite, $-A$, becomes small.

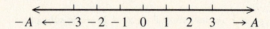

Therefore we can treat the problem of minimizing the function $C = 7x + 3y$ as one of maximizing $-C = -7x - 3y$. Thus, condition (1) is satisfied.

Next, we must transform the constraints so that they are $\leq$ inequalities rather than $\geq$ inequalities. We multiply both sides of each inequality by -1. $2x + 5y \geq 20$ becomes $-2x - 5y \leq -20$ and $x + 6y \geq 17$ becomes $-x - 6y \leq -17$. Condition (2) is now satisfied.

We assume $x_i \geq 0$ for all problems in this text, therefore condition (3) is satisfied.

While the inequalities are now in the form $a_1x_1 + a_2x_2 + \ldots a_nx_n \leq b$, we do not have $b \geq 0$, and this requirement is not easily overcome. The simplex method will need some refinements. At this stage, however, we set up a tableau in the usual manner to maximize $-C = -7x - 3y$ subject to $-2x - 5y \leq -20$ and $-x - 6y \leq -17$.

	x	y	x_1	x_2	
x_1	-2	-5	1	0	-20
x_2	-1	-6	0	1	-17
$-C$	-7	-3	0	0	0

Before we begin using the simplex method, we must eliminate the negative numbers from the righthand column. The procedure is similar to the simplex method itself. It is summarized in the table that follows.

To eliminate negative numbers from the righthand column of a tableau:

1. Within the rectangle, locate the bottommost row containing a negative entry in the righthand column.
2. Choose as the pivotal column any other column with a negative entry in this row.
   a. If there are no positive entries below it in the rectangle, then this entry is used as the pivot.
   b. If there are positive entries below it in the rectangle, then choose the pivot by dividing each of these positive entries into the rightmost entry of the same row. The lowest ratio determines the pivot.
3. Divide each entry of the pivotal row by the pivot.
4. Replace the entries above and below the pivot by 0's as in the simplex method.
5. Replace the variable on the lefthand side of the pivotal row by the variable at the top of the pivotal column.
6. Repeat steps 1–5, if necessary, until all of the values in the rightmost column are positive.

After these steps are done, the tableau will satisfy all four conditions required to use the simplex method.

We now return to the example. The tableau is shown again.

	x	y	x_1	x_2		
x_1	-2	-5	1	0	-20	**(1)**
x_2	$\boxed{-1}$	-6	0	1	-17	**(2)**
$-C$	-7	-3	0	0	0	**(3)**

1. Within the rectangle, row (2) is the bottommost row with a negative entry in the rightmost column.
2. We arbitrarily choose the first column as the pivotal column (we could have chosen the second column, since it, too, has a negative entry). Since there are no positive entries below this one, it will be used as the pivot.
3.–5. These three steps are shown in the tableau that follows.

	x	y	x_1	x_2			
x_1	0	7	1	-2	14	Row (1) + 2 × pivotal row	**(5)**
x	1	6	0	-1	17	Pivotal row	**(4)**
$-C$	0	39	0	-7	119	Row (3) + 7 × pivotal row	**(6)**

Now that all of the entries in the righthand column are positive, the conditions required to use the simplex method are all satisfied. This tableau, then, is the initial tableau.

The first tableau would be:

	x	y	x_1	x_2			
y	0	1	1/7	$-2/7$	2	Pivotal row	(7)
x	1	0	$-6/7$	⑤/7	5	Row (4) + -6 × pivotal row	(8)
$-C$	0	0	$-39/7$	29/7	41	Row (6) + -39 × pivotal row	(9)

The second tableau would be:

	x	y	x_1	x_2			
y	2/5	1	$-1/5$	0	4	Row (7) + (2/7) × pivotal row	
x_1	7/5	0	$-6/5$	1	7	Pivotal row	
$-C$	$-29/5$	0	$-3/5$	0	12	Row (9) + $(-29/7)$ × pivotal row	

Now that all of the entries in the last row are negative (except the last), we read the solution from the tableau. We have $y = 4$, $x_1 = 7$, $x = 0$, and $x_2 = 0$. Since the maximum value for $-C$ is -12, the minimum value for C is 12. As a shortcut, we drop the negative sign in front of C on the lefthand side of the tableau. Then it is not necessary to change the sign of the objective variable as it was for maximization problems.

To solve a linear programming minimization problem:

1. Multiply the objective function by -1 so it will be a maximization problem.
2. Multiply both sides of each $\geq$ inequality by -1.
3. Eliminate the negative entries from the rightmost column using the procedure described in this section.
4. Solve using the simplex method.

EXAMPLE 1

Find values of x and y to minimize $C = x + 2y$ subject to

$$3x + \ y \geq 10$$
$$x + \ y \geq 8$$
$$2x + 5y \geq 25.$$

Solution The problem is equivalent to: Find values of x and y to maximize $-C = -x - 2y$ subject to

$$-3x - y \leq -10$$
$$-x - y \leq -8$$
$$-2x - 5y \leq -25.$$

This information is displayed in a tableau:

	x	y	x_1	x_2	x_3		
x_1	-3	-1	1	0	0	-10	**(1)**
x_2	-1	-1	0	1	0	-8	**(2)**
x_3	-2	(-5)	0	0	1	-25	**(3)**
C	-1	-2	0	0	0	0	**(4)**

We now eliminate the negative entries from the rightmost column. Row (3) is the bottommost row with a negative entry in the rightmost column. Either of the first two columns could be chosen as the pivotal column. We choose the second, making -5 the pivot. The entries of Row (3), then, are divided by -5.

	x	y	x_1	x_2	x_3			
x_1	$-13/5$	0	1	0	$-1/5$	-5	Row (1) + 1 × pivotal row	**(5)**
x_2	$-3/5$	0	0	1	$-1/5$	-3	Row (2) + 1 × pivotal row	**(6)**
y	$(2/5)$	1	0	0	$-1/5$	5	Pivotal row	**(4)**
C	$-1/5$	0	0	0	$-2/5$	10	Row (4) + 2 × pivotal row	**(7)**

There are still negative entries in the rightmost column, so we continue to eliminate these. Row (6) is now the bottommost row with a negative entry in the rightmost column. Either the first or fifth column could be used as the pivotal column. We choose the first. The positive entry below $-3/5$ will serve as the pivot.

	x	y	x_1	x_2	x_3			
x_1	0	$13/2$	1	0	$-3/2$	$55/2$	Row (5) + (13/5) × pivotal	**(9)**
x_2	0	$(3/2)$	0	1	$-1/2$	$9/2$	Row (6) + (3/5) × pivotal	**(10)**
x	1	$5/2$	0	0	$-1/2$	$25/2$	Pivotal row	**(8)**
C	0	$1/2$	0	0	$-1/2$	$25/2$	Row (7) + (1/5) × pivotal	**(11)**

Since this tableau has all positive entries in the rightmost column, we now begin to apply the simplex method.

	x	y	x_1	x_2	x_3		
x_1	0	0	1	$-13/3$	2/3	8	Row (9) + $(-13/2) \times$ pivotal row
y	0	1	0	2/3	$-1/3$	3	Pivotal row
x	1	0	0	$-5/3$	1/3	5	Row (8) + $(-5/2) \times$ pivotal row
C	0	0	0	$-1/3$	$-1/3$	11	Row (11) + $(-1/2) \times$ pivotal row

Since all entries in the bottom row (except the last) are negative, we read the solution from the tableau: $x_1 = 8$, $y = 3$, $x = 5$, $C = 11$, $x_2 = 0$, and $x_3 = 0$. ∎

The method presented in this section is the one most often used to solve minimization problems with computers. One popular method involves converting columns into rows. But linear programming problems usually have more variables than constraints. A tableau with many columns is easier to handle by machine than one with many rows.

There are also situations that we have not covered. If there is more than one way to choose the pivotal column in the simplex method, there is more than one solution. In preparing a tableau for the simplex method, there may be a negative entry in the rightmost column but no other negative entry in that row. In the simplex method, there may be a positive entry in the last row, but no other positive entry in that column. Such problems have no solution.

9.7 EXERCISES

For problems 1–6, do problems 7–12 of Section 9.1 Exercises by the simplex method. See Example 1.

7. Find values of x, y, and z to minimize $C = x + 2y + 3z$ subject to

$$x + 2y + z \geq 3$$
$$x + y + 2z \geq 5$$
$$x + 3y + z \geq 7.$$

8. Find values of x, y, and z to minimize $C = 3x + 2y + z$ subject to

$$x + 2y + z \geq 3$$
$$x + y + 2z \geq 5$$
$$x + 3y + z \geq 7.$$

9. Find values of x, y, and z to minimize $C = x + y + 2z$ subject to

$$x + 2y + z \geq 5$$
$$3x + y + z \geq 7$$
$$2x + y + z \geq 9.$$

10. Find values of x, y, and z to minimize $C = x + 3y + z$ subject to

$$x + y + 2z \geq 3$$
$$2x + y + z \geq 7$$
$$x + y + 3z \geq 11.$$

11. Find values of x, y, and z to minimize $C = 3x + 2y + 7z$ subject to

$$4x + 2y + 8z \geq 7$$
$$x + y + z \geq 3$$
$$2x + y + 3z \geq 4.$$

✹**12.** Find values of x and y to maximize $P = 3x + y$ subject to

$$x + 3y \leq 27$$
$$7x + 3y \geq 63.$$

✹**13.** Find values of x and y to minimize $C = 2x + 3y$ subject to

$$x + y \geq 6$$
$$x + y \leq 10.$$

CHAPTER REVIEW

The restrictions on the variables of a linear programming problem are called
_____.

The _____ function is a formula for the quantity that is to be
maximized or minimized.

The region of the plane in which the points satisfy all of the constraints is called
the _____ region.

The solution to a linear programming problem always lies at a
_____ of the feasible region.

To solve a linear programming problem with four variables, the
_____ method is used.

To rewrite each constraint as an equation, a variable x_i, called a
_____ variable, is added to the smaller side of the inequality.

A table showing the value of each variable in a linear programming problem is
called a _____.

CHAPTER TEST

Consider the following problem: A delicatessen sells a regular ham and cheese
sandwich and a jumbo ham and cheese sandwich. A regular sandwich requires 6 slices
of ham and 2 slices of cheese. A jumbo sandwich requires 8 slices of ham and 3 slices
of cheese. The profit on a regular sandwich is $.50 and on a jumbo sandwich it is
$.60. If there is enough ham to make 1000 slices and enough cheese to make 360
slices, how many sandwiches of each type should be made to get the maximum profit?
You may assume that all sandwiches made can be sold.

1. What is the objective?

2. Identify the variables.

3. Write the objective function.

4. Write the constraints.

5. Are all variables greater than or equal to zero?

Sketch the graph of each inequality.

6. $y > 2x - 4$ **7.** $y < \frac{1}{2}x + 7$ **8.** $x + 2y \leq 5$

9. $3x - 2y \geq 4$ **10.** $x + 1 > 3y$

Use a graph to find the values of x and y that will maximize $P = 3x + y$ subject to the given constraints.

11. $x \geq 0, y \geq 0$ **12.** $x \geq 0, y \geq 0$
  $x \leq 4, y \leq 6$ $x + y \leq 6$
  $2x + y \leq 10$ $x + 3y \leq 12$

Use a graph to find the values of x and y that will minimize $C = x + 3y$ subject to the given constraints.

13. $x \geq 0, y \geq 0$ **14.** $x \geq 0, y \geq 0$
  $x \leq 4, y \leq 6$ $x + y \geq 6$
  $2x + y \geq 4$ $x + 3y \leq 12$

15. Use a graph to find the values of x and y that will maximize $P = 5x + 7y$ subject to the constraints
  $x \geq 0, y \geq 0,$
  $x \leq 6, y \leq 4$
  $x + 3y \leq 15.$

16. Use a graph to solve the problem used for problems 1–5.

Problems 17–19 refer to the following problem: A delicatessen sells a regular chicken salad sandwich and a diet chicken salad sandwich. Each regular sandwich requires 1/2 cup of chicken and 1/4 cup of relish. Each diet sandwich requires 1/3 cup of chicken and 1/8 cup of relish. There are at least 30 cups of chicken that must be used and at least 13 3/4 cups of relish that must be used. It costs $.80 to make a regular sandwich and $.90 (for extra preparation time trimming fat, and so on) to make a diet sandwich. How many of each type should be made in order to minimize cost?

17. State the mathematical formulation of the problem.

18. Sketch the graph of the feasible region.

19. Solve the problem.

Solve the following problems by the simplex method. You may assume that x, y, and z are all greater than or equal to zero.

20. Find the values of x and y that will maximize $P = 3x + 2y$ subject to
  $x + 2y \leq 8$
  $-3x + 2y \leq 0.$

21. Find the values of x and y that will maximize $P = 5x + 2y$ subject to
  $x + y \leq 6$
  $y - x \leq 2.$

22. Find the values of x and y that will maximize $P = 2x + 3y$ subject to
$$2x + 3y \leq 7$$
$$3x + 4y \leq 10.$$

23. Find the values of x, y, and z that will maximize $P = 2x + y + 3z$ subject to
$$x + 2y + z \leq 3$$
$$x + y + 2z \leq 5$$
$$x + 3y + z \leq 7.$$

24. Find the values of x, y, and z that will minimize $C = 2x + 7y + 6z$ subject to
$$x + 2y + 3z \geq 5$$
$$2x + 5y + 9z \geq 7$$
$$x + 3y + 4z \geq 9.$$

FOR FURTHER READING

Halmos, Paul. "The Legend of John von Neumann," *The American Mathematical Monthly,* Vol. 80 (1973), 382–394.

Kolata, Gina. "A Fast Way to Solve Hard Problems," *Science,* (September 21, 1984), 1379–1380. (Karmarkar's algorithm is discussed.)

Lee, Kil S. and Wayne Marx. "Demonstrating the Efficiency of Linear Programming," *The Mathematics Teacher,* Vol. 76 No. 9 (December 1983), 664–666. (Solutions to a problem are compared with and without linear programming techniques.)

Semmes, Pat. "A Geometric Approach to Linear Programming in the Two-Year College," *The Two-Year College Mathematics Journal,* Vol. 5 No. 1 (Winter 1974), 37–40. (A geometric model is described.)

Biography   John von Neumann

John von Neumann, a banker's child, was born December 28, 1903, in Hungary. At the age of six, he could divide two eight-digit numbers in his head. He studied chemistry in Berlin and Zurich and received a PhD in math in 1926.

He taught at Princeton and then at the Institute for Advanced Study. He married Marietta Kövesi. They had one daughter and were divorced in 1937. He later married Klára Dán.

During the 1930s and 40s he worked in the areas of quantum mechanics, statistics, hydrodynamics, ballistics, meteorology, game theory, linear programming, and computer design. There are many stories about him that have circulated through the years. A scientist worked until 4:30 one morning with a desk calculator on five cases of a difficult problem. When he presented it to von Neumann, the latter began to calculate the result mentally. He threw back his head, mumbling to himself, and within minutes had solved the first four cases. Before he had finished the fifth case, the scientist, who remembered the answer, blurted it out. Von Neumann gave him a funny look, continued his calculations for about 30 seconds, and was astounded that the scientist had gotten the correct answer first!

In 1944, he met Herman Goldstine, and they discussed problems involved in their work. Von Neumann was working on the hydrogen bomb and hoped that the high-speed calculations Goldstine projected for ENIAC would be of help to him. He became interested in computers, and in 1945, wrote a paper suggesting the concept of the stored program, which allows the instructions to be held internally until they are executed.

In 1955, von Neumann fell and was injured. While treating him, the doctors discovered that he had bone cancer. He died February 8, 1957.

Chapter Ten
Probability and Statistics

For a group of 25 people, the probability is greater than $\frac{1}{2}$ that at least two of them have the same birthday. The same mathematical ideas used to compute this probability are used in the study of computer errors caused by the failure of memory chips.

10.1   Probability

If you bought a jar of mixed nuts that said ''less than 80 percent peanuts'' (which are relatively cheap) and found that there were 90 percent peanuts, would you buy that brand again? Would you be certain that you made the right decision? Is it possible that you got the only can on the shelf with a disproportionate number of peanuts? It is impossible, however, to open all of the cans and weigh the nuts. You must make a decision based on incomplete data. Many situations in business and industry are similar. Inferential statistics is the branch of mathematics that deals with drawing conclusions about a set and evaluating those conclusions when only part of the set can be studied.

Since predictions based on data from only part of a set cannot be made with absolute certainty, the laws of probability are used. The word *experiment* is often used in probability to refer to an action that is under study. Tossing a coin, rolling a die, or drawing a card from a shuffled deck of 52 playing cards are examples of experiments. Associated with each experiment is the result or outcome.

> **Definition**
> The *sample space* for an experiment is the set of all possible outcomes. Each element in the sample space is called a *sample point*.

EXAMPLE 1

What is the sample space for tossing a coin?

Solution   Since the coin can show heads or tails, we say that {Heads, Tails} or {*H, T*}) is the sample space. ■

EXAMPLE 2

What is the sample space for tossing three coins?

Solution   Since each coin can be either heads or tails, we want to list every possible arrangement. We could have any of the following arrangements: {*HHH, HHT, HTH, HTT, THH, THT, TTH, TTT*}. (The pattern is a lot like listing T's and F's in a truth table.) ■

If each sample point is just as apt to occur as any other, we say that the sample points are *equally likely*. In example 2, we considered *HHT, HTH,* and *THH* to be different outcomes, so that all of the sample points in the sample space were equally likely. If we were to list the sample space as 1 head, 2 heads, or 3 heads (that is, {1*H*, 2*H*, 3*H*}), the events would not be equally likely. It is a good idea to list a sample space so that it contains equally likely events, if possible.

The probability of an event is a fraction that indicates how likely the event is to occur. We say that an experiment is successful if the event under study does occur. Consider tossing a coin. If we want the probability of getting heads, we say that "getting heads" is the event under study. If we are studying the defective parts coming off of an assembly line, we would consider the experiment successful if we found a defective part.

> **Definition**
> The *probability* of an event A (denoted $P(A)$) is the number of successful outcomes in the sample space divided by the total number of outcomes in the sample space, if all outcomes are equally likely.

EXAMPLE 3

What is the probability of tossing heads on a single toss of a fair coin?

Solution   To say that a coin is "fair" means that the possible outcomes are equally likely. Since there are two sample points in the sample space $\{H, T\}$, and one of them (H) is successful, the probability of getting heads is 1/2 or 0.5. ∎

EXAMPLE 4

What is the probability of getting at least one head if three fair coins are tossed?

Solution   The sample space was given in Example 2. Since there are eight outcomes and seven of them have at least one H, the probability of at least one head is 7/8 or 0.875. ∎

EXAMPLE 5

What is the probability of getting exactly one head if three fair coins are tossed?

Solution   Referring to the sample space listed in Example 2, we see that there are three outcomes that have exactly one H. They are HTT, THT, and TTH. Since there were a total of eight sample points in the sample space, the probability of exactly one head is 3/8 or 0.375. ∎

The smallest probability that can occur is 0. The probability of an event is 0 if there are no successful outcomes in the sample space. The largest probability that can occur is 1. The probability of an event is 1 if every sample point in the sample space is a successful outcome. The sum of the probabilities of all of the sample points in the sample space is 1. If the probability that event A occurs is given by $P(A)$, then the probability that A does *not* occur is given by $1 - P(A)$.

EXAMPLE 6

Find the probability:

(a)   Of rolling a 7 on one roll of a single fair die.

(b)   Of rolling a number less than 7 on one roll of a single fair die.

(c)   Of not rolling a 4 on one roll of a single fair die.

Solution   The sample space is {1, 2, 3, 4, 5, 6}.

(a)   There are no successful outcomes in the sample space. The probability of rolling a 7 is 0/6 = 0.

(b)   There are six successful outcomes in the sample space. The probability of rolling a number less than 7 is 6/6 = 1.

(c)   The probability of rolling a 4 is 1/6, so the probability of *not* rolling a 4 is 1 − 1/6 = 5/6. ∎

Suppose we draw a card from a shuffled deck of 52 playing cards. To find the probability of drawing a jack or a heart, we could examine the entire sample space. There are 13 hearts and 3 more jacks besides the jack of hearts (which we counted in the 13 hearts). That is, there are 16 successful outcomes, so the probability is 16/52 = 4/13. If we are trying to find the probability of event A or B it may be easier, however, to base the answer on the individual probabilities of the events rather than to examine the entire sample space. In this example, the probability of getting a heart is 13/52, and the probability of getting a jack is 4/52. Drawing the jack of hearts was counted as a successful outcome in computing both probabilities. We compensate for this "double counting" in the final answer: P(jack or heart) = 13/52 + 4/52 − 1/52.

> **Probability Rule 1**
> If A and B are events in the same sample space, then $P(A$ or $B) = P(A) + P(B) - P(A$ and $B)$.

EXAMPLE 7

Find the probability of drawing a red card or a face card from a shuffled deck of 52 playing cards.

Solution   There are 26 red cards, so P(Red) = 26/52. There are 12 face cards (J, Q, and K of each suit), so P(face card) = 12/52. There are 6 red face cards, so P(red and face card) = 6/52. Therefore P(red or face card) = 26/52 + 12/52 − 6/52 = 32/52 = 8/13. ∎

EXAMPLE 8

Find the probability of rolling a number divisible by 2 or 3 in one roll of a single fair die.

Solution   The numbers divisible by 2 are 2, 4, and 6, so the probability of rolling a number divisible by 2 is 3/6. The numbers divisible by 3 are 3 and 6, so the probability of rolling a number divisible by 3 is 2/6. The only number divisible by both 2 and 3 is 6, and the probability of rolling a 6 is 1/6. The probability of rolling a number divisible by 2 or 3 is 3/6 + 2/6 − 1/6 or 4/6 = 2/3. ∎

EXAMPLE 9   Find the probability of drawing a diamond or a club in a single draw from a shuffled deck of 52 playing cards.

Solution   There are 13 diamonds, so P(diamonds) = 13/52. There are 13 clubs, so P(clubs) = 13/52. It is impossible to draw both a diamond and a club in a single draw, so P(diamonds and clubs) = 0. The probability of drawing a club or a diamond is $13/52 + 13/52 - 0 = 26/52 = 1/2$ or 0.5. ■

If it is impossible for two events A and B to both occur, we say that A and B are mutually exclusive, and $P(A$ and $B) = 0$.

10.1 EXERCISES

Specify the sample space for each experiment. See Examples 1 and 2.

1. Rolling a fair die.

2. Drawing a marble from a box of six red and five white marbles if you are blindfolded.

3. Drawing a marble from a box of three red, four white, and six blue marbles if you are blindfolded.

4. Drawing a card from a partial deck containing only the diamonds.

5. Drawing a card from a partial deck containing only the face cards (do not count aces).

6. Having a family of two children (specify the sex of each child).

7. Having a family of three children (specify the sex of each child).

8. Drawing a coin from a box containing a penny, a nickel, and a dime.

9. Drawing two coins from a box containing a penny, a nickel, and a dime. (Hint: There are six possible outcomes.)

10. Drawing two coins from a box containing a penny, a nickel, a dime, and a quarter. (Hint: There are twelve possible outcomes.)

Find the probability of each event. See Examples 3, 4, and 5.

11. Rolling a 4 in a single roll of a fair die.

12. Drawing a red marble from a box containing five red and four blue marbles if you are blindfolded.

13. Drawing a red marble from a box containing four red, three white, and two blue marbles if you are blindfolded.

14. Drawing a face card from a shuffled deck of 52 cards.

15. Drawing a diamond from a shuffled deck of 52 cards.

16. Drawing the ace of spades from a shuffled deck of 52 cards.

17. Drawing a dime from a box containing a penny, a nickel, and a dime.

18. Drawing a penny and a nickel from a box containing a penny, a nickel, and a dime. (It doesn't matter which is first.)

19. Drawing a nickel and a dime from a box containing a penny, a nickel, and a dime. (It doesn't matter which is first.)

20. Drawing a dime and a quarter from a box containing a penny, a nickel, a dime, and a quarter. (It doesn't matter which is first.)

Find the probability of each event. See Example 6.

21. Drawing a black marble from a box containing six red and five white marbles if you are blindfolded.

22. Drawing a red card from a partial deck containing only the diamonds.

23. Drawing at least 5¢ from a box containing a nickel, a dime, and a quarter.

24. Drawing at least 25¢ from a box containing a penny, a nickel, and a dime.

25. Not drawing the quarter from a box containing a penny, a nickel, a dime, and a quarter.

Find the probability of each event. See Examples 7, 8, and 9.

26. Rolling an even number or a number greater than 4 on one roll of a single fair die.

27. Rolling an odd number or a prime number on one roll of a single fair die.

28. Rolling an even number or a prime number on one roll of a single fair die.

29. Drawing a diamond or a queen from a shuffled deck of 52 cards.

30. Drawing a black card or a king from a shuffled deck of 52 cards.

31. Drawing a red king or a black queen from a shuffled deck of 52 cards.

32. Drawing a 7 or a face card from a shuffled deck of 52 cards.

✳33. Your friend has three cards: one red on both sides, one blue on both sides, and one that is red on one side and blue on the other. He says that if you draw one of the cards (without looking) from a box and lay it on the table, he will guess what color the other side is. If he is right, you owe him a candy bar; if he is wrong, he owes you a candy bar. He claims that since the other side is either red or blue, the probability that he is correct is only 0.50. Is his analysis correct?

✳34. The *odds* in favor of an event are given by the number of successful outcomes possible divided by the number of unsuccessful outcomes possible.
    (a) If the odds for an event are given as 1/1, what is the probability of the event?
    (b) if the probability of an event is 3/4, what are the odds against it?

✳35. Find the odds for guessing the correct color in problem 33 if your friend always guesses the same color that is showing.

10.2   Counting Principles

Since computing probabilities requires knowing how many possible outcomes there are for an event, it is helpful to know some rules called "counting principles." Suppose you toss a coin and roll a die. There are twelve possible outcomes for this experiment. They are $\{H1, H2, H3, H4, H5, H6, T1, T2, T3, T4, T5, T6\}$. There are two outcomes for tossing the coin, six outcomes for rolling the die, and $2 \cdot 6$ outcomes for this experiment. We generalize this result to obtain the following formula.

> **Fundamental Counting Principle**
> If one operation can be done in m ways and a second operation can be done in n ways, then they can be done together in $m \cdot n$ ways.

EXAMPLE 1

In how many ways can two letters be put together if the first must be a consonant and the second a vowel (consider y to be both).

Solution   There are 21 consonants and 6 vowels. Therefore, there are $21 \cdot 6 = 126$ possibilities. ■

EXAMPLE 2

How many outcomes are there if two dice are rolled?

Solution   The first die can be rolled in six ways, and the second can be rolled in six ways. Therefore there are $6 \cdot 6 = 36$ ways to roll two dice. ■

EXAMPLE 3

In how many ways can a code consisting of two letters followed by three digits be formed, if digits may repeat, but letters may not?

Solution   The first letter can be chosen in 26 ways, but once chosen, cannot be repeated. There are, then, only 25 choices left for the second letter. The first digit can be 0–9, so there are 10 choices. There are also 10 choices for the second digit, and 10 for the third, since repetitions are allowed. Thus, there are $26 \cdot 25 \cdot 10 \cdot 10 \cdot 10 = 650,000$ ways to form the code. ■

Probability problems often involve the possible arrangements of objects. In mathematics, the arrangements are called *permutations*. Consider the letters *A, B,* and *C*. We can arrange them in the following ways: *ABC, ACB, BAC, BCA, CAB,* or *CBA*. We say there are six permutations. We can write six as $3 \cdot 2 \cdot 1$. That is, there are three ways we could choose the first letter in the arrangement, but once we have chosen it, there are only two choices left for the second letter. Once it is chosen, there is only one choice left for the third letter. The ! symbol is called a *factorial* symbol. The expression $n!$ (where n is a positive integer) indicates the product of all positive integers less than or equal to n. Thus 3! (read *three factorial*) means $3 \cdot 2 \cdot 1$ or 6.

> **Definition**
> The number of permutations of n distinct objects is $n!$.

EXAMPLE 4

In how many ways can five runners finish a race, if there are no ties?

Solution   The number of arrangements of the five runners in order is $5! = 5 \cdot 4 \cdot 3 \cdot 2 \cdot 1 = 120$. ■

EXAMPLE 5

In how many ways can six different books be arranged on a shelf?

Solution   The number of arrangements for six books is $6! = 6 \cdot 5 \cdot 4 \cdot 3 \cdot 2 \cdot 1 = 720$. ■

Suppose you only have room for three of the six books of Example 5. Then you could choose any one of the six as the first one to go on the shelf. There would be five left from which to choose the second one. The third book would be chosen from the four remaining books. That is, there are $6 \cdot 5 \cdot 4 = 120$ possible arrangements. Since we used only three of the six books, we have found the number of permutations of six objects *taken three at a time*. Instead of multiplying all of the numbers less than six, we only multiplied three of them. We generalize to obtain a formula.

> **Definition**
> The number of permutations of n distinct objects taken r at a time (denoted $_nP_r$) is $n!/(n - r)!$.

Writing $n!/(n - r)!$ means that the last $(n - r)$ factors of the product in the numerator will divide out, so only r factors are left. Mathematicians say that $0! = 1$, so that the formula is correct for all situations.

EXAMPLE 6

How many three-character codes can be made from the letters A–E, if repetitions are not allowed?

Solution   There are five letters to be taken three at a time, so $_5P_3 = 5 \cdot 4 \cdot 3 = 60$. ■

EXAMPLE 7

If five people are on a committee, how many ways are there to choose a chairman and a secretary?

Solution   There are five people to be taken two at a time, so $_5P_2 = 5 \cdot 4 = 20$. ∎

If we list all permutations of the letters A, B, and C taken two at a time, we have AB, AC, BA, BC, CA, and CB. We treat AB and BA as different arrangements (as we do AC and CA, or BC and CB). There are only three distinct pairings in this example if order is not a consideration. The arrangements of n distinct objects taken r at a time *without regard to order* are called *combinations*.

> **Definition**
> The number of combinations of n distinct objects taken r at a time (denoted $_nC_r$) is given by $n!/((n - r)!r!)$.

The formula is easier to use in practice than it appears to be. The $(n - r)!$ will divide out the last $n - r$ factors of the product in the numerator, leaving only r factors. The $r!$ will also give r factors in the denominator. It is easy to write the numerator by starting with n and multiplying successively smaller numbers until there are r factors and to write the denominator by starting with 1 and multiplying successively larger factors until there are r factors.

EXAMPLE 8

In how many ways can a committee of three be chosen from a group of five?

Solution   Since a committee does not depend on the order in which the members are chosen, we find $_5C_3 = (5 \cdot 4 \cdot 3)/(1 \cdot 2 \cdot 3) = 5 \cdot 2 = 10$. ∎

EXAMPLE 9

In how many ways can you choose two books out of seven to lend to a friend?

Solution   The order in which the books are chosen is not important. We compute $_7C_2 = (7 \cdot 6)/(1 \cdot 2) = 21$. ∎

The counting rules are often helpful in computing probabilities.

EXAMPLE 10

If a password for access to a computer follows the pattern of letter-digit-letter-digit (as in R2D2), what is the probability of guessing it on the first try?

Solution   The fundamental counting principle tells us that there are $26 \cdot 10 \cdot 26 \cdot 10$ possible passwords, so the probability of guessing it on the first try is $1/67,600$. ∎

EXAMPLE 11   If there are 500 batteries on an assembly line, and two of them are bad, what is the probability that an inspector choosing two of them to test will pick the two bad ones?

Solution   The order in which the batteries are tested is not important. Therefore there are $_{500}C_2 = (500 \cdot 499)/(1 \cdot 2) = 124{,}750$ ways to choose two batteries out of 500. Only one of these combinations consists of the two bad ones. The probability of getting the two bad ones is 1/124,750.   ■

EXAMPLE 12   If a combination lock has a three-digit combination, and you can only remember that none of the digits are larger than 5 and that they are all different, what is the probability that you can open the lock on the first try?

Solution   Since the digits must be chosen from 0–5 and order is important, we consider that there are $_6P_3 = 6 \cdot 5 \cdot 4 = 120$ possibilities. The probability of guessing the correct one on the first try is 1/120.   ■

10.2 EXERCISES

Use the fundamental counting principle to determine the total number of ways each event can occur. See Examples 1–3.

1. Rolling three dice.

2. Tossing five coins.

3. Writing a three-digit number (1st digit $\neq$ 0).

4. Writing a three-digit number if digits cannot be repeated.

5. Writing an even three-digit number (1st digit $\neq$ 0, but digits can be repeated).

6. Writing a code of two letters followed by three digits if no character is repeated and 0 can be used in any of the three positions.

7. Writing a code of two letters followed by three digits if repetitions are allowed.

8. Writing a three-digit number less than 400.

9. Writing a three-digit number less than 400 and divisible by five.

10. Writing a code of three letters using *A–H* if repetitions are allowed.

Determine the total number of ways each event can occur. See Examples 4 and 5.

11. Forming a number from the digits {1, 2, 3, 4} with no repetitions.

12. Arranging four books on a shelf.

13. Parking three cars in three parking places.

14. Seating five people in a row.

15. Arranging the letters *A–F*.

16. Arranging a buffet table with seven items in a row.

17. Arranging six bottles in a row at a bar.

18. Assigning five classes to five classrooms.

19. Arranging four cans of fruit on a shelf.

20. Arranging the program with six speakers at a convention.

Determine the total number of ways each event can occur. See Examples 6 and 7.

21. Forming a three-digit number from {1, 2, 3, 4} without repetition.

22. Arranging two books out of four on a shelf.

23. Parking a car in one of three parking places.

24. Seating four people out of five.

25. Arranging the letters *A–F* three at a time.

26. Arranging four items out of seven in a row on a buffet table.

27. Arranging three bottles out of six in a row at a bar.

28. Arranging your schedule to take two classes out of five.

29. Arranging two cans out of four on a shelf.

30. Arranging a program with four speakers out of six for a convention.

Determine the total number of ways that each event can occur. See Examples 8 and 9.

31. Choosing two letters from the set {a, b, c, d, e, f}.

32. Choosing three coins from a penny, a nickel, a dime, and a quarter.

33. Arranging a tennis tournament with eight players if each one plays all the others.

34. Arranging for ten teams to play a season where each team plays the others once.

35. Choosing three people out of eight for a committee.

36. Choosing three people out of seven for a committee.

37. Drawing two socks out of a drawer with ten socks.

38. Choosing three candy bars from a machine with nine.

39. Buying two lamps out of four that you like.

40. Choosing three toppings out of eight for a pizza.

➡41. Use pseudocode or a flowchart to show an algorithm for finding $n!$.

✳42. Find the number of arrangements for seating four people around a round table.

✳43. The number of arrangements of n objects of which m are indistinguishable is given by $n!/m!$. For example, the number of different arrangements of the letters in the word *Mississippi* is given by $11!/(4!\ 4!\ 2!)$. Explain why this rule works.

✳44. How many divisors does 2520 have?

✳45. How many zeroes occur at the end of 80!?

✳46. What does the formula for $_nC_3$ suggest about the product of three consecutive positive integers?

10.3   Independent Events and Conditional Probability

If you toss a coin, the probability of getting heads is 1/2. If you toss that coin a second time, the probability of getting heads the second time is also 1/2. The coin has no mechanism for keeping track of its tosses; each toss represents a fresh start with equally likely outcomes of heads and tails. We say that the two events are independent.

> **Definition**
> Two events are independent if the outcome of the first has no influence on the outcome of the second.

Independent events may occur together or in sequence. The probability of getting two heads is the same whether we toss two coins at the same time or toss one coin twice in a row. The sample space for either can be represented as {*HH, HT, TH, TT*}, and the probability of getting two heads is 1/4. The probability that two independent events will *both* happen depends on their individual probabilities.

> **Probability Rule 2**
> The probability that two independent events A and B will both happen (either simultaneously or successively) is given by the formula:
> $P(A \text{ and } B) = P(A) \cdot P(B)$.

EXAMPLE 1   Find the probability of rolling doubles (i.e., two faces alike) on a single roll of a pair of fair dice.

Solution   It doesn't matter what the first die shows, so the probability of a successful outcome is 1. There is, however, only one way in six for the second die to match it. The probability of success is 1/6. The probability that both events happen, then, is $1 \cdot 1/6 = 1/6$. ■

EXAMPLE 2   What is the probability of getting five heads in succession when tossing a coin?

Solution   Each toss is independent of the others, and $P(\text{heads}) = 1/2$ for each. Therefore $P(5 \text{ heads}) = (1/2)^5 = 1/32$. ■

EXAMPLE 3   What is the probability of drawing two aces in succession from a shuffled deck of 52 playing cards if the first card is returned to the deck before the second draw?

Solution P(ace) $= 4/52 = 1/13$ for each draw.
P(two aces) $= (1/13)(1/13) = 1/169$. ■

If the first card were not returned to the deck, the second draw would be affected by the first. If the first card were an ace, there would be only three aces left among the remaining 51 cards, so P(ace) $= 3/51$ on the second draw. On the other hand, if the first card were not an ace, then P(ace) $= 4/51$ for the second draw. This example illustrates the concept of conditional probability. We calculate the probability of the two events based on the condition that the first outcome is successful. The probability is denoted by $P(B|A)$ and is read, "the probability of B, given A."

> **Probability Rule 3**
> The probability that two dependent events A and B will both happen (either simultaneously or successively) is given by the formula:
> $P(A \text{ and } B) = P(A) \cdot P(B|A)$.

EXAMPLE 4

What is the probability of drawing two aces from a shuffled deck of 52 cards if the first card is not replaced before the second draw?

Solution   On the first draw, P(ace) $= 4/52 = 1/13$. On the second draw, assuming that an ace was drawn the first time, P(ace) $= 3/51 = 1/17$, so P(two aces) $= (1/13)(1/17) = 1/221$. ■

EXAMPLE 5

If you wash four black socks, six blue ones, and two brown ones:

(a)   what is the probability that the first two you take out of the washing machine will both be brown?

(b)   what is the probability that the first two you take out of the washing machine will both be blue?

Solution

(a)   Since there are two brown socks among twelve, P(brown) $= 2/12 = 1/6$ for the first sock. Assuming it is brown, then there is only one brown sock left among eleven, so P(brown) $= 1/11$ for the second sock. The probability that both socks are brown is $(1/6)(1/11) = 1/66$.

(b)   Since there are six blue socks among twelve, P(blue) $= 6/12 = 1/2$ for the first sock. Assuming it is blue, then there are five blue socks left among eleven, so P(blue) $= 5/11$ for the second sock. The probability that both socks are blue is $(1/2)(5/11) = 5/22$. ■

If you have tossed a coin ten times and have gotten nine heads, the probability that you get heads on the next toss is 1/2. Remember that each toss, including the tenth, is an independent event having probability of 1/2. Many people mistakenly think that there is a higher probability that tails would occur, since in an experiment of tossing a coin ten times, the probability of getting ten heads in a row is small. The probability that the first nine tosses are heads is 1/512. We would expect the nine heads to be followed by heads about half of the time and by tails about half of the time. The law of averages (also called the law of large numbers and Bernoulli's law) says that if we repeated this experiment (tossing a coin ten times) very many times, we would get ten heads on approximately 1/1024 of the trials. That is, the law of averages allows us to predict the average outcome if an experiment is performed very many times. It does not allow us to predict the outcome of a single experiment.

Insurance companies are able to write policies based on this principle. If you are a 35-year-old male who wants to take out a $100,000 life insurance policy, the company cannot predict whether *you* will die in the next ten years or not. But they can predict the percent of 35-year-old males that will die in the next ten years. They insure many 35-year-old males, and they base their charges on statistics.

10.3 EXERCISES

Find the probability of each event. See Examples 1, 2, and 3.

1. Rolling two odd numbers on one roll of a pair of fair dice.

2. Rolling a sum of 12 on one roll of a pair of fair dice.

3. Rolling a 2 on the first die and 3 on the second when rolling a pair of fair dice.

4. Tossing heads followed by two tails when tossing a fair coin three times.

5. Tossing two tails followed by two heads when tossing a fair coin four times.

6. Tossing alternating heads and tails (starting with heads) when tossing a fair coin four times.

7. Having three boys in a row if the probability that a baby will be a boy is 0.51.

8. Having three girls in a row if the probability that a baby will be a girl is 0.49.

9. Having a boy followed by a girl if the probability of having a boy is 0.51 and the probability of having a girl is 0.49.

10. Drawing two red cards in succession from a shuffled deck of 52 if the first card is replaced.

11. Drawing a ten from a shuffled deck of 52 followed by something larger (*J*, *Q*, *K*, or *A*) if the first card is replaced.

12. Drawing a club followed by a king from a shuffled deck of 52 if the first card is replaced.

13. Pulling two blue socks out of a drawer (without looking) containing two blue, four black, and six brown socks if the first sock is replaced.

14. Pulling two brown socks out of a drawer (without looking) containing two blue, four black, and six brown socks if the first sock is replaced.

15. Pulling a blue followed by a brown sock out of a drawer (without looking) containing two blue, four black, and six brown socks if the first sock is replaced.

Find the probability of each event. See Examples 4 and 5.

16. Drawing two red cards in succession from a shuffled deck of 52 if the first card is not replaced.

17. Drawing a ten followed by something larger (*J*, *Q*, *K*, or *A*) from a shuffled deck of 52 if the first card is not replaced.

18. Drawing a club followed by a spade from a shuffled deck of 52 if the first card is not replaced.

19. Pulling two blue socks out of a drawer (without looking) containing two blue, four black, and six brown socks if the first sock is not replaced.

20. Pulling two brown socks out of a drawer (without looking) containing two blue, four black, and six brown socks if the first sock is not replaced.

21. Pulling a blue sock followed by a brown sock from a drawer (without looking) containing two blue, four black, and six brown socks if the first sock is not replaced.

22. Your winning 1st place and your friend winning 2nd place in a raffle if you each buy one ticket out of 5000.

23. Your winning both 1st and 2nd places in a raffle if you buy two tickets out of 5000.

24. Your winning 1st place and your friend winning 2nd place in a raffle if you each buy five tickets out of 5000.

25. Your winning 2nd place and your friend winning 3rd place in a raffle if you each buy one ticket out of 5000. (Hint: Neither of you may win first place.)

26. Pulling out two bad batteries on the 1st and 2nd draw from five batteries, if two of them are bad.

27. Pulling out two bad batteries on the 2nd and the 3rd draw from five batteries, if two of them are bad. (Hint: The first must be good.)

28. Pulling out two bad batteries on the 3rd and 4th draw from five batteries, if two of them are bad. (Hint: The 1st and 2nd must be good.)

29. Pulling out two bad batteries on the 1st and 4th draw from five batteries, if two of them are bad.

30. Pulling out two bad batteries on the 1st and 5th draw from five batteries, if two of them are bad.

Find the probability of each event.

✳31. (a) Rolling a sum of 3 or 11 on one roll of a pair of fair dice.

  (b) Rolling a sum less than 11 on one roll of a pair of fair dice.

✳32. Drawing a king followed by a club from a shuffled deck of 52 if the first card is not replaced.

✳**33.** Drawing at least two kings out of three draws from a shuffled deck of 52 cards (without replacement).

✳**34.** Tossing more heads than tails on four tosses of a fair coin.

✳**35.** In a group of r people, the probability that at least two of them have the same birthday is calculated by: $1 - P$ (all the birthdays are different).

    (a) Calculate the probability that at least two people in a group of three have the same birthday.

    (b) For large groups, the probability, p, can be approximated by the formula $p = 1 - e^{-r(r-1)/(2(365))}$, in which computation is easier. What is the probability that in a group of thirty people, at least two have the same birthday?

✳**36.**† In computer programming a "word" is the number of bits in the largest data value that can be transferred in one operation. Depending on the computer, a word may be from 8 to 60 bits. Occasionally a bit is transferred incorrectly. To detect such errors, one or more extra bits are included as "parity" bits in the coded representation of a character. For even parity, the extra bits are chosen so that the sum of the bits in any character is even. A single error is then easily detectable. If two or more errors occur in the same word, however, the error is not so easily detectable. If r errors are made, the probability that at least two of them occur in the same word is computed in the same way as the birthday probability (see problem 35). Instead of 365 days, we use the total number of words possible in memory.

    (a) How many 32-bit words are possible in a one megabyte (2^{20}) memory?

    (b) Modify the formula of problem 35 for the conditions described in this problem.

    (c) If there are 500 errors, what is the probability that at least two of them occur in the same word?

    (d) If an error occurs on the average of one every 35.7 days, and an average of 642 errors are required to produce two errors in the same word, how long would you expect a computer to operate before two errors occur in the same word?

10.4 Measures of Central Tendency

If your grades in math are 85, 93, 87, 89, and 86 for the term, your teacher would probably say your grade for the term is 88, which is the *average* of these scores. Some of the grades are higher than 88 and some are lower than 88; 88 is a typical grade that represents the entire list, since they cluster around 88. The average is a *measure of central tendency*. We will consider three different

†This problem is based on information in the article, "The Reliability of Computer Memories," by Robert J. McEliece, *Scientific American*, Vol. 252 No. 1 (January 1985), 88–95.

averages or measures of central tendency. The one used in the example above is the arithmetic mean and is the one most often used in statistical studies.

The set of all objects having a certain characteristic is called a *population*. If the characteristic under study can be measured, the measurement for each object in the population is called a *variate*. The set of all variates for members of a population is called a *statistical population*.

> **Definition**
> The *arithmetic mean*, denoted μ, for a statistical population is the sum of the variates divided by the number of variates.

EXAMPLE 1

Find the arithmetic mean of the statistical population: 34, 23, 42, 27, 41, and 13.

Solution　The sum of the numbers is $34 + 23 + 42 + 27 + 41 + 13 = 180$. Since there are six variates in the population, $\mu = 180/6 = 30$. ∎

EXAMPLE 2

Find the arithmetic mean of the statistical population: 101, 203, 491, 321, and 114.

Solution　The sum of the numbers is 1230. Since there are five variates in the population, $\mu = 1230/5 = 246$. ∎

Suppose that five people have annual incomes of $20,000, $23,000, $19,000, $18,000, and $500,000 respectively. The arithmetic mean is $116,000 for this population. The arithmetic mean, however, does not seem very representative of the numbers in this population. Since the arithmetic mean gives equal weight to all of the variates, one or two extreme values (such as $500,000) may result in misleading conclusions about the distribution. In this example, a different measure of central tendency called the *median* gives a more accurate picture. Half of the numbers in a statistical population will be greater than or equal to the median and half will be less than or equal to the median.

> **Definition**
> For a statistical population containing an odd number of variates, the *median* is the number in the middle of the list when the variates are arranged in ascending (or descending) order. For a statistical population containing an even number of variates, the median is the arithmetic mean of the two middle numbers.

The median income for the five people in the example above would be the middle number in the distribution when the variates are ordered: {$18,000, $19,000, $20,000, $23,000, $500,000}. The median is $20,000.

EXAMPLE 3

Find the median for the statistical population: 53, 22, 47, 61, 65, 35, 47.

Solution   We first list the variates in order: 22, 35, 47, 47, 53, 61, 65. The number in the middle is 47, so 47 is the median. ■

EXAMPLE 4

Find the median for the statistical population: 68, 51, 73, 62, 70, 81.

Solution   We first list the variates in order: 51, 62, 68, 70, 73, 81. Since there are six numbers, we find the arithmetic mean of the two variates in the middle. The median is $(68 + 70)/2 = 138/2 = 69$. ■

If the manager of a produce market decides to package apples, he or she might watch sales of apples for a few days to determine how many a typical customer buys at once. Suppose sales of apples for one day are as follows: 1, 1, 2, 2, 2, 2, 2, 3, 4, 4, 4, 4, 5, 7, 9. The arithmetic mean is 52/15 = 3 7/15. The manager certainly would not package apples with 3 7/15 apples in a package. The median is 3; yet only one person bought three apples. The manager might very well decide to package apples in groups of two, since more people bought two than any other number. This measure of central tendency is called the *mode*.

Definition
The *mode* of a statistical population is the variate that appears most frequently in the population.

A statistical population may have more than one mode. If each number in the population occurs only once, we say there is no mode.

EXAMPLE 5

Find the mode of the statistical population: 53, 22, 47, 61, 65, 35, 47.

Solution   We list the variates in order: 22, 35, 47, 47, 53, 61, 65. Since 47 occurs twice, and each other variate occurs only once, 47 is the mode. ■

EXAMPLE 6

Find the mode of the statistical population: 3, 4, 4, 5, 6, 6, 7.

Solution   There are two modes: 4 and 6. ■

We have seen three different measures of central tendency. Sometimes one type of measurement is more appropriate than the others. The mean is the best choice when every variate is to be given equal weight. The median is the best choice when there are one or two extreme values. The mode is the best choice for something that can only be measured in certain increments, such as clothing sizes.

EXAMPLE 7

Determine which type of average is most appropriate for each statistical population:

(a) $76,000, $45,000, $72,000, $74,000, and $73,000, where the variates are the prices of homes in one neighborhood.

(b) 4, 5, 5 1/2, 6, 6, 6 1/2, 6 1/2, 6 1/2, 7, 7, 7 1/2, 8, where the variates are sizes of shoes sold by a salesman.

Solution

(a) The median is the most appropriate average, because it is not affected by the one relatively small value.

(b) The mode is the most appropriate average, because shoe sizes occur only in whole and half sizes. ■

EXAMPLE 8

Determine which type of average is most appropriate for each statistical population:

(a) 92, 93, 94, 95, 100, where the variates are scores of students on a test.

(b) $22,000, $20,000, $28,000, $19,000, and $4,500, where the variates are the annual incomes of five people.

Solution

(a) The arithmetic mean is the most appropriate average, because it is affected by every score.

(b) The median is the most appropriate average, because it is not affected by the one relatively small value. ■

Averages, by themselves, often give an incomplete picture from which erroneous conclusions are drawn. For example, in 1984, it was widely reported that the average price of a new home exceeded $100,000. Many people concluded that it would be impossible for them to buy an average home. At the time, however, interest rates were high. Builders were constructing larger, more expensive homes that would appeal to affluent buyers for whom interest rate was not a primary consideration. The median price for resale homes was $73,300 for the same period of time.

Another example is that in 1955, average pay for women was 64 percent of the average pay for men. The Equal Pay Act of 1963 made it illegal for employers to pay women less for equal work, but in 1985, the average pay for women was 69 percent of the average pay for men. Some people concluded that discrimination against women was not much better than in 1955. The average, however, fails to consider that women tend to be concentrated in clerical and service areas, in which wages are low. Also, women in many of the higher-paid jobs traditionally held by men are young women who do not have as much experience as the men who have held those jobs for many years.

When you read reports that cite ''average'' figures, you should ask yourself, which average was used—mean, median, or mode? Also consider whether there might be factors that are not mentioned in the report that could have affected the average.

10.4 EXERCISES

Find the arithmetic mean of each statistical population. See Examples 1 and 2.

1. {7, 15, 12, 19, 13, 8, 10}   **2.** {32, 19, 42, 30, 19, 40, 28}

3. {41, 32, 50, 40, 55, 43, 32, 18, 58}   **4.** {14, 5, 17, 12, 10, 5, 17, 8}

5. {23, 10, 8, 10, 2, 25}   **6.** {15, 2, 30, 11, 2, 30, 10, 16}

7. {1, 1, 2, 3, 5, 8, 13, 21}   **8.** {2, 3, 5, 7, 11, 13, 17, 19}

9. {2, 4, 8, 16, 32}   **10.** {11, 3, 22, 7, 1, 16, 3}

Find the median of each statistical population. See Examples 3 and 4.

11. {7, 15, 12, 19, 13, 8, 10}   **12.** {32, 19, 42, 30, 19, 40, 28}

13. {41, 32, 50, 40, 55, 43, 32, 18, 58}   **14.** {14, 5, 17, 12, 10, 5, 17, 8}

15. {23, 10, 8, 10, 2, 25}   **16.** {15, 2, 30, 11, 2, 30, 10, 16}

17. {1, 1, 2, 3, 5, 8, 13, 21}   **18.** {2, 3, 5, 7, 11, 13, 17, 19}

19. {2, 4, 8, 16, 32}   **20.** {11, 3, 22, 7, 1, 16, 3}

Find the mode of each statistical population. See Examples 5 and 6.

21. {7, 15, 12, 19, 13, 8, 10}   **22.** {32, 19, 42, 30, 19, 40, 28}

23. {41, 32, 50, 40, 55, 43, 32, 18, 58}   **24.** {14, 5, 17, 12, 10, 5, 17, 8}

25. {23, 10, 8, 10, 2, 25}   **26.** {15, 2, 30, 11, 2, 30, 10, 16}

27. {1, 1, 2, 3, 5, 8, 13, 21}   **28.** {2, 3, 5, 7, 11, 13, 17, 19}

29. {2, 4, 8, 16, 32}   **30.** {11, 3, 22, 7, 1, 16, 3}

Determine the most appropriate average to use in each case. See Examples 7 and 8.

31. {$3, $4, $3.50, $4.25, $20} where each amount is the price of a pound of chocolate candy.

32. {103, 96, 104, 136, 106, 107, 108, 101} where each number is an IQ.

33. {2.3, 1.9, 2.1, 1.8, 1.8, 2.0} where each number is the run time in seconds for a computer program written as a homework assignment by students.

34. {6 5/8, 6 7/8, 7 1/4, 7 1/2, 6 7/8} where each number is a hat size.

35. {3, 7, 19, 22, 25, 45, 67, 72, 75} where each number is the age of a patient admitted to the hospital.

36. {1, 2, 2, 3, 4, 4, 107} where each number is the length of a hospital stay in days.

37. {110, 120, 120, 130, 130, 135, 140, 160} where each number is the systolic blood pressure of a patient in the hospital.

38. {1, 2, 2, 2, 2, 3, 3, 4, 4} where each number is the number of students sharing an apartment.

39. {2, 2, 2, 3, 3, 4, 4, 4, 4, 4, 4} where each number is the number of doors on automobiles sold by a certain dealer.

40. {12,000, 15,000, 22,000, 24,000, 110,000} where each number is the number of miles on used cars sold by a certain dealer.

✳**41.** Four gymnasts have scored an average of 9.6 on the vault. How high a score will the fifth gymnast need to raise the team average to 9.7?

✳**42.** What happens to the mean, median, and mode if each variate in the statistical population is

(**a**) increased by 4?

(**b**) multiplied by 4?

✳**43.** After averaging 27 test scores, a teacher decided to check the average, but inadvertently included the average with the 27 scores, and found the average of the 28 scores. What effect would this mistake have on the result?

10.5   Measures of Dispersion

A measure of central tendency is useful for summarizing the data in a statistical population, but it does not give a complete picture. Consider the two populations {39, 40, 41, 42, 43} and {1, 21, 41, 61, 81}. Each has a mean of 41, but the values in the first population are clustered more closely around that mean. The values in the second population are widely scattered on both sides of the mean. There are ways to measure this scatter or *dispersion*.

Definition
The *range* is the difference between the largest and smallest numbers in a statistical population.

EXAMPLE 1

Find the mean and the range of each statistical population.

(**a**)   {39, 40, 41, 42, 43}

(**b**)   {1, 21, 41, 61, 81}

Solution

(**a**) $\mu = 41$, and the range is $43 - 39 = 4$.

(**b**) $\mu = 41$, and the range is $81 - 1 = 80$. ∎

EXAMPLE 2

Find the mean and the range of each statistical population.

(a) {47, 48, 49, 50, 51, 52, 53}

(b) {1, 48, 49, 50, 51, 52, 99}

Solution

(a) $\mu = 50$, and the range is $53 - 47 = 6$.

(b) $\mu = 50$, and the range is $99 - 1 = 98$. ■

In Example 2, the two distributions had the same mean, and most of the values in both populations were clustered around that mean. The two extreme values of 1 and 99 in the second population, however, produce a large range. A mean of 50 and a range of 98 could also describe the distribution {1, 2, 3, 50, 97, 98, 99}. In this distribution, the values tend to be clustered near the largest and smallest values in the population.

A measure of dispersion that takes into account how much each score deviates from the mean would be more useful than the range, which is affected only by the largest and smallest values in the distribution. The *variance* is one such measure of dispersion.

Definition
The *deviation* is the difference between a single number in a statistical population and the mean of that population. The *variance* is the arithmetic mean of the squared deviations for all numbers in a distribution.

EXAMPLE 3

Find the variance of each statistical population.

(a) {39, 40, 41, 42, 43}

(b) {1, 21, 41, 61, 81}

Solution

(a) The mean is 41. The deviations are {39 − 41, 40 − 41, 41 − 41, 42 − 41, 43 − 41} or {−2, −1, 0, 1, and 2}. The squares of the deviations are {4, 1, 0, 1, 4}. The variance is $(4 + 1 + 0 + 1 + 4)/5 = 10/5 = 2$.

(b) The mean is 41. The deviations are {1 − 41, 21 − 41, 41 − 41, 61 − 41, 81 − 41} or {−40, −20, 0, 20, 40}. The squares of the deviations are {1600, 400, 0, 400, 1600}. The variance is $(1600 + 400 + 0 + 400 + 1600)/5 = 4000/5 = 800$. ■

EXAMPLE 4

Find the variance of each distribution.

(a) {47, 48, 49, 50, 51, 52, 53}
(b) {1, 48, 49, 50, 51, 52, 99}
(c) {1, 2, 3, 50, 97, 98, 99}

Solution

(a) The mean is 50. The deviations are $\{-3, -2, -1, 0, 1, 2, 3\}$. The squared deviations are {9, 4, 1, 0, 1, 4, 9}. The variance is $(9 + 4 + 1 + 0 + 1 + 4 + 9)/7 = 28/7 = 4$.

(b) The mean is 50. The deviations are $\{-49, -2, -1, 0, 1, 2, 49\}$. The squared deviations are {2401, 4, 1, 0, 1, 4, 2401}. The variance is $(2401 + 4 + 1 + 0 + 1 + 4 + 2401)/7 = 4812/7$, which is about 687.

(c) The mean is 50. The deviations are $\{-49, -48, -47, 0, 47, 48, 49\}$. The squared deviations are {2401, 2304, 2209, 0, 2209, 2304, 2401}. The variance is $(2401 + 2304 + 2209 + 0 + 2209 + 2304 + 2401)/7$, which is about 1975. ■

In Examples 4(b) and 4(c), the mean is 50 and the range is 98 for both populations. The variance indicates that the values in the second population are more widely scattered.

One difficulty in using the variance as a measure of dispersion is that squaring the deviations also squares any units that are being used. For example, if our population is a group of college students and the distribution is their heights, the deviations will be expressed in inches, but the variance will be expressed in square inches, making comparison with the mean (expressed in inches) difficult. The most widely used measure of dispersion, the *standard deviation*, avoids this problem.

> **Definition**
> The *standard deviation* for a statistical population, denoted σ, is the square root of the variance.

EXAMPLE 5

Find the standard deviation of each statistical population.

(a) {39, 40, 41, 42, 43}
(b) {1, 21, 41, 61, 81}

Solution

(a) The variance was calculated in Example 3. Since the variance is 2, the standard deviation is $\sqrt{2}$, or approximately 1.41.

(b) The variance was calculated in Example 3. Since the variance was 800, the standard deviation is $\sqrt{800}$, or approximately 28.28. ■

EXAMPLE 6   Find the standard deviation of the statistical population $\{1, 6, 11, 16, 21, 26, 31\}$.

Solution   The mean is $112/7 = 16$. The variance is $(225 + 100 + 25 + 0 + 25 + 100 + 225)/7 = 700/7 = 100$. Therefore $\sigma = \sqrt{100} = 10$. ■

A computer program to find the standard deviation with the definition we have used would require two passes through the data. The first time through, the sum of the variates would be stored in an accumulator so that the mean could be computed. The second time through, the mean would be subtracted from each variate, and the sum of the squares of these deviations would be stored in an accumulator so that the definition could be used.

Standard deviation is sometimes computed by using the formula $\sigma = \sqrt{\text{mean of (variates)}^2 - (\text{mean of variates})^2}$. With this formula, one accumulator is used to store the sum of the squares of the variates and another is used to store the sum of the variates themselves. After one pass through the data, the formula could be used. Round-off error, however, is more serious with this formula.

There is an interesting relationship between the range and the standard deviation. The standard deviation is always less than or equal to one-half of the range. You may wish to use this relationship to help detect arithmetic errors made in calculating σ. It will not show every error, but will show some of the worst.

10.5 EXERCISES

Find the mean and the range of each statistical population. See Examples 1 and 2.

1. $\{9, 10, 12, 13, 16\}$ **2.** $\{11, 24, 25, 30, 35\}$

3. $\{0, 1, 12, 23, 24\}$ **4.** $\{4, 23, 25, 35, 38\}$

5. $\{26, 41, 42, 44, 46, 59\}$ **6.** $\{32, 36, 40, 44, 48, 52\}$

7. $\{29, 30, 31, 55, 56, 57\}$ **8.** $\{45, 46, 51, 53, 57, 61, 72\}$

9. $\{25, 47, 50, 53, 56, 62, 92\}$ **10.** $\{11, 12, 13, 85, 87, 88, 89\}$

Find the variance for each statistical population. See Examples 3 and 4.

11. $\{9, 10, 12, 13, 16\}$ **12.** $\{11, 24, 25, 30, 35\}$

13. $\{0, 1, 12, 23, 24\}$ **14.** $\{4, 23, 25, 35, 38\}$

15. $\{26, 41, 42, 44, 46, 59\}$ **16.** $\{32, 36, 40, 44, 48, 52\}$

17. $\{29, 30, 31, 55, 56, 57\}$ **18.** $\{45, 46, 51, 53, 57, 61, 72\}$

19. $\{25, 47, 50, 53, 56, 62, 92\}$ **20.** $\{11, 12, 13, 85, 87, 88, 89\}$

Find the standard deviation for each statistical population. See Examples 5 and 6.

21. $\{4, 6, 8, 10\}$ **22.** $\{7, 10, 16, 19\}$

23. $\{1, 2, 12, 13\}$ **24.** $\{10, 13, 14, 15\}$

25. {19, 21, 22, 23, 25}   **26.** {1, 4, 22, 40, 43}

27. {3, 5, 10, 12, 17, 19}   **28.** {1, 5, 8, 12, 15, 19}

29. {2, 3, 5, 7, 11, 14}   **30.** {1, 2, 5, 7, 9, 10, 11, 11}

31. Use pseudocode or a flowchart to show the algorithm for finding the standard deviation using the formula $\sigma = \sqrt{\text{variance}}$, given a statistical population.

32. Use pseudocode or a flowchart to show the algorithm for finding the standard deviation using the formula $\sigma = \sqrt{\text{mean of (variates)}^2 - (\text{mean of variates})^2}$, given a statistical population.

33. Work Example 6 using the formula in problem 32.

34. What happens to the range and standard deviation if

   (a) each variate is increased by 4?

   (b) each variate is multiplied by 4?

10.6   Histograms and the Normal Curve

The data used in statistics is often classified as either discrete or continuous. The data is discrete if measurements can be made only in certain increments. For example, the number of television sets per household is a discrete variable, because, if a household increases the number of television sets, they must increase it by 1, or 2, or so on, not by 1/2 or 3/4. Height, on the other hand, is a continuous variable, because height does not have to be measured in intervals of 1″, but can assume any value between whole inches, such as 5′ 4 1/2″.

A graphical interpretation of the data in a statistical population is often useful. One type of graph is called a *frequency distribution*. The horizontal axis shows the possible values for the variable, and the vertical axis indicates the number of times a value occurs in the population. Figure 10.1 shows a frequency distribution for the number of heads obtained when five coins were tossed 64 times.

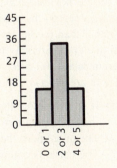

Figure 10.1

The horizontal axis is divided into intervals called *class intervals*. For continuous variables, the center for an interval may be chosen arbitrarily. Each class interval, however, must have the same width. For discrete variables, each class interval is usually centered on one of the variables. For example, an interval centered at 2 would run from 1.5 to 2.5. A rectangle is drawn with the class interval as the base. The height of the rectangle is the number on the vertical axis that indicates how many variates fall within that class interval. Such a graph is called a *histogram*.

The histogram in Figure 10.1 illustrates the frequency with which different numbers of heads occurred when five coins were tossed 64 times. The class intervals are each two units wide. If we show the same data on a histogram with class intervals of only one unit (see Figure 10.2), the histogram begins to approach a shape called the *normal curve*. See Figure 10.3.

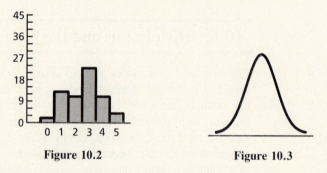

Figure 10.2   Figure 10.3

The normal curve, sometimes called a *bell curve*, is often familiar to students, as the scores on many standardized tests are distributed according to this curve. At one time it was thought that any data taken from nature, such as height or weight, would be normally distributed. Although many exceptions have been found, the normal curve still has important applications in statistical theory.

The normal curve is like a histogram for continuous variables, where each point on the horizontal axis represents a class interval. The vertical axis of a histogram does not have to show a scale. The graph can be drawn so that the area (rather than the height) of each rectangle represents the number of variates within that class interval. When drawn in this manner, the area under the entire histogram totals 100 percent.

There are several important characteristics of the normal curve.

(1) It is symmetric about the center. That is, the lefthand side is a mirror image of the righthand side.

(2) The area under the curve represents 100 percent of the population.

(3) The curve is completely specified by the mean and standard deviation of the population.

We say that the curve is completely specified by the mean and standard deviation of the population, because the curve is centered at the mean and approximately 68 percent of the area under the curve lies within 1 standard deviation on either side of the mean. Approximately 95 percent of the area under the curve lies within 2 standard deviations on either side of the mean. Approximately 99 percent of the area under the curve lies within 3 standard deviations of the mean. By placing the center of the normal curve at the mean and locating the points along the horizontal axis that are ± 1, ± 2, and ± 3 standard deviations from the mean, we can draw the normal curve.

In order to use the normal curve for statistical analysis, it is necessary to convert the units of the horizontal axis (called *raw scores*) to units called *z-scores,* or standard scores. A *z*-score is a measure of how many standard deviations a number is from the mean. Thus, the mean, μ, has a *z*-score of 0. A *z*-score of 2 represents two standard deviations to the right of the mean. Negative *z*-scores indicate measurements to the left of the mean.

To convert a raw score x to a standard score z, use the formula:

$$z = \frac{x - \mu}{\sigma}.$$

EXAMPLE 1

If a normally distributed population has a mean of 50 and a standard deviation of 5, find the *z*-score for each measurement.

(a) $x = 53$

(b) $x = 42$

Solution

(a) $z = (53 - 50)/5 = 3/5 = 0.6$

(b) $z = (42 - 50)/5 = -8/5 = -1.6$ ■

EXAMPLE 2

A cereal company packages cereal in boxes. If the weights of the boxes are normally distributed with a mean of 12 oz and a standard deviation of 0.3 oz, by how many standard deviations is a box weighing 11.5 oz underweight?

Solution $z = (11.5 - 12)/0.3 = -0.5/0.3 = -5/3 \approx -1.67$ (The symbol $\approx$ means "is approximately equal to.") ■

Table 10.1 shows what percent (written as a decimal) of the area under the normal curve lies between the mean and $|z|$ for various values of $|z|$. It is possible to use this table to answer many questions about a particular normal distribution.

Table 10.1

To find the percent of area under the normal curve between the mean and a particular z-score, locate the number in the lefthand column that has the same units and tenths digits as the z-score. Locate the entry in this row that is in the column headed by the hundredths digit of the z-score. Change this decimal entry to a percent.

	0	1	2	3	4	5	6	7	8	9
0.0	0.000	0.004	0.008	0.012	0.016	0.020	0.024	0.028	0.032	0.036
0.1	0.040	0.044	0.048	0.052	0.056	0.060	0.064	0.068	0.071	0.075
0.2	0.079	0.083	0.087	0.091	0.095	0.099	0.103	0.106	0.110	0.114
0.3	0.118	0.122	0.126	0.129	0.133	0.137	0.141	0.144	0.148	0.152
0.4	0.155	0.159	0.163	0.166	0.170	0.174	0.177	0.181	0.184	0.188
0.5	0.192	0.195	0.199	0.202	0.205	0.209	0.212	0.216	0.219	0.222
0.6	0.226	0.229	0.232	0.236	0.239	0.242	0.245	0.249	0.252	0.255
0.7	0.258	0.261	0.264	0.267	0.270	0.273	0.276	0.279	0.282	0.285
0.8	0.288	0.291	0.294	0.297	0.300	0.302	0.305	0.308	0.311	0.313
0.9	0.316	0.319	0.321	0.324	0.326	0.329	0.332	0.334	0.337	0.339
1.0	0.341	0.344	0.346	0.349	0.351	0.353	0.355	0.358	0.360	0.362
1.1	0.364	0.367	0.369	0.371	0.373	0.375	0.377	0.379	0.381	0.383
1.2	0.385	0.387	0.389	0.391	0.393	0.394	0.396	0.398	0.400	0.402
1.3	0.403	0.405	0.407	0.408	0.410	0.412	0.413	0.415	0.416	0.418
1.4	0.419	0.421	0.422	0.424	0.425	0.427	0.428	0.429	0.431	0.432
1.5	0.433	0.435	0.436	0.437	0.438	0.439	0.441	0.442	0.443	0.444
1.6	0.445	0.446	0.447	0.448	0.450	0.451	0.452	0.453	0.454	0.455
1.7	0.455	0.456	0.457	0.458	0.459	0.460	0.461	0.462	0.463	0.463
1.8	0.464	0.465	0.466	0.466	0.467	0.468	0.469	0.469	0.470	0.471
1.9	0.471	0.472	0.473	0.473	0.474	0.474	0.475	0.476	0.476	0.477
2.0	0.477	0.478	0.478	0.479	0.479	0.480	0.480	0.481	0.481	0.482
2.1	0.482	0.483	0.483	0.483	0.484	0.484	0.485	0.485	0.485	0.486
2.2	0.486	0.486	0.487	0.487	0.488	0.488	0.488	0.488	0.489	0.489
2.3	0.489	0.490	0.490	0.490	0.490	0.491	0.491	0.491	0.491	0.492
2.4	0.492	0.492	0.492	0.493	0.493	0.493	0.493	0.493	0.493	0.494
2.5	0.494	0.494	0.494	0.494	0.495	0.495	0.495	0.495	0.495	0.495
2.6	0.495	0.496	0.496	0.496	0.496	0.496	0.496	0.496	0.496	0.496
2.7	0.497	0.497	0.497	0.497	0.497	0.497	0.497	0.497	0.497	0.497
2.8	0.497	0.498	0.498	0.498	0.498	0.498	0.498	0.498	0.498	0.498
2.9	0.498	0.498	0.498	0.498	0.498	0.498	0.499	0.499	0.499	0.499
3.0	0.499	0.499	0.499	0.499	0.499	0.499	0.499	0.499	0.499	0.499
3.1	0.499	0.499	0.499	0.499	0.499	0.499	0.499	0.499	0.499	0.499
3.2	0.499	0.499	0.499	0.499	0.499	0.499	0.499	0.500	0.500	0.500

We might ask what percent of the boxes of cereal (in Example 2) weigh more than 12.7 oz? We convert 12.7 oz to a z-score: $z = (12.7 - 12)/0.3 = 0.7/0.3 = 7/3 \approx 2.33$ We are asking what percent of the area lies to the right of this z-score. See Figure 10.4.

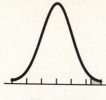

Figure 10.4

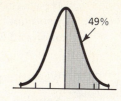

Figure 10.5

Table 10.1 tells us that 0.49 or 49 percent of the area under the curve lies between 0 and $z = 2.33$ (that is, between 12 and 12.7). See Figure 10.5. Since 50 percent of the area lies to the right of the mean, the area we want is $0.50 - 0.49 = 0.01$. Therefore, 1 percent of the boxes weigh more than 12.7 oz.

> **To find the percent of variates in a population that satisfy a condition:**
> 1. Convert raw scores to standard scores.
> 2. Sketch the normal curve and shade in the area that satisfies the condition. The mean is at the center of the normal curve.
> 3. Use Table 10.1 to determine the area under the curve between the mean and the boundary of the shaded area, and sketch the normal curve with this area shaded.
> 4. Compare the two sketches from step 2 and step 3. Remember that the area from the mean to the far right (or left) of the normal curve is 50 percent of the area under the curve.
> 5. Add or subtract as necessary to determine the area shaded in step 2.

EXAMPLE 3

What percent of the cereal boxes (see Example 2) weigh between 11.8 oz and 12.3 oz?

Solution We want the area between 11.8 and 12.3. The z-score associated with 11.8 is $(11.8 - 12)/0.3 = -0.2/0.3 \approx -.67$. The z-score associated with 12.3 is $(12.3 - 12)/0.3 = 0.3/0.3 \approx 1.00$. The diagrams that follow illustrate the area we want and the area given by Table 10.1. Table 10.1 gives the area between the mean and $|z| = 0.67$ as 0.249 or 24.9 percent. The area between the mean and $|z| = 1.00$ is .341 or 34.1 percent.

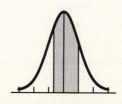

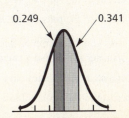

We conclude that the area between -0.67 and 1.00 is 24.9 percent + 34.1 percent or 59.0 percent. That is, 59 percent of the boxes weigh between 11.8 oz and 12.3 oz. ■

EXAMPLE 4

What percent of the cereal boxes weigh less than 11.5 oz?

Solution   The z-score associated with 11.5 is $(11.5 - 12)/0.3 = -5/3 \approx -1.67$. Table 10.1 gives the area between the mean and $|z| = 1.67$ as 0.453 or 45.3 percent. The diagrams that follow illustrate the area we want and the area given by Table 10.1.

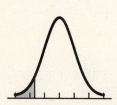 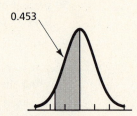

0.453

The percent of boxes that weigh less than 11.5 oz is 50 percent $-$ 45.3 percent $= 4.7$ percent. ■

10.6 EXERCISES

If a normally distributed population has a mean of 45 and a standard deviation of 4, find the *z*-score for each measurement (to the nearest hundredth). See Example 1.

1. $x = 51$ **2.** $x = 39$ **3.** $x = 55$ **4.** $x = 35$ **5.** $x = 34$

If a normally distributed population has a mean of 32 and a standard deviation of 7, find the *z*-score for each measurement (to the nearest hundredth). See Example 1.

6. $x = 42$ **7.** $x = 22$ **8.** $x = 15$ **9.** $x = 49$ **10.** $x = 52$

Answer each question. See Examples 3 and 4.

A company packages printer paper so that the lengths of the rolls are normally distributed with an average (mean) length of 100 feet with a standard deviation of 3 feet.

11. What percent of the rolls are over 105′ long?

12. What percent of the rolls are less than 98′ long?

13. What percent of the rolls are between 96′ and 104′ long?

A bakery makes chocolate chip cookies whose diameters are normally distributed with an average diameter of 12″ and a standard deviation of 0.5″.

14. What percent of the cookies have a diameter greater than 13″?

15. What percent of the cookies have a diameter less than 11.25″?

16. What percent of the cookies are between 11.75″ and 12.25″ in diameter?

The times it takes groups of four to eat at a certain restaurant are normally distributed with a mean of 1 hour 15 minutes with a standard deviation of 7 minutes.

17. What percent of the groups take longer than 1 hour 25 minutes?

18. What percent of the groups take less than 1 hour?

19. What percent of the groups take between 1 hour 12 minutes and 1 hour 18 minutes?

The typing speeds of the secretaries in a typing pool are normally distributed with a mean of 80 words per minute and a standard deviation of 10.

20. What percent of them type between 75 and 90 words per minute?

21. What percent of them type between 72 and 82 words per minute?

22. What percent of them type between 68 and 88 words per minute?

A candymaker finds that the weights of boxes of chocolates are normally distributed with an average weight of 16 oz and a standard deviation of 2 oz.

23. What percent of the boxes weigh between 11 oz and 19 oz?

24. What percent of the boxes weigh between 15 oz and 18 oz?

25. What percent of the boxes weigh between 13 oz and 17 oz?

The scores for a test are normally distributed with a mean of 100 and a standard deviation of 10.

26. What percent of the scores are between 90 and 105?

27. What percent of the scores are between 98 and 107?

28. What percent of the scores are between 83 and 113?

10.7 Estimating the Mean of a Population

Thus far, every time we have computed a mean, median, mode, or standard deviation, we have been given a list of numbers that represents a population, and our computations have been based on every number in the list. Even to compute the median or the range using at most two numbers from the set, we had to consider the entire list to determine which numbers to use.

It is not always feasible (or even possible) to study an entire population. When a company that bakes bread wants to know the mean lifetime of its loaves, they cannot possibly let every loaf get stale to see how long it takes. Instead, they choose some loaves from the population, and if those loaves are representative of the entire population, the mean can be estimated. Care must be taken to select a truly representative sample, but we will consider only the computational aspects of the problem.

Definition

A part of a population is called a *sample,* and the mean of the sample (called the sample mean) is denoted by $\bar{x}$.

Consider as an example a population whose variates are {1, 3, 5, 7, 9}. The population mean, μ, is (1 + 3 + 5 + 7 + 9)/5 = 25/5 = 5. If we were to choose a sample of size 3, there would be ten possibilities: {1, 3, 5}, {1, 3, 7}, {1, 3, 9}, {1, 5, 7}, {1, 5, 9}, {1, 7, 9}, {3, 5, 7}, {3, 5, 9}, {3, 7, 9}, and {5, 7, 9}. The sample means are, respectively, 3, 11/3, 13/3, 13/3, 5, 17/3, 5, 17/3, 19/3, and 7. From this data, you can see that if we consider the sample means themselves to be a population, the mean of the population is 1/10(3 + 11/3 + 13/3 + 13/3 + 5 + 17/3 + 5 + 17/3 + 19/3 + 7) = 1/10 (50) = 5.

It is not a coincidence that the mean of the sample means is equal to the population mean. The theory of statistical sampling is based on an important principle called the *central limit theorem.*

The Central Limit Theorem

For any population, if all samples of size *n* are considered, the mean of the sample means will equal the population mean.

For samples of size *n* (where $n \geq 30$), the population of sample means (formed by taking the mean of each possible sample of size *n*) will be approximately normally distributed with a standard deviation (denoted *s*) of $\sigma/\sqrt{n}$ (where σ is the standard deviation of the original population). The quantity $\sigma/\sqrt{n}$ is called the *standard error.*

It is important to the theory of sampling that these sample means are approximately normally distributed, even if the original population was not. The properties of the normal curve can be used to study the population of sample means for samples that contain 30 or more variates. For the remainder of the chapter, we will assume samples of this size.

For example, we know that approximately 68 percent of the sample means fall within 1 standard deviation on either side of the mean of the sample means. But since the mean of the sample means is equal to the population mean, we can say that about 68 percent of the sample means will fall within one standard deviation on either side of the population mean. Therefore, if we pick one sample and calculate its mean, the probability is about 0.68 that it is within ± 1 standard deviation = $(\sigma/\sqrt{n})$ of the population mean.

The interval within which the population mean is expected to fall, based on the sample mean, is called a *confidence interval.* With a 0.95 confidence interval, the probability is 0.95 that μ falls within the given interval. With a 0.99 confidence interval, the probability is 0.99 that μ falls within the given interval. The 0.95 and 0.99 confidence interval estimates are commonly used.

EXAMPLE 1

On a placement test, a sample of 36 students from a college showed a mean score of $\bar{x} = 100$ with a standard deviation of $s = 12$. Establish a 0.95 confidence interval estimate for the mean score for all students at the college.

Solution   We must determine an interval within which 95 percent of all sample means would fall. We do not know either μ or σ, the mean and standard deviation of the population of students at the college. If $\mu = \bar{x}$, then for the population of sample means, we could say that because one-half of 95 percent is 47.5 percent, a z-score of 1.96 corresponds to 0.475. Thus 95 percent of the area under the normal curve falls between μ and $\pm 1.96 \, \sigma/\sqrt{n}$ (remember that $\sigma/\sqrt{n}$ is the standard deviation for the population of sample means). See the following figure.

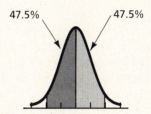

If $\mu \neq \bar{x}$, then we can still claim that 95 percent of the time, μ will fall within $\pm 1.96 \, \sigma/\sqrt{n}$ of $\bar{x}$. That is, if 95 percent of the $\bar{x}$'s are within ± 1.96 $\sigma/\sqrt{n}$ of μ, then μ is within $\pm 1.96 \, \sigma/\sqrt{n}$ of $\bar{x}$ for 95 percent of the $\bar{x}$'s. Thus there is a 0.95 probability that μ falls within $\pm 1.96 \, \sigma/\sqrt{n}$ of the mean for this sample. Using $s = 12$ as an estimate for σ, we have $\sigma/\sqrt{n} = 12/6 = 2$, which can be used as the standard error. The 0.95 confidence interval estimate for μ, then, is $100 \pm 1.96(2) = 100 \pm 3.92 = 96.08$ to 103.92. ∎

The procedure that we used in working out this problem is summarized in the following table.

> **To determine a p percent confidence interval estimate for the population mean μ, based on a sample mean, $\bar{x}$:**
> 1. Find the z-score that corresponds to $(1/2)p$ percent of the area under the normal curve.
> 2. Estimate the standard error (i.e., the standard deviation for the population of sample means) as $s/\sqrt{n}$, where s is the standard deviation of the sample.†
> 3. The confidence interval estimate for μ is $\bar{x} \pm z(s/\sqrt{n})$.

†When the standard deviation of a sample is calculated, the sum of the squared deviations is divided by $n - 1$ (rather than n) before the square root is taken.

EXAMPLE 2

Establish a 0.99 confidence interval estimate for the average annual salary of employees of a large corporation if a sample of 100 workers shows an annual salary of $15,000 with a standard deviation of $500.

Solution We calculate a z-score corresponding to 99 percent of the area under the normal curve as shown in the following figures.

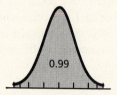
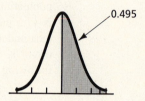

Corresponding to an area of 1/2 of 99 percent = $0.5 \times 0.99 = 0.495$, we have $z = 2.58$. The standard error can be estimated as $500/\sqrt{100} = 500/10 = 50$. Thus, 99 percent of the sample means fall within $15,000 \pm 2.58(50) = 15,000 \pm 129$. The confidence interval estimate for μ is $14,871 to $15,129. ■

When statistics are cited in newspaper and magazine articles, the confidence level is not always stated. Sometimes, however, enough information is given to calculate it.

EXAMPLE 3

A hospital claims that based on a sample of 49 babies, the mean weight of babies born there is between 7 lbs 5 oz and 7 lbs 9 oz, with a standard deviation of 7 oz. Determine the confidence level for this claim.

Solution The value for $\bar{x}$ is in the center of the interval, so $\bar{x} = 7$ lbs 7 oz and $\mu = 7$ lbs 7 oz ± 2 oz. From the formula for a confidence interval estimate $\mu = \bar{x} \pm z(s/\sqrt{n})$, we know that $z(s/\sqrt{n}) = 2$ oz or $z(7/\sqrt{49}) = 2$, or $z = 2$. Since a z-score of 2 corresponds to an area of 0.477 or 47.7 percent of the area between the mean μ and $+2$ standard deviations, 95.4 percent of the area falls between ± 2 standard deviations, and the confidence level must be 0.954. ■

> **To determine the confidence level given the confidence interval estimate of $\mu = \bar{x} \pm y$, the sample standard deviation s, and the sample size n:**
> 1. Determine $\bar{x}$ by finding the value at the center of the confidence interval given.
> 2. Obtain the z-score by solving the equation $y = z(s/\sqrt{n})$ for z.
> 3. Using Table 10.1, determine the percent of area under the normal curve associated with the z-score obtained in step 2.
> 4. Double the area and change to a decimal to get the confidence level.

EXAMPLE 4   A hospital claims that based on a sample of 81 babies, the mean weight of babies born there is between 7 lbs 5 oz and 7 lbs 9 oz, with a standard deviation of 7 oz. Determine the confidence level for this claim.

Solution   The value for $\bar{x}$ is in the center of the interval, so $\bar{x} = 7$ lbs 7 oz and $\mu = 7$ lbs 7 oz ± 2 oz. From the formula for a confidence interval estimate $\mu = \bar{x} \pm z(s/\sqrt{n})$, we know that $z(s/\sqrt{n}) = 2$ oz or $z(7/\sqrt{81}) = 2$, or $z = 2.57$. Since a z-score of 2.57 corresponds to an area of 0.495 or 49.5 percent of the area between the mean μ and $+2$ standard deviations, 99 percent of the area falls between ± 2 standard deviations, and the confidence level must be 0.99. ■

10.7 EXERCISES

For each sample below, establish a 0.95 confidence interval estimate for μ and a 0.99 confidence interval estimate for μ. See Examples 1 and 2.

1. A group of 100 15-year-old girls averages 10 minutes 8 seconds to run a mile, and the standard deviation of the sample is 16 seconds.

2. A group of 64 15-year-old girls averages 10 minutes 8 seconds to run a mile, and the standard deviation of the sample is 16 seconds.

3. A group of 81 17-year-old girls averages 10 minutes 37 seconds to run a mile, and the standard deviation of the sample is 27 seconds.

4. A group of 49 male college graduates, age 23, have an average expected lifetime earnings of $1,200,000 with a standard deviation of $70,000.

5. A group of 49 male high school graduates, age 23, have an average expected lifetime earnings of $860,000 with a standard deviation of $61,600.

6. A group of 49 female high school drop-outs, age 23, have an average expected lifetime earnings of $360,000 with a standard deviation of $77,000.

7. A group of 100 recently graduated PhD's earned an average of $23,700 with a standard deviation of $4000.

8. A group of 81 recently graduated BA's in computer science earned an average of $26,000 with a standard deviation of $4000.

9. A group of 64 recently graduated BA's in education earned an average of $14,500 with a standard deviation of $1200.

Determine the confidence level for each average given. See Examples 3 and 4.

10. A real estate agent claims that based on a sample of 36 homes sold in the city, average price is between $84,040 and $87,960 with a standard deviation of $6000.

11. A real estate agent in a different city claims that based on a sample of 64 homes sold in the city, average price is between $89,420 and $94,580 with a standard deviation of $8000.

12. A group of 36 secretaries could type an average of between 54 and 66 words per minute with a standard deviation of 14 words per minute.

13. A group of 49 secretaries could type an average of between 56 and 64 words per minute with a standard deviation of 14 words per minute.

14. A group of 100 people watch an average of between 1.87 and 2.13 hours of television daily with a standard deviation of 0.5 hour.

15. A group of 81 people watch an average of between 1.9 and 2.1 hours of television daily with a standard deviation of 20 minutes (1/3 hour).

✳**16.**†The characteristic under study for a population may not be measurable. For example, each student in a class has a marital status. Each is single, married, divorced, or widowed. To summarize the data, we use the *proportion* or percent of the population in each group. The letter π represents a population proportion while p represents a sample proportion. If all possible samples of size n (where n is less than 10 percent of the population, but fairly large) are selected from a population, then the population of sample proportions approximates a normal distribution. The average of this population is π and the standard deviation is $\sqrt{\pi(1 - \pi)/n}$ (where π is written as a decimal). Thus, confidence interval estimates for π can be made, based on p, in a manner similar to the confidence interval estimates for μ.

(a) If the Nielsen ratings report that 20 percent of the American homes (based on a sample of 993 homes) were tuned in to a certain t.v. show, find the 0.95 confidence interval estimate for π.

(b) If the Nielsen ratings report that 20% ± 1.3% of American homes (based on a sample of 993 homes) watched a certain t.v. show, what is the confidence level?

CHAPTER REVIEW

The set of all possible outcomes for an experiment is called the _____ _____ _____.

If each sample point is just as apt to occur as any other, we say that the sample points are _____ _____.

The smallest probability that can occur is _____, and the largest is _____.

$n!$ is read as n _____.

$_nP_r$ is read, "the number of _____ of _____ objects taken _____ at a time."

$_nC_r$ is read, "the number of _____ of _____ objects taken _____ at a time."

†This problem is based on information in the article, "Can You Believe the Ratings," by David Chagall, *T.V. Guide*, June 24, 1978, 2–13.

Two events are _____ if the outcome of the first has no influence on the outcome of the second.

If the characteristic under study can be measured, the measurement for each object in the population is called a _____.

The mean, median, and mode are measures of _____ _____.

The range, variance, and standard deviation are measures of _____ _____.

A part of a population is called a _____.

CHAPTER TEST

Specify the sample space for each experiment.

1. Drawing a marble from a box containing five red and four green marbles, if you are blindfolded.

2. Rolling a tetrahedral die (a die that has four triangular sides).

Find the probability of each event.

3. Drawing a red marble from a box containing five red and four green marbles, if you are blindfolded.

4. Drawing a blue marble from a box containing five red and four green marbles, if you are blindfolded.

5. Rolling a number greater than 2 or less than 5 on one roll of a single fair (six-sided) die.

Determine the total number of ways each event can occur.

6. Writing a code word in the form letter-digit-letter-digit if repetitions of letters are not allowed, but repetitions of digits are.

7. Forming a number from the digits {1, 2, 3} with no repetitions.

8. Forming a two-letter code from the letters {A, F, G, K} without repetition.

9. Choosing two people out of five to attend a conference.

10. Setting two switches out of seven to the "off" position.

Find the probability of each event.

11. Rolling two prime numbers on one roll of a pair of fair dice.

12. Rolling three 6's in a row if one fair die is rolled three times.

13. Drawing two face cards (K, Q, J) from a shuffled deck of 52, if the first card is replaced.

14. Drawing two face cards (K, Q, J) from a shuffled deck of 52, if the first card is not replaced.

15. Drawing two cards of the same suit from a shuffled deck of 52, if the first card is not replaced.

16. Find the arithmetic mean: $\{4, 9, 11, 4, 7\}$.

17. Find the median: $\{4, 9, 11, 4, 7\}$.

18. Find the mode: $\{4, 9, 11, 4, 7\}$.

Determine the most appropriate average to use.

19. $\{\$25, \$28, \$30, \$32, \$75\}$ where each amount is the price of dinner for two at a local restaurant.

20. $\{4, 3, 2, 4, 2\}$ where each number is the number of hours spent by students on a homework assignment.

21. Find the mean and the range: $\{27, 40, 43, 43, 47, 58\}$.

Find the variance.

22. $\{5, 22, 26, 34, 38\}$

23. $\{26, 41, 42, 44, 45, 60\}$

Find the standard deviation.

24. $\{2, 5, 23, 41, 44\}$

25. $\{0, 1, 3, 6, 9, 9, 10, 10\}$

If a normally distributed population has a mean of 38 and a standard deviation of 6, find the z-score for each raw score x.

26. $x = 29$

27. $x = 52$

Assume that potato chips are packaged so that the weights of the bags are normally distributed with an average weight of 12 oz and a standard deviation of 0.5 oz.

28. What percent of the bags weigh more than 13.5 oz?

29. What percent of the bags weigh less than 11.0 oz?

30. What percent of the bags weigh between 11.75 and 12.75 oz?

For each sample, establish a 0.95 confidence interval estimate for μ.

31. On an IQ test, a sample of 49 college students shows a mean of 110 and a standard deviation of 7.

32. The mean height of 36 players in a basketball conference is 78″ with a standard deviation of 3″.

33. The mean weight of 64 male employees of a company is 165 with a standard deviation of 16 pounds.

Determine the confidence level for each average given.

34. A group of 81 cars on a used-car lot has a mean age of between 6.57 and 7.63 years with a standard deviation of 2 years.

35. A group of 64 outpatients at a clinic had a mean temperature of between 98.0 and 99.2 with a standard deviation of 2 degrees.

36. The following flowchart shows an algorithm to calculate $_nP_r$. Modify it to find $_nC_r$.

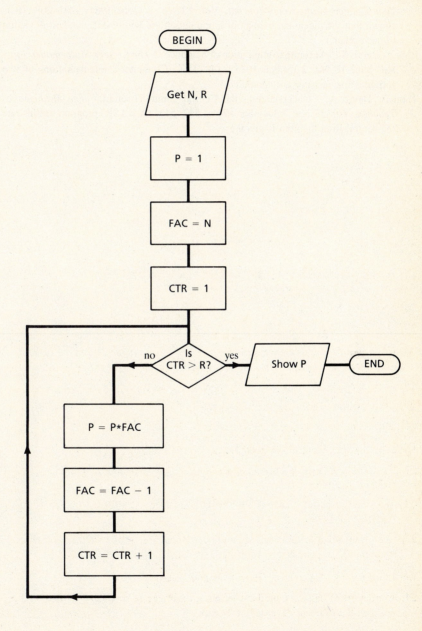

FOR FURTHER READING

Book, Stephen A. "Why $n - 1$ in the Formula for the Sample Standard Deviation?," *The Two-Year College Mathematics Journal*, Vol. 10 No. 5 (November 1979), 330–333.

Falk, Ruma. "Another Look at the Mean, Median, and Standard Deviation," *The Two-Year College Mathematics Journal*, Vol. 12 No. 3 (June 1981), 207–208. (It is shown that the mean and median of a population are within one standard deviation of each other.)

Roberts, Jerry A. "Accurate Computation of Variance," *The College Mathematics Journal*, Vol. 16 No. 2 (March 1985), 149–150. (Computer implementations of two formulas for variance are compared.)

Thomas, David A. "Understanding the Central Limit Theorem," *The Mathematics Teacher*, Vol. 77 No. 7 (October 1984), 542–543. (A BASIC program to illustrate the central limit theorem is given.)

Biography   Abraham DeMoivre

Abraham DeMoivre was born May 26, 1667, in France where his father was a surgeon. His parents were Huguenots, and when he was 18, the family moved to London seeking tolerance of their religious beliefs. DeMoivre supported himself by private tutoring and by solving mathematical puzzles. He bought a copy of Newton's book on calculus and tore out the pages so that he could carry one or two in his pocket to study between appointments. He hoped to secure a position at a university, but even though he had the help of influential friends, he was denied a position since he was not born in England.

In 1711, he published a paper called *The Laws of Chance*. In 1716, he expanded this work and entitled it *The Doctrine of Chances "A Method of Calculating the Probability of Events in Play."* It was one of the first books to deal exclusively with probability. Annuities, mortality statistics, permutations, and combinations were among the topics covered.

In 1733, he wrote a seven-page pamphlet, which he presented to his friends. There are two copies still in existence. This paper is important because it develops the concept of the normal curve. The curve is given by the equation $y = (1/\sqrt{2\pi})e^{(-x^2/2)}$.

A bachelor all of his life, he spent a great deal of time solving math problems in the coffee houses of London. It is said that Newton would send for him at one of the coffee houses each evening so that they could discuss philosophy.

As he grew old, both his sight and hearing failed. He died November 27, 1754, at the age of 87. There is a legend that he began to sleep 10 or 15 minutes more each night than he had the previous night. When he had reached the point at which he was to sleep 24 hours, he did not wake up.

Appendix   Signed Numbers

To discuss arithmetic using signed numbers, we begin with a definition.

> **Definition**
> The absolute value of a real number x is x if $x \geq 0$ and the absolute value of a real number x is $-x$ if $x < 0$. The absolute value of x is denoted $|x|$.

Thus, if $x = 3$, we have $|x| = x$ since $x \geq 0$. That is, $|3| = 3$. But if $x = -3$, we have $|x| = -x$ since $x < 0$. That is, $|-3| = -(-3) = 3$. This example shows that the absolute value of a real number is positive. It is the magnitude of the number without regard to sign. The rules for performing arithmetic operations using signed numbers are listed below.

> **To add two signed numbers:**
> (a)   If the signs of the two numbers are alike, add the absolute values of the numbers. The sum has the same sign as the numbers.
> (b)   If the signs of the two numbers are different, subtract the smaller absolute value from the larger absolute value. The difference has the same sign as the number with larger absolute value.

EXAMPLE 1

Perform the following additions:

(a) $3 + 5$

(b) $-3 + 5$

(c) $3 + (-5)$

(d) $-3 + (-5)$

Solution

(a) $3 + 5 = 8$

(b) $-3 + 5 = 2$

(c) $3 + (-5) = -2$

(d) $-3 + (-5) = -8$ ■

> **To subtract two signed numbers:**
> Change the sign of the subtrahend (the number being subtracted) and add to the minuend using the addition rule for signed numbers.

EXAMPLE 2

Perform the following subtractions:

(a) $3 - 5$

(b) $-3 - 5$

(c) $3 - (-5)$

(d) $-3 - (-5)$

Solution

(a) $3 - 5 = 3 + (-5) = -2$

(b) $-3 - 5 = -3 + (-5) = -8$

(c) $3 - (-5) = 3 + 5 = 8$

(d) $-3 - (-5) = -3 + 5 = 2$ ■

> **To multiply two signed numbers:**
> (a)   If the signs of the two numbers are alike, multiply their absolute values. The product is positive.
> (b)   If the signs of the two numbers are different, multiply their absolute values. The product is negative.

EXAMPLE 3

Perform the following multiplications:

(a) $8(4)$

(b) $8(-4)$

(c) $-8(4)$

(d) $-8(-4)$

Solution

(a) $8(4) = 32$

(b) $8(-4) = -32$

(c) $-8(4) = -32$

(d) $-8(-4) = 32$ ■

> **To divide two signed numbers:**
> (a)   If the signs of the two numbers are alike, divide their absolute values. The quotient is positive.
> (b)   If the signs of the two numbers are different, divide their absolute values. The quotient is negative.

EXAMPLE 4   Perform the following divisions:

(a) $8 \div 4$

(b) $8 \div (-4)$

(c) $-8 \div 4$

(d) $-8 \div (-4)$

Solution

(a) $8 \div 4 = 2$

(b) $8 \div (-4) = -2$

(c) $-8 \div 4 = -2$

(d) $-8 \div (-4) = 2$ ∎

APPENDIX EXERCISES

Perform the indicated operation. See Examples 1, 2, 3, and 4.

1. $7 + 12$ **2.** $-8 + 11$ **3.** $6 + (-10)$ **4.** $-9 + (-8)$

5. $-11 + 8$ **6.** $13 + 12$ **7.** $-10 + (-14)$ **8.** $15 + (-13)$

9. $7 - 12$ **10.** $-8 - 11$ **11.** $6 - (-10)$ **12.** $-9 - (-8)$

13. $-11 - 8$ **14.** $13 - 12$ **15.** $-10 - (-14)$ **16.** $15 - (-13)$

17. $7 (12)$ **18.** $6(-11)$ **19.** $-5(13)$ **20.** $-4(-14)$

21. $-8(10)$ **22.** $9(9)$ **23.** $15(-11)$ **24.** $-14(-12)$

25. $24 \div (-4)$ **26.** $25 \div 5$ **27.** $-26 \div (-2)$ **28.** $-27 \div 3$

29. $32 \div 8$ **30.** $-33 \div 11$ **31.** $34 \div (-2)$ **32.** $-36 \div (-6)$

Answers to Selected Exercises

Chapter 1

Section 1.1   (page 5):   **(1)** 45045   **(3)** 23023   **(5)** N = 73, D = 73
(7) X = 4, Y = 3, Z = 5   **(9)** X = 6, Y = 8, Z = 10   **(11)** 120
(13) LEFT = 2, MID = 1, RIGHT = 3   **(15)** LEFT = 8, MID = 4, RIGHT = 7

Section 1.2   (page 9):

(1) $89.25, $505.75
1. Get LIST, RATE.
2. Multiply RATE by 0.01 and assign to RATE.
3. Multiply LIST by RATE to get DISC.
4. Subtract DISC from LIST to get SALE.
5. Show DISC, SALE.
(3) $5, $18.54, $481.46
1. Get B, R, and D.
2. Multiply R by 0.01 and divide by 12: Assign to R.
3. Multiply R by B to get I.
4. Subtract I from D to get P.
5. Subtract P from B and assign to B.
6. Show I, P, and B.
(5) $400
1. Get S, R, A, and M.
2. Multiply R by 0.01 and assign to R.
3. If M is greater than A then subtract A from M and multiply by R to get BONUS.
   Otherwise, assign 0 to BONUS.
4. Add BONUS to S to get P.
5. Show P.
(7) $337
1. get BAL.
2. Get AMT.
3. While AMT is not 0, do steps 4–7.
4. Get nature of transaction.
5. If AMT is a deposit, then add AMT to BAL.
   Otherwise, subtract AMT from BAL.
6. Get next AMT.
7. Show BAL.
(9) Four months
1. Get BAL, MIN.
2. Assign CTR a value of 0.
3. While BAL is greater than MIN do steps 4–5.
4. Subtract half of BAL from BAL.
5. Add 1 to CTR.
6. Show CTR.

Section 1.3 (page 18):

(1) Assign BAL a value of 0.
Assign CTR a value of 1.
While CTR is less than 13,
    Add 100 to BAL.
    Multiply BAL by 0.01 to get INT.
    Add INT to BAL (and assign to BAL).
    Show BAL.
    Add 1 to CTR.
    ENDWHILE.

(3) Get AMT, NUM, RATE.
Assign BAL a value of 0.
Divide RATE by 12, multiply by 0.01, and assign to RATE.
Assign CTR a value of 1.
While CTR is less than NUM + 1,
    Add AMT to BAL.
    Multiply BAL by RATE to get INT.
    Add INT to BAL.
    Show BAL.
    Add 1 to CTR.
    ENDWHILE.

(5) Get BAL, PURCHASE, PAY.
Subtract PAY from BAL.
If BAL is greater than 0, multiply BAL by 0.01 to get CHARGE.
    Otherwise, assign CHARGE a value of 0.
    ENDIF.
Add BAL, PURCHASE, and CHARGE and assign to BAL.

(7) Not structured   **(9)**   **(11)**

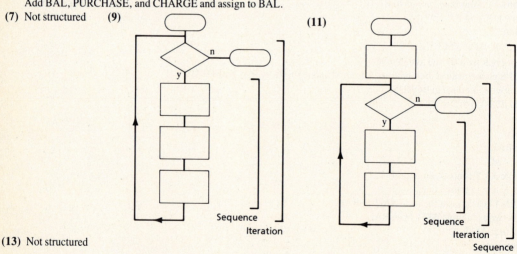

(13) Not structured

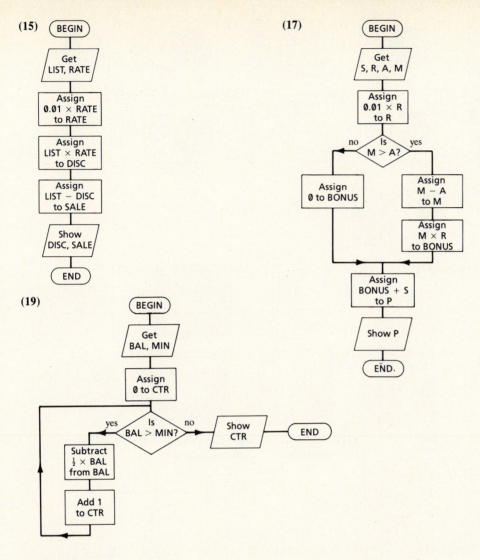

(15)

BEGIN

Get
LIST, RATE

Assign
0.01 × RATE
to RATE

Assign
LIST × RATE
to DISC

Assign
LIST − DISC
to SALE

Show
DISC, SALE

END

(17)

BEGIN

Get
S, R, A, M

Assign
0.01 × R
to R

Is
M > A?   no / yes

Assign
0 to BONUS

Assign
M − A
to M

Assign
M × R
to BONUS

Assign
BONUS + S
to P

Show P

END.

(19)

BEGIN

Get
BAL, MIN

Assign
0 to CTR

Is
BAL > MIN?   yes / no

Subtract
½ × BAL
from BAL

Add 1
to CTR

Show
CTR

END

Chapter 2

Section 2.1   (page 28):

(1) a. {3}   b. {2, 3, 5, 6, 7, 9}   **(3)** a. {1, 4, 8, 10}   b. {1, 4, 8, 10}
(5) a. {6, 9}   b. {2, 5, 7}   **(7)** a. { }   b. {5, 7, 9, 11}
(9) a. {5, 7}   b. {1, 3, 9, 11}   **(11)** a. {1, 3, 5, 7, 9, 11}   b. {1, 3}   **(13)** {4}   **(15)** {4}   **(17)** {4}
(19) {2, 4, 6}   **(21)** {4}   **(23)** {8, 10}   **(25)** {4, 8, 10}   **(27)** $W' \cap S$ **(29)** S' **(31)** $W' \cap S'$
(33) $W \cap O'$ **(35)** $S' \cap O$ **(37)** $W \cap (S' \cap O')$ **(39)** $W' \cap (S \cup O)$ **(41)** $F \cap M'$ **(43)** $T \cap R$
(45) M' **(47)** $T' \cap F'$

Section 2.2   (page 33):

(1)

(3)

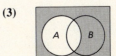

(5)

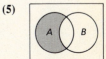

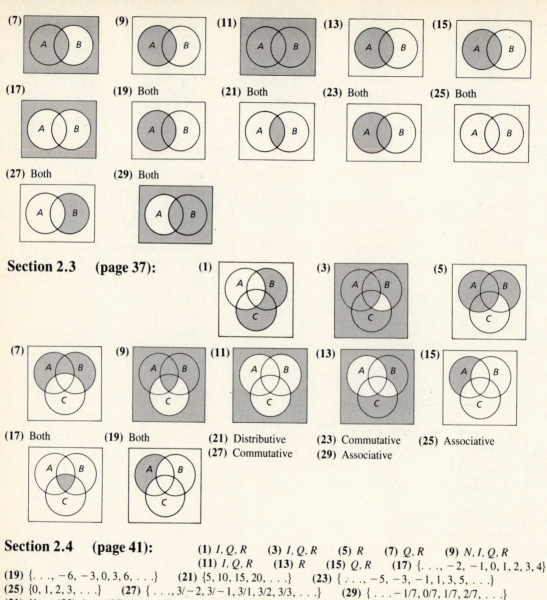

(7) **(9)** **(11)** **(13)** **(15)**

(17)   **(19)** Both   **(21)** Both   **(23)** Both   **(25)** Both

(27) Both   **(29)** Both

Section 2.3   (page 37):

(1)   **(3)**   **(5)**

(7)   **(9)**   **(11)**   **(13)**   **(15)**

(17) Both   **(19)** Both   **(21)** Distributive   **(23)** Commutative   **(25)** Associative

(27) Commutative   **(29)** Associative

Section 2.4   (page 41):
(1) I, Q, R **(3)** I, Q, R **(5)** R **(7)** Q, R **(9)** N, I, Q, R
(11) I, Q, R **(13)** R **(15)** Q, R **(17)** $\{\ldots, -2, -1, 0, 1, 2, 3, 4\}$
(19) $\{\ldots, -6, -3, 0, 3, 6, \ldots\}$ **(21)** $\{5, 10, 15, 20, \ldots\}$ **(23)** $\{\ldots, -5, -3, -1, 1, 3, 5, \ldots\}$
(25) $\{0, 1, 2, 3, \ldots\}$ **(27)** $\{\ldots, 3/-2, 3/-1, 3/1, 3/2, 3/3, \ldots\}$ **(29)** $\{\ldots -1/7, 0/7, 1/7, 2/7, \ldots\}$
(31) N **(33)** I **(35)** N **(37)** R **(39)** R **(41)** R

Section 2.5   (page 47):
(1) 21.27   **(3)** 19.74   **(5)** 142   **(7)** 3   **(9)** 4   **(11)** 1
(13) 4   **(15)** 2   **(17)** 2   **(19)** 3   **(21)** 9.7   **(23)** 0.1
(25) 60,000   **(27)** 5.8×10^1 **(29)** 4.09×10^2 **(31)** 8.732×10^3 **(33)** 6×10^{-3} **(35)** 4.278×10^{-1}
(37) 234   **(39)** 0.508   **(41)** 820   **(43)** 3452   **(45)** 0.657

Section 2.6   (page 52):
(1) 14   **(3)** 2   **(5)** 98   **(7)** −1   **(9)** −4
(11) 16   **(13)** 486   **(15)** −56   **(17)** 36   **(19)** 18
(21) 4*Y+2*X   **(23)** (X+2*Y)*(2*X−Y)   **(25)** X−2*Y+3   **(27)** 3*(X+2*(Y−1))
(29) (4*A*B)/(5*C*D) or 4*A*B/(5*C*D)   **(31)** (4+X)/(Y−5)   **(33)** (A+3)/(B+2)
(35) 3*(X*(Y*(Z−2)−3)−4)   **(37)** 2*2*2   **(39)** X*Y*Y

Section 2.7   (page 57):
(1) yes, yes   **(3)** yes, no   **(5)** no, yes   **(7)** no, no   **(9)** no, yes
(11) yes, yes   **(13)** no, no   **(15)** no, yes   **(17)** no, no
(19) no, yes   **(21)** yes, no   **(23)** no, no   **(25)** no, no   **(27)** no, no   **(29)** no, yes   **(31)** Associative
for addition   **(33)** Commutative for addition   **(35)** Identity for addition   **(37)** Identity for multiplication
(39) Commutative and associative for multiplication   **(41)** Inverse for addition

Section 2.8   (page 61):
(1) 2 2/5, 2   **(3)** 3 1/2, 3   **(5)** 2, 0   **(7)** 16, 0   **(9)** 9, 9
(11) 4, 0   **(13)** 6 2/5, 6   **(15)** 6 2/5, 0   **(17)** 1, 0   **(19)** 2 1/2, 2
(21) 0   **(23)** 3   **(25)** 0   **(27)** 3   **(29)** 3   **(31)** 4   **(33)** 5   **(35)** 2

Chapter Test 2   (page 63):
(1) a. {3, 5},   b. {1, 3, 4, 5, 6, 7}   **(2)** a. {4},   b. {1, 2, 4, 5, 7}
(3) a. {4, 6}   b. {1, 2, 3, 5, 6, 7}   **(4)** a. C ∪ F
b. F ∩ B   **(5)** a. B ∩ F′   b. (C ∪ F) ∩ B′   **(6)** a. (B ∪ C)′   b. (B ∩ C) ∩ F′   **(7)**

(8)   **(9)** Both: **(10)** Both: **(11)**   **(12)**

(13) Both: **(14)** Both:

(15) a. Distributive   b. Associative   c. Commutative   **(16)** a. N, I, Q, R   b. Q, R   c. R   d. I, Q, R
e. Q, R   f. N, I, Q, R   **(17)** {5, 6, 7, 8, 9, . . .}   **(18)** {. . . −3, −2, −1, 0, 1}   **(19)** R   **(20)** I
(21) a. 47.4   b. 8.08   **(22)** a. 2   b. 3   c. 2   **(23)** a. 0.001   b. 10 to 2 significant digits
(24) a. 8.35×10^2 b. 5.1×10^{-3} **(25)** a. 0.0793   b. 6400   **(26)** 81   **(27)** 1   **(28)** 125
(29) 3*(X−3*(Y−1))   **(30)** (3*A + B)/(4*C*D)   **(31)** yes, yes   **(32)** no, yes   **(33)** no, no
(34) a. Associative   b. Commutative   c. Inverse for addition   **(35)** a. Distributive   b. Identity for multiplication   c. Inverse for multiplication   **(36)** 1, 0   **(37)** 1, 0   **(38)** 5   **(39)** 0   **(40)** 2

(41)

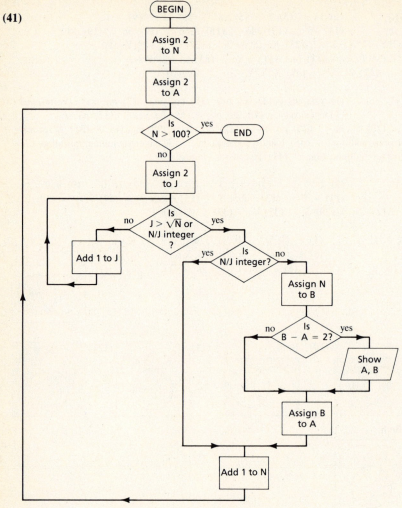

Chapter 3

Section 3.1   (page 72):

(1) $c \vee f$ **(3)** $c \wedge f$ **(5)** $(c \wedge b) \vee p$ **(7)** $\sim m \wedge \sim a$
(9) $(m \wedge a) \wedge \sim v$ **(11)** $(\sim m \wedge a) \wedge \sim t$ **(13)** $o \wedge r$ **(15)** $(o \wedge r)$ $\wedge d$ **(17)** $s \vee (o \wedge d)$ **(19)** $a \wedge r$ **(21)** $f \wedge \sim(w \vee s)$ **(23)** $\sim(w \vee s)$ **(25)** $t \wedge \sim z$ **(27)** $f \vee \sim z$ **(29)** $f \wedge \sim z$

Section 3.2   (page 80):

(1) TFFF   **(3)** FT   **(5)** FFFT   **(7)** TTFT   **(9)** FTFF
(11) FFFT for both   **(13)** $\sim r \vee \sim s$ **(15)** $p \wedge \sim q$ **(17)** $r \wedge s$
(19) $\sim(p \vee q)$ **(21)** yes   **(23)** no

Section 3.3   (page 84):

(1) $(p \wedge (\sim q)) \vee r$ **(3)** $p \vee ((\sim q) \wedge r)$ **(5)** $((\sim p) \wedge q) \wedge (\sim r)$
(7) $(p \wedge q) \vee (r \wedge s)$ **(9)** $((\sim p) \vee q) \vee (r \wedge (\sim s))$
(11) $((p \wedge (\sim q)) \wedge r) \vee (\sim s)$ **(13)** TTTTTFTT   **(15)** TTTTFFTF   **(17)** TFTTTTTT   **(19)** FFFFFFTF
(21) FFFTTTTT   **(23)** no   **(25)** yes   **(27)** yes   **(29)** no

Section 3.4   (page 88):

(1) $f \to p$ **(3)** $s \to e$ **(5)** $l \to e$ **(7)** $d \to e$ **(9)** $(h \lor m) \to d$ **(11)** $\sim d \to \sim (b \lor h)$ **(13)** $p \to f$ **(15)** $h \to l$ **(17)** $c \to m$ **(19)** $t \to n$ **(21)** $(p \lor (\sim q)) \to r$ **(23)** $p \to (q \land r)$ **(25)** $((\sim p) \land q) \to r$ **(27)** $(p \land r) \to (q \lor s)$ **(29)** $((p \land q) \lor s) \to r$ **(31)** If your program is easy to follow, then it is structured. If your program is not structured, then it is not easy to follow. If your program is not easy to follow, then it is not structured. It is not the case that if your program is structured, then it is easy to follow.   **(33)** If your programs are easy to maintain, then you document them. If you do not document your programs, then they will not be easy to maintain. If your programs are not easy to maintain, then you do not document them. It is not the case that if you document your programs, then they will be easy to maintain.   **(35)** If you can log onto the computer, then you know the password. If you do not know the password, then you cannot log onto the computer. If you cannot log onto the computer, then you do not know the password. It is not the case that if you know the password, then you can log onto the computer.

Section 3.5   (page 96):

(1) TTTF   **(3)** FTTT   **(5)** TFTT for both   **(7)** FTTT for both   **(9)** TFTTTTTT   **(11)** TTTTTFTT   **(13)** FTTTTTTT   **(15)** TTFTTTTT   **(17)** TTFTTTTT   **(19)** valid   **(21)** invalid   **(23)** valid

Section 3.6   (page 99):

(1) $(A < 3)$ OR $(B > 5)$ **(3)** $((A > 0)$ OR $(A < 5))$ OR $(C > -3)$ **(5)** $(A > 0)$ OR $((C > -3)$ AND $(C < 5))$ **(7)** $(X <> 0)$ OR $(Y <> 0)$ **(9)** $((C = 3)$ AND $(X <> 0))$ OR $(Y <> 0)$ **(11)** $\{1, 2\}$ **(13)** $\{\ \}$ **(15)** $\{0, 7, 8, 9, \ldots\}$ **(17)** $\{1, 2, 6, 7, 8, \ldots\}$ **(19)** $\{1, 2\}$ **(21)** action 2   **(23)** action 1   **(25)** action 1   **(27)** action 1   **(29)** action 1

Section 3.7   (page 104):

(1) Commutative: TFFF   **(3)** Associative: TFFFFFFF

for both   for both

(5) Distributive: TTTTTFFF   **(7)** Idempotent: TF   **(9)** Law of complementation: FF

for both   for both   for both

(11) De Morgan's law: FTTT   **(13)** Law of absorption: TTFF   **(15)** $A + 1 = 1$: TT

for both   for both   for both

(17) $1' = 0$: F   **(19)** $A \cdot 1 = A$: TF

for both   for both

Section 3.8   (page 110):

(1) $AB + C$ **(3)** $A(B + B')$ **(5)** $(AB') + (BA')$ **(7)** $A(B + C)$ **(9)** $(B + C') + (B'C)$ **(11)**

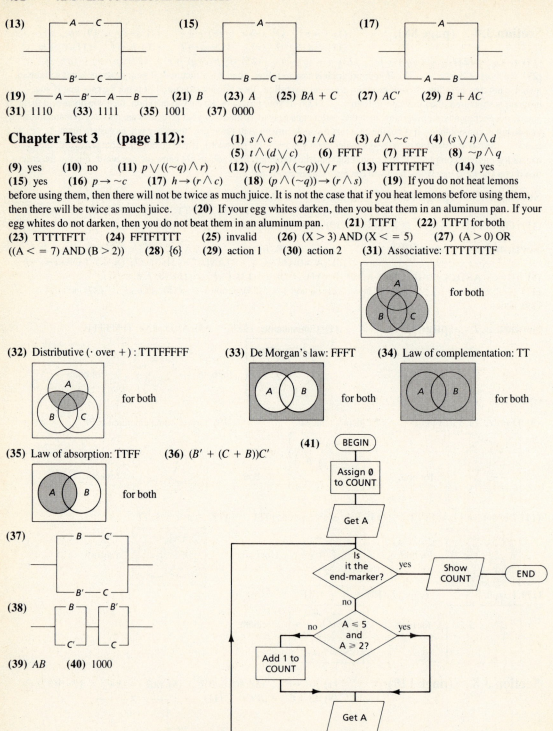

(13)   **(15)**   **(17)**

(19) — A — B' — A — B —   **(21)** B **(23)** A **(25)** $BA + C$ **(27)** AC' **(29)** $B + AC$
(31) 1110   **(33)** 1111   **(35)** 1001   **(37)** 0000

Chapter Test 3   (page 112):

(1) $s \wedge c$ **(2)** $t \wedge d$ **(3)** $d \wedge \sim c$ **(4)** $(s \vee t) \wedge d$
(5) $t \wedge (d \vee c)$ **(6)** FFTF   **(7)** FFTF   **(8)** $\sim p \wedge q$
(9) yes   **(10)** no   **(11)** $p \vee ((\sim q) \wedge r)$ **(12)** $((\sim p) \wedge (\sim q)) \vee r$ **(13)** FTTTFTFT   **(14)** yes
(15) yes   **(16)** $p \to \sim c$ **(17)** $h \to (r \wedge c)$ **(18)** $(p \wedge (\sim q)) \to (r \wedge s)$ **(19)** If you do not heat lemons before using them, then there will not be twice as much juice. It is not the case that if you heat lemons before using them, then there will be twice as much juice.   **(20)** If your egg whites darken, then you beat them in an aluminum pan. If your egg whites do not darken, then you do not beat them in an aluminum pan.   **(21)** TTFT   **(22)** TTFT for both
(23) TTTTTFTT   **(24)** FFTFTTTT   **(25)** invalid   **(26)** $(X > 3)$ AND $(X < = 5)$ **(27)** $(A > 0)$ OR $((A < = 7)$ AND $(B > 2))$ **(28)** {6}   **(29)** action 1   **(30)** action 2   **(31)** Associative: TTTTTTTT

for both

(32) Distributive ($\cdot$ over $+$) : TTTFFFFF   for both   **(33)** De Morgan's law: FFFT   for both   **(34)** Law of complementation: TT   for both

(35) Law of absorption: TTFF   for both   **(36)** $(B' + (C + B))C'$

(41)

(37)

(38)

(39) AB **(40)** 1000

Chapter 4

Section 4.1   (page 123):   **(1)** 6   **(3)** 7   **(5)** 11   **(7)** 12   **(9)** 55   **(11)** 58   **(13)** 35
(15) 150   **(17)** 189   **(19)** 187   **(21)** 1001   **(23)** 10010
(25) 100011   **(27)** 101111   **(29)** 110101   **(31)** 111111   **(33)** 1000010   **(35)** 1010001
(37) 1101110   **(39)** 11000110

Section 4.2   (page 129):   **(1)** 3.375   **(3)** 3.1875   **(5)** 6.4375   **(7)** 6.53125
(9) 0.01$\overline{001}$ **(11)** 0.010$\overline{110}$ **(13)** 0.0011$\overline{1100}$ **(15)** 1000.011
(17) 55   **(19)** 45   **(21)** 36   **(23)** 100010   **(25)** 1001111   **(27)** 10111001

Section 4.3   (page 136):   **(1)** 10010   **(3)** 1001.101   **(5)** 1001.111   **(7)** 10100
(9) 101101.001   **(11)** 1001   **(13)** 1.0101   **(15)** 1.10
(17) 101.010   **(19)** 10.11   **(21)** 10101   **(23)** 1.111   **(25)** 0.101   **(27)** 11.0111   **(29)** 1000.1111
(31) 100   **(33)** 101   **(35)** 111 R 1   **(37)** 11 R 1   **(39)** 110 R 101

Section 4.4   (page 141):   **(1)** 10110110   **(3)** 10011010   **(5)** 11111111   **(7)** 01001011
(9) 00011001   **(11)** 01101101   **(13)** 00100000   **(15)** 00110011
(17) 00101011   **(19)** 01011100   **(21)** 01101101, same   **(23)** 00100000, same   **(25)** 00110011, same
(27) 43, same   **(29)** 92, same

Section 4.5   (page 145):   **(1)** 00100101   **(3)** 00101101   **(5)** 00110100   **(7)** 01001110
(9) 01011011   **(11)** 01111011   **(13)** 45   **(15)** 54   **(17)** 85
(19) -103 **(21)** -86 **(23)** -1 **(25)** 00001000   **(27)** 11000100   **(29)** 01000010   **(31)** 00011000
(33) 00001111   **(35)** 11001111

Section 4.6   (page 149):   **(1)** 58   **(3)** 132   **(5)** 9.140625   **(7)** 34.765625
(9) 520.126953125   **(11)** 324   **(13)** 267   **(15)** 26.75
(17) 5.625   **(19)** 483.25390625   **(21)** 42   **(23)** 73.2   **(25)** 1144   **(27)** 120.3   **(29)** 1057
(31) 22   **(33)** 3B.4   **(35)** 264   **(37)** 50.6   **(39)** 22F

Section 4.7   (page 154):   **(1)** 56   **(3)** 132.7   **(5)** 153.4   **(7)** 12.54   **(9)** 110010
(11) 10010.011   **(13)** 11.0001   **(15)** 1010.100111101   **(17)** 2E
(19) 5A.E   **(21)** 6B.8   **(23)** A.B   **(25)** 10010011   **(27)** 10100011.1   **(29)** 10.1101111
(31) 11011.00001111   **(33)** 22   **(35)** A.9C   **(37)** 3.64   **(39)** 12.126

Section 4.8   (page 162):   **(1)** 722   **(3)** 66.35   **(5)** 12.756   **(7)** 5176   **(9)** 136.32
(11) 564   **(13)** 60.55   **(15)** 7.131   **(17)** 3F2   **(19)** A.43
(21) 578   **(23)** 6CC.C   **(25)** EB.DFC   **(27)** 3A8   **(29)** 4.C3

Chapter Test 4   (page 163):   **(1)** 229   **(2)** 157   **(3)** 110111   **(4)** 10011001
(5) 11001001   **(6)** 5.5625   **(7)** 3.34375   **(8)** 7.8125
(9) 0.0011   **(10)** 111.00$\overline{1001}$ **(11)** 1001.011   **(12)** 11001   **(13)** 100.001   **(14)** 111.0011
(15) 1000 R 11   **(16)** a. 11010011   b. 00100101   **(17)** 00111110   **(18)** 00100100   **(19)** 00111110
(20) 36   **(21)** a. 00111111   b. 11000001   **(22)** a. 01100111   b. 10011001   **(23)** a. 23   b. -26
(24) 00001111   **(25)** 11001011   **(26)** 307   **(27)** 700   **(28)** 43.75   **(29)** 323.6   **(30)** D3.C
(31) 307.1   **(32)** 101111.0111   **(33)** C7.2   **(34)** 110100111.1110011   **(35)** 322.37   **(36)** 160.44
(37) 63.14   **(38)** 30.51   **(39)** 4.70D   **(40)** 24F

(41)

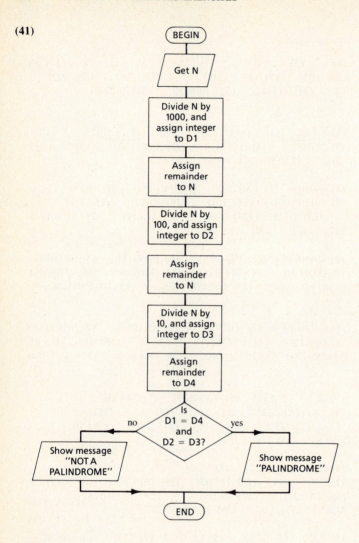

Chapter 5

Section 5.1   (page 172):

(1) $6x^5$ **(3)** $14y^7$ **(5)** $48x^9$ **(7)** $2xy^2$ **(9)** $y^2/2$ **(11)** 1   **(13)** 1/16   **(15)** 1/1000   **(17)** $56/x$ **(19)** x^3y^3z **(21)** $2x/y$ **(23)** $4y^2/z$ **(25)** $1/(2a^4)$ **(27)** $4y^3/x^4$ **(29)** $2y/(3x)$

Section 5.2   (page 177):

(1) a^6 **(3)** 2^{12} or 4096   **(5)** 5^9 **(7)** $9a^2$ **(9)** $16b^2$ **(11)** $8/x^3$ **(13)** $y^5/6^5$ **(15)** a^4/b^4 **(17)** not a real number   **(19)** 3   **(21)** -3 **(23)** 4   **(25)** not a real number   **(27)** 9   **(29)** $\sqrt[4]{125}$ **(31)** x^8y^4 **(33)** $3x^6/y^3$ **(35)** $1/(5xy)$ **(37)** $9b^2/16$ **(39)** $64b^6/125$ **(41)** y^3/x **(43)** $3a/(5b^2)$ **(45)** $b^3c^6/(8a^3)$ **(47)** $16a^2/9$ **(49)** $a^3b^3/6$ **(51)** 4/5   **(53)** $x^3/8$ **(55)** $1/(27a^3)$

Section 5.3   (page 183):

(1) 2 factors   **(3)** 2 terms   **(5)** 3 terms   **(7)** 2 factors
(9) 2 terms   **(11)** $3x + 2y$ **(13)** $3x - 3 + y$ **(15)** $4x^2 - x + 1$
(17) $3a^2 + 5a - 3$ **(19)** $x + 3$ **(21)** $2x^2 - 2x$ **(23)** $3x^3 - 3x^2 + 12x$ **(25)** $2x^2 - x - 3$
(27) $15y^2 - 23y + 4$ **(29)** $x^2 - 7xy + 10y^2$ **(31)** $6x^3 - 13x^2 + 8x - 3$ **(33)** $x^3 + 27$ **(35)** $3a^3 - 2a^2$
$- 13a + 10$ **(37)** $y^4 + y^3 - 7y^2 + 7y - 2$ **(39)** $8x^3 + 12x^2 + 6x + 1$ **(41)** $x^2 + x - 1$ **(43)** $x^2 - x - 7$
(45) $x^2 - 4$ **(47)** $6x^2 - 6$

Section 5.4   (page 188):

(1) $2x(x + 1)$ **(3)** $3m(m^2 - m + 2)$ **(5)** $(x + 2)(x + 3)$
(7) $(z - 3)(z + 2)$ **(9)** $(p + 4)(p + 6)$ **(11)** $(y - 3)(y - 8)$
(13) $(2z - 3)(3z - 2)$ **(15)** $(5x + 1)(x + 5)$ **(17)** $(3p + 4)(4p - 3)$ **(19)** $3(x - 2)(x - 2)$
(21) $(z + 8)(z - 3)$ **(23)** $4(m + 2)(m - 2)$ **(25)** prime   **(27)** $4y^2(y + 2)$ **(29)** $(p + 4)(p + 5)$
(31) $(m - 3)(m - 5)$ **(33)** $2(a - 10)(a - 2)$ **(35)** $(2a + 1)(3a - 5)$ **(37)** $(2m + 3)(m - 3)$
(39) prime   **(41)** $(3x - 2)(3x + 2)$

Section 5.5   (page 193):

(1) 2/3   **(3)** 4/15   **(5)** 1/3   **(7)** $\dfrac{y + 3}{y - 3}$ **(9)** $\dfrac{x + 2}{x + 3}$

(11) $\dfrac{z - 4}{z - 5}$ **(13)** $\dfrac{y + 2}{y + 1}$ **(15)** 18/49   **(17)** 1/6   **(19)** 1

(21) $\dfrac{y - 2}{y + 3}$ **(23)** $\dfrac{m + 5}{m - 3}$ **(25)** $\dfrac{a^2 - 4a + 3}{a^2 + 4a + 3}$ **(27)** 1   **(29)** $\dfrac{x - 3}{x + 4}$ **(31)** 3/4   **(33)** m^2

(35) $\dfrac{x + 1}{x - 2}$

Section 5.6   (page 199):

(1) $\dfrac{a^2 + b^2}{ab}$ **(3)** $\dfrac{p^3 + q^2}{p^2q}$ **(5)** $\dfrac{y^2 + 1}{(y - 1)^2(y + 1)}$

(7) $\dfrac{-x^2 + 6x + 9}{x(2x - 3)(2x + 3)}$ **(9)** $\dfrac{2y^2 + 2}{y^2 - 1}$ **(11)** $\dfrac{p^2 + 1}{2p(p^2 - 1)}$

(13) $\dfrac{8m + 6}{m^2 - 4}$ **(15)** $\dfrac{-x - 1}{(3x + 1)^2}$ **(17)** $\dfrac{-z + 5}{z^2 + z}$ **(19)** $\dfrac{a^2 + 4a + 2}{a^2 - 1}$ **(21)** $\dfrac{-m^2 + 9m + 13}{m^2 - 2m - 8}$

(23) $\dfrac{z + 1}{z}$

Section 5.7   (page 203):

(1) X*(X − 7) + 2   **(3)** X*(X*4 + 3) − 1
(5) X*(X*(X + 4) − 3) + 2   **(7)** X*(X*(X*8 − 6) + 4) − 3
(9) X*(X*(X − 2) + 0) + 3   **(11)** X*(X*(X*(X + 3) + 2) − 1) − 1   **(13)** X*(X*(X*(X*3 + 2) − 1) + 7) − 5
(15) X*(X*(X*(X − 1) + 1) − 1) + 1   **(17)** 3, 2   **(19)** 2, 2   **(21)** 6, 3   **(23)** 5, 3   **(25)** 3, 2
(27) 10, 4   **(29)** 10, 4

Chapter Test 5   (page 204):

(1) $35x^4y^3$ **(2)** $\dfrac{35}{a^2b}$ **(3)** 16   **(4)** $9mn$ **(5)** $\dfrac{15x^3z^2}{yw^2}$
(6) $8a^3b^6$ **(7)** $3a^6/b^4$ **(8)** $8x^3/27$ **(9)** $9z^2/4$ **(10)** 25
(11) $5x^2 + x + 3$ **(12)** $y^2 - 4y + 4$ **(13)** $6m^2 - 7m - 3$ **(14)** $8p^3 - 4p + 1$ **(15)** $z^2 - 5z + 11$
(16) $5x(x - 2)$ **(17)** $(z - 4)(z + 3)$ **(18)** $(a + 4)(a + 7)$ **(19)** $(3x - 1)(2x + 3)$ **(20)** $y(3y - 1)(3y + 1)$
(21) 2/3   **(22)** $\dfrac{2x + 1}{x + 2}$ **(23)** 4/15   **(24)** $\dfrac{m + 1}{m + 3}$ **(25)** $\dfrac{p + 1}{p - 3}$ **(26)** $\dfrac{x^3 + y^2}{x^2y}$ **(27)** $\dfrac{3a^2 + 2a}{a^2 - 4}$
(28) $\dfrac{z^2 - z - 1}{2z^2 + z - 3}$ **(29)** $\dfrac{2p - 1}{p^2 - p - 6}$ **(30)** $\dfrac{-z^2 + 6z + 9}{3z^2}$ **(31)** X*(X*3 + 2) − 1
(32) X*(X*(X*3 − 2) + 5) − 4   **(33)** X*(X*(X*(X + 2) − 3) + 4) − 5   **(34)** 6, 4   **(35)** 6, 3

(36)

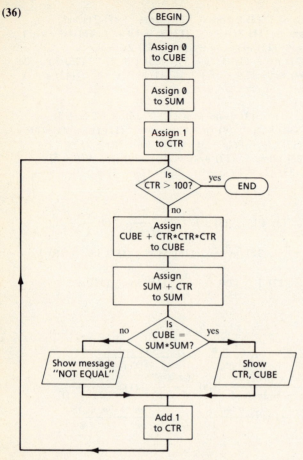

Chapter 6

Section 6.1 (page 213):

(1) $x = 4/5$ (3) $z = -2/3$ (5) $x = 25$ (7) $m = 9$
(9) $x = 9/2$ (11) $p = -14/5$ (13) $x = -1/9$ (15) $x = 0$
(17) $p = 7/8$ (19) $m = 9$ (21) $x = -2$ (23) $x = 5$ (25) $z = 2$ (27) $y \le 7/2$ (29) $x < 9/4$
(31) $p > 6$ (33) $y \ge 2/5$ (35) $m < -2/3$

Section 6.2 (page 217):

(1) $m = F/a$ (3) $g = \dfrac{\omega}{nmh}$ (5) $L = \dfrac{P - 2W}{2}$

(7) $a = \dfrac{v^2 - v_0^2}{2(x - x_0)}$ (9) $F = Eq$ (11) $V = h\pi r^2$ (13) $G = \dfrac{Fr^2}{M_1 M_2}$

(15) $q = F/E$ (17) $x_0 = x - v_0 t - \dfrac{1}{2} at^2$ (19) $t_2 = \dfrac{s_2 - s_1 + vt_1}{v}$ (21) $e = \dfrac{Itm}{Mz}$ (23) $s = \dfrac{pE}{E - p}$

(25) $r = \dfrac{2ak^2 m}{v^2 a - k^2 m}$ (27) $i = \dfrac{of}{o - f}$ (29) $m_2 = \dfrac{4a^3\pi^2 - p^2 k^2 m_1}{p^2 k^2}$

Section 6.3 (page 223):

(1) 99 (3) 32 ft/sec^2 (5) 15 percent (7) 2 liters (9) $500
at 8 percent, $1500 at 9 percent (11) must sell $1,000,000, must list
$2,000,000 (13) 2 hours 12 minutes (15) 3 mph and 3.5 mph (17) $120 (19) 150 (21) 1 lb.
(23) about 4 hours 46 minutes (25) about 47 oz

Section 6.4   (page 228):

(1) 1, 4   **(3)** $-3, 2$ **(5)** $-6, 4$ **(7)** $2, -1$ **(9)** 1, 8
(11) 3/2, 2/3   **(13)** 1/2, $-3/2$ **(15)** 1/3   **(17)** $0, -1$
(19) $1, -1$ **(21)** $4, -4$ **(23)** $6/5, -6/5$ **(25)** $5/2, -1/3$ **(27)** $-3/2, 1/3$ **(29)** $5, -2$ **(31)** 3, 5
(33) $2/3, -5$ **(35)** $2, -2$

Section 6.5   (page 234):

(1) $2 \pm \sqrt{2}$ **(3)** $\dfrac{5 \pm \sqrt{13}}{2}$ **(5)** $\dfrac{-5 \pm \sqrt{33}}{4}$ **(7)** $\pm\sqrt{21}$

(9) 0, 7/2   **(11)** $\dfrac{1 \pm \sqrt{13}}{6}$ **(13)** $7, -1$ **(15)** $-4 \pm \sqrt{15}$

(17) $\dfrac{-2 \pm \sqrt{6}}{2}$ **(19)** $\dfrac{-5 \pm \sqrt{17}}{2}$ **(21)** $\dfrac{-1 \pm \sqrt{13}}{6}$ **(23)** 0, 3

Section 6.6   (page 239):

(1) 0, 2, 3   **(3)** $\pm 1, \pm 3$ **(5)** $\pm 1, \pm 2$ **(7)** ± 1 **(9)** 0, 1/2, 1
(11) 1.73   **(13)** 1.62   **(15)** 1.82   **(17)** 2.4   **(19)** 2.3
(21) 4.64   **(23)** 4.05

Section 6.7   (page 244):

(1) $x = -9/10$ **(3)** $x = 2$ **(5)** $z = 1/4$ **(7)** no solution
(9) $x = 33/13$ **(11)** $x = 6, -25/4$ **(13)** $a = 1$
(15) $y = -1/3$ **(17)** $a = 1$ **(19)** no solution   **(21)** no solution   **(23)** $x = 1/4$

Section 6.8   (page 248):

(1) X = 5   **(3)** TC = 10   **(5)** X = 4, Y = 4   **(7)** A = 2,
B = 5   **(9)** X = 2, Y = 2   **(11)** A = 5, B = 4, C = 4
(13) A = 2, B = 3   **(15)** X = -1, Y = 4   **(17)** A = 5, B = 4, C = 8

Chapter Test 6   (page 249):

(1) $x = 3$ **(2)** $p = 50$
(3) $z = -1$ **(4)** $x = 1$
(5) $y \le 3$ **(6)** $x < -1$ **(7)** $R = D/T$ **(8)** $W = A/L$
(9) $C = \dfrac{5F - 160}{9}$ **(10)** $a = L - (n - 1)d$ **(11)** $d = \dfrac{L - a}{n - 1}$
(12) $a = S(1 - r)$ **(13)** \$30,000   **(14)** 6 liters   **(15)** 50 mph and 55 mph
(16) 14, 1   **(17)** 3, 7   **(18)** $-5/2, -1$ **(19)** $-5, 1$ **(20)** $5, -3$
(21) $\dfrac{-5 \pm \sqrt{13}}{2}$ **(22)** $-1 \pm \sqrt{2}$ **(23)** $2, -1/2$ **(24)** $\dfrac{2 \pm \sqrt{2}}{2}$
(25) $\dfrac{1 \pm \sqrt{7}}{3}$ **(26)** 0, 1, 2   **(27)** $0, -2, 1$
(28) $0, \pm 1, \pm 2$ **(29)** 2.93   **(30)** 3.07
(31) $x = 5/3$ **(32)** $a = -3$
(33) no solution   **(34)** $x = -6$
(35) no solution   **(36)** A = 2, B = 6
(37) B = 9   **(38)** A = 18, B = 6
(39) X = 1, Y = 4, Z = 4
(40) X = 16, Y = 4

(41)

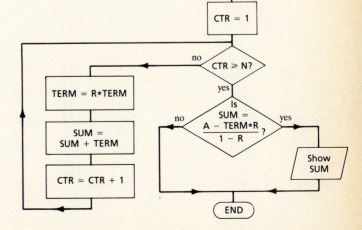

Chapter 7

Section 7.1 (page 258):
(1) yes (3) yes (5) yes (7) yes (9) no (11) 0
(13) 0 (15) 5 (17) −1 (19) 19 (21) 1 (23) 6
(25) 5 (27) no (29) yes, $\{x \mid x \in R, x \neq 1\}$ (31) yes, $\{x \mid x \in R, x \geq 4\}$ (33) yes, R (35) yes, R

Section 7.2 (page 263):
(1)
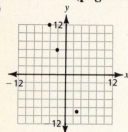

(3)
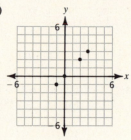

(5)
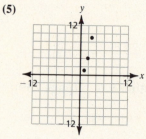

(7)
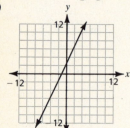

(9)
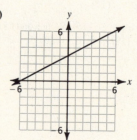

(11) no (13) no (15) yes

(17) yes (19) no

Section 7.3 (page 270):
(1)
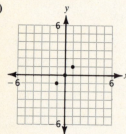

(3)
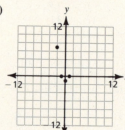

(5)
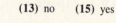

(7)
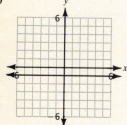

(9)
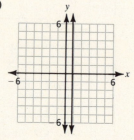

(11)
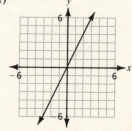

(13) $m = 3, b = -2$ **(15)** $m = -2, b = 5$ **(17)** $m = 2, b = -4$ **(19)** $m = 1/2, b = -1/2$

(21) $m = 3, b = -7$ **(23)** $m = 1/2, b = 0$ **(25)** $m = 0, b = 4$

(27)
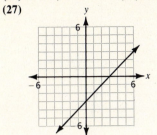

(29)
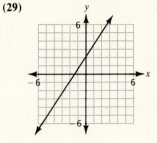

(31)
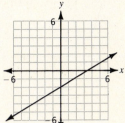

(33)
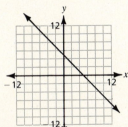

(35)
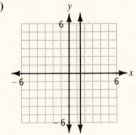

Section 7.4   (page 274):

(1) $(-7/2, -1/4)$
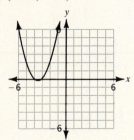

(3) $(2, -8)$
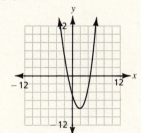

(5) $(0, -1)$
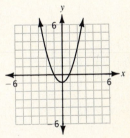

(7) $(2, 4)$
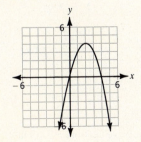

(9) $(1/2, 0)$
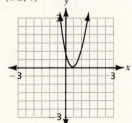

(11) $(1/2, -1/4)$
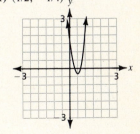

(13) $(-1, 2)$

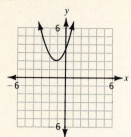

(15) $(-3/4, -1/8)$

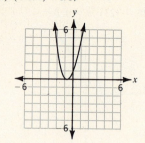

(17) $(-3/4, 1/8)$

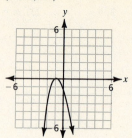

(19) $(1/2, -3/4)$

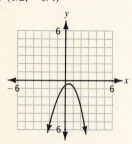

Section 7.5 (page 280):

(1)

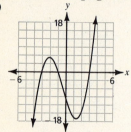

(3)

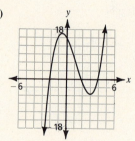

(5)

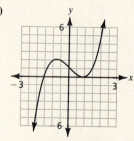

(7)

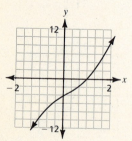

(9)

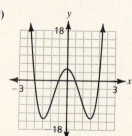

(11)

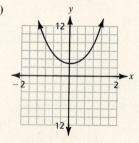

(13) Plot x-axis from $(0, 80)$ to $(279, 80)$ and y-axis from $(140, 0)$ to $(140, 159)$; let x take on values from -7 to 7 in increments of 0.05; $y = x^2 - 5x + 6$; plot $(20x + 140, 20y + 80)$ **(15)** Plot x-axis from $(0, 80)$ to $(279, 80)$ and y-axis from $(140, 0)$ to $(140, 159)$; let x take on values from -14 to 14 in increments of 0.1; $y = x^3 - 5x + 6$; plot $(10x + 140, 20y + 80)$ **(17)** Plot x-axis from $(0, 80)$ to $(279, 80)$ and y-axis from $(140, 0)$ to $(140, 159)$; let x take on values from -3.5 to 3.5 in increments of 0.025; $y = x^2 - 9$; plot $(40x + 140, 20y + 80)$

Section 7.6 (page 284):

(1) $\ln 1.649 = 0.5$ **(3)** $\ln 12.182 = 2.5$ **(5)** $\ln 90.017 = 4.5$
(7) $\ln 665.142 = 6.5$ **(9)** $e^{0.405} = 1.5$ **(11)** $e^{1.253} = 3.5$
(13) $e^{1.705} = 5.5$ **(15)** $e^{3\ln 2}$ **(17)** $e^{5\ln 4}$ **(19)** $e^{9\ln 8.3}$ **(21)** $e^{4.1\ln 7.8}$ **(23)** $e^{-3\ln 7.45}$ **(25)** $e^{-8.1\ln 101.3}$ **(27)** EXP(0.5)
(29) EXP(-3.5)   **(31)** EXP(-6.5)   **(33)** LN(7.3)   **(35)** EXP(4*LN(3))   **(37)** LN(0.5)   **(39)** LN(5.5)
(41) EXP(360*LN(1 + R))

Section 7.7 (page 290):

(1) 2.3   **(3)** 7   **(5)** 5.7   **(7)** 6.33   **(9)** yes   **(11)** no
(13) no   **(15)** 2   **(17)** -6 **(19)** 2   **(21)** -20 **(23)** 2
(25) -19 **(27)** 2   **(29)** -20 **(31)** 4.21   **(33)** 3.17   **(35)** -2.73

Chapter Test 7 (page 292):

(1) yes   **(2)** 5   **(3)** -8 **(4)** no   **(5)** yes, $\{x \mid x \in R, x \neq 1\}$
(6) yes, R

(7)

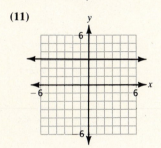

(8)

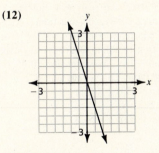

(9) no   **(10)** yes

(11)

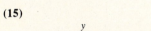

(12)

(13) $m = -1, b = 4$

(14) $m = 1/2, b = 3/2$

(15)

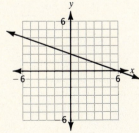

(16) $(0, -16)$

(17) $(5/2, 25/4)$

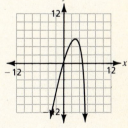

(18) $(-2, 0)$

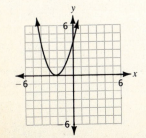

(19) $(-1/2, 5/2)$

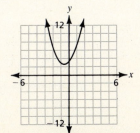

(20) $(-1/6, 13/12)$

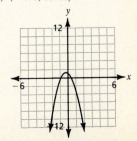

(21)

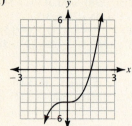

(22)

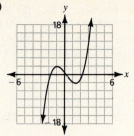

(23)

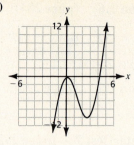

(24)

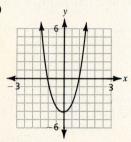

(25)

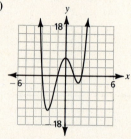

(26) Plot x-axis from $(0, 80)$ to $(279, 80)$ and y-axis from $(140, 0)$ to $(140, 159)$; let x take on values from -7 to 7 in increments of 0.05; $y = x^2 - 4x + 4$; plot $(20x + 140, 20y + 80)$ **(27)** Plot x-axis from $(0, 80)$ to $(279, 80)$ and y-axis from $(140, 0)$ to $(140, 159)$; let x take on values from -7 to 7 in increments of 0.05; $y = 2x^2 - 4$; plot $(20x + 140, 20y + 80)$ **(28)** Plot x-axis from $(0, 80)$ to $(279, 80)$ and y-axis from $(140, 0)$ to $(140, 159)$; let x take on values from -7 to 7 in increments of 0.05; $y = x^2$; plot $(20x + 140, 20y + 80)$ **(29)** $\ln 1808.04 = 7.5$ **(30)** $e^{2.01} = 7.5$
(31) $e^{2.7\ln 56} = 52{,}493.76$ **(32)** EXP(7.5)   **(33)** LN(7.5)
(34) a. 16.1 b. 16.1 **(35)** a. 4 b. -5
(36) a. 2 b. 2 c. -3 **(37)** no   **(38)** $x = 5.73$

(39)

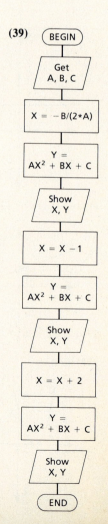

Chapter 8

Section 8.1   (page 302):   **(1)** $(3, 5)$ **(3)** $(1/5, -1/5)$ **(5)** $(1/5, 2/5)$ **(7)** inconsistent
(9) dependent   **(11)** $(3, 5)$ **(13)** $(1/5, -1/5)$ **(15)** $(1/5, 2/5)$
(17) inconsistent   **(19)** dependent   **(21)** $(-2, 3)$ **(23)** $(0, 0)$

Section 8.2   (page 307):   **(1)** $(3, 2, -1)$ **(3)** $(1, 1, 1)$ **(5)** $(3, -3, -3)$ **(7)** $(1/2,$
$-1/2, 0)$ **(9)** $(2/3, -1/3, 2)$ **(11)** $(2, -1, 1)$

Section 8.3   (page 313):   **(1)** $(3, 2, -1)$ **(3)** $(1, 1, 1)$ **(5)** $(3, -3, -3)$
(7) $(1/2, -1/2, 0)$ **(9)** $(2/3, -1/3, 2)$ **(11)** $(2, -1, 1)$

Section 8.4   (page 317):   **(1)** $(2.60, 0.20)$ **(3)** $(3.89, -0.33)$ **(5)** does not converge
(7) $(2.14, 1.71)$ **(9)** does not converge   **(11)** $(1.60, 2.20)$
(13) $(2.70, 0.10)$ **(15)** does not converge

Section 8.5   (page 324):   **(1)** -5 **(3)** -2 **(5)** -2 **(7)** 14 **(9)** 1 **(11)** $(3, 5)$
(13) $(1/5, 2/5)$ **(15)** dependent   **(17)** $(18/17, -27/17)$
(19) $(1/2, -1/2, 0)$ **(21)** inconsistent   **(23)** $(2, -1, 0)$

Section 8.6   (page 331):
(1) $\begin{bmatrix} 2 & 2 \\ 10 & -5 \end{bmatrix}$ **(3)** $\begin{bmatrix} 10 & -8 & 0 \\ 6 & -2 & -2 \\ 0 & 12 & -7 \end{bmatrix}$ **(5)** $\begin{bmatrix} 9 & -1 \\ 0 & 3 \\ 4 & 1 \end{bmatrix}$ **(7)** undefined

(9) $\begin{bmatrix} 38 & 53 \\ 22 & -3 \end{bmatrix}$ **(11)** $\begin{bmatrix} 1 & 0 \\ 0 & 1 \end{bmatrix}$ **(13)** $\begin{bmatrix} 12 \\ 14 \\ 28 \end{bmatrix}$ **(15)** $\begin{bmatrix} 1 & 2 \\ 2 & 13 \end{bmatrix}$

(17) $\begin{bmatrix} 21 & 5 \\ 5 & 10 \end{bmatrix}$ **(19)** $\begin{bmatrix} 2 & -1 & -1 \\ -3 & 4 & -1 \\ 3 & -5 & 2 \end{bmatrix}$ **(21)** $\begin{bmatrix} 5 & 2 & -4 \\ -3 & 0 & 1 \\ -2 & -2 & 3 \end{bmatrix}$ **(23)** $\begin{bmatrix} 1 & 0 & 0 \\ 0 & 1 & 0 \\ 0 & 0 & 1 \end{bmatrix}$

Section 8.7   (page 338):
(1) $\begin{bmatrix} 5 & -4 \\ -6 & 5 \end{bmatrix}$ **(3)** $\begin{bmatrix} 3 & 1 \\ -5 & -2 \end{bmatrix}$ **(5)** $\begin{bmatrix} -1 & 1 \\ 2/3 & -1/3 \end{bmatrix}$ **(7)** $\begin{bmatrix} 1/2 & 2 \\ 0 & -1 \end{bmatrix}$

(9) $\begin{bmatrix} 3/2 & -1/2 & 0 \\ -3/2 & -1/2 & 1 \\ 1 & 1 & -1 \end{bmatrix}$ **(11)** $\begin{bmatrix} 0 & 3/2 & -1 \\ 1 & -1/2 & 0 \\ -1 & -1/2 & 1 \end{bmatrix}$ **(13)** $\begin{bmatrix} 13 & -3 & -4 \\ -7 & 2 & 2 \\ -2 & 0 & 1 \end{bmatrix}$ **(15)** $\begin{bmatrix} -1/3 & 2/3 & 0 \\ -2 & 5 & -1 \\ 14/3 & -31/3 & 2 \end{bmatrix}$

(17) $(2/3, 1/2)$ **(19)** $(1, -1)$ **(21)** $(1/2, 1/3)$ **(23)** $(2, 1, -1)$ **(25)** $(2, 1, -1)$

Chapter Test 8   (page 340):   **(1)** $(4, 2)$ **(2)** $(-3, -5)$ **(3)** $(-3, -5)$ **(4)** inconsistent
(5) $(2, 4)$ **(6)** $(1, 1, -1)$ **(7)** $(-2, -2, -4)$ **(8)** $(1, 2, 3)$
(9) dependent   **(10)** $(2, 1, -2)$ **(11)** $(1, 1, -1)$ **(12)** $(-2, -2, -4)$ **(13)** $(1, 2, 3)$ **(14)** dependent
(15) $(2, 1, -2)$ **(16)** $(5.30, 1.10)$ **(17)** $(2.30, 1.70)$ **(18)** $(3.75, 1.25)$ **(19)** $(1.25, 2.25)$
(20) $(-5.50, -6.50)$ **(21)** -5 **(22)** 11 **(23)** -47 **(24)** $(5/4, 9/4)$ **(25)** $(1, 2, 3)$
(26) $\begin{bmatrix} 3 & -3 \\ 2 & 1 \end{bmatrix}$ **(27)** $\begin{bmatrix} 0 & 1 \\ -2 & -7 \\ 4 & 10 \end{bmatrix}$ **(28)** undefined   **(29)** $\begin{bmatrix} 2 & -5 \\ 18 & -15 \end{bmatrix}$ **(30)** $\begin{bmatrix} 9 & -4 & -3 \\ 9 & 10 & 18 \\ 7 & -9 & -4 \end{bmatrix}$

(31) $\begin{bmatrix} 3 & -5 \\ -1 & 2 \end{bmatrix}$ **(32)** $\begin{bmatrix} 1 & -3 & 1 \\ 2 & -7 & 3 \\ -5 & 19 & -8 \end{bmatrix}$ **(33)** $(-1, 2)$ **(34)** $(2, 0)$ **(35)** $(1, -1, 2)$

(36)

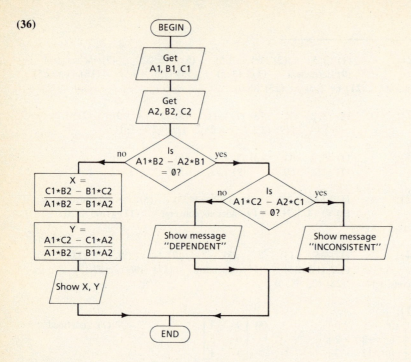

Chapter 9

Section 9.1   (page 347):

(1) Maximize profit; let x = no. boxes of first type, let y = no. boxes of second type, let P = profit; $P = 3x + 2y$; $10x + 20y \leq 110$, $24x + 8y \leq 104$, $x \geq 0$, $y \geq 0$ **(3)** Maximize profit; let x = no. regular boxes, let y = no. premium boxes, let P = profit; $P = 0.20x + 0.50y$; $(3/4)x + (1/2)y \leq 6000$, $(1/2)x + y \leq 4000$, $x \geq 0$, $y \geq 0$ **(5)** Maximize profit; let x = no. 5×7 frames, let y = no. 8×10 frames, let P = profit; $P = x + 2y$; $(1/2)x + (3/4)y \leq 480$, $(1/3)x + y \leq 480$, $x \geq 0$, $y \geq 0$ **(7)** Minimize cost; let x = no. oz. cheddar, let y = no. oz. Swiss, let C = cost; $C = 0.25x + 0.20y$; $210x + 270y \geq 1170$, $300x + 330y \geq 1560$, $x \geq 0$, $y \geq 0$;   **(9)** Minimize cost; let x = no. servings chocolate, let y = no. servings peanuts, let C = cost; $C = 0.30x + 0.20y$; $104x + 26y \geq 234$, $x + 2y \geq 4$, $x \geq 0$, $y \geq 0$ **(11)** Minimize cost; let x = no. cups wheat germ, let y = no. cups dry milk, let C = cost; $C = 0.41x + 0.34y$; $(3/2)x + (1/2)y \geq 1$, $(1/2)x + (3/2)y \geq 1$, $x \geq 0$, $y \geq 0$

Section 9.2   (page 352):

(1)

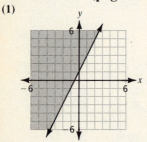

(3)

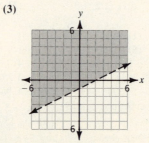

(5)

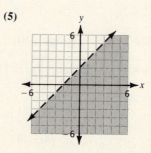

(7)

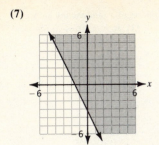

(9)

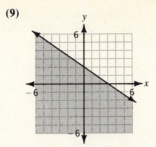

(11)

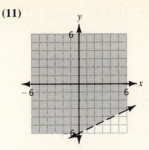

(13)

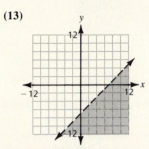

(15)

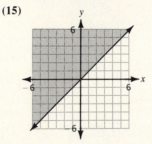

(17)

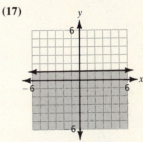

(19)

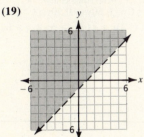

(21)

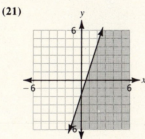

(23)

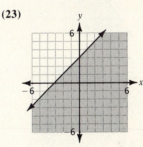

Section 9.3   (page 357):

(1) $x = 2, y = 4$ **(3)** $x = 2, y = 6$ **(5)** $x = 18, y = 0$
(7) $x = 4, y = 0$ **(9)** $x = 2, y = 2$ **(11)** $x = 3, y = 1$

(13) $x = 3, y = 0$ **(15)** $x = 3, y = 3$

Section 9.4   (page 361):

(1) $x = 3, y = 4$ **(3)** $x = 0, y = 4000$ **(5)** $x = 480, y = 320$
(7) $x = 0, y = 4\,8/11$ **(9)** $x = 2, y = 1$ **(11)** $x = 1/2, y = 1/2$

Section 9.5   (page 367):

(1) $x = 2, y = 4, (x_1 = 0, x_2 = 0, P = 22)$ **(3)** $x = 8, y = 0, (x_1 = 0,$
$x_2 = 24, P = 16)$ **(5)** $x = 8, y = 8, (x_1 = 0, x_2 = 0, x_3 = 4, P = 24)$
(7) 8 servings of spaghetti, 0 servings of pizza $(x_1 = 0, x_2 = 1.25, P = 280g)$ **(9)** 8 style 1, 6 style 2, and 6 style 3
$(x_1 = 0, x_2 = 0, x_3 = 0, P = \$460)$

Section 9.6   (page 371):

(1) $x = 3, y = 4 \, (x_1 = 0, x_2 = 0, P = \$17)$ **(3)** $x = 0, y = 4000$
$(x_1 = 4000, x_2 = 0, P = \$2000)$ **(5)** $x = 480, y = 320 \, (x_1 = 0, x_2 = 0,$
$P = \$1120)$ **(7)** $x = 3, y = 0, z = 0, (x_1 = 0, x_2 = 2, x_3 = 4, P = 9)$ **(9)** $x = 4, y = 7, z = 0, (x_1 = 9, x_2 = 0,$
$x_3 = 0, P = 18)$ **(11)** $x = 5/3, y = 0, z = 4/3, (x_1 = 0, x_2 = 0, x_3 = 11/3, P = 14/3)$

Section 9.7   (page 376):

(1) 0 oz. cheddar, 52/11 oz. Swiss $(x_1 = 1170/11, x_2 = 0, C = 52/55$ or
$\$0.95)$ **(3)** 2 servings chocolate, 1 serving peanuts, $(x_1 = 0, x_2 = 0,$
$C = \$0.80)$ **(5)** 1/2 cup wheat germ, 1/2 cup dry milk, $(x_1 = 0, x_2 = 0, C = 3/8$ or $\$0.38)$ **(7)** $x = 4, y = 1,$
$z = 0, (x_1 = 3, x_2 = 0, x_3 = 0, C = 6)$ **(9)** $x = 13/3, y = 1/3, z = 0, (x_1 = 0, x_2 = 19/3, x_3 = 0, C = 14/3)$
(11) $x = 1, y = 2, z = 0, (x_1 = 1, x_2 = 0, x_3 = 0, C = 7)$

Chapter Test 9 (page 377):

(1) Maximize profit **(2)** Let x = no. regular sandwiches, let y = no. jumbo sandwiches, let P = profit **(3)** $P = 0.50x + 0.60y$

(4) $6x + 8y \leq 1000, 2x + 3y \leq 360, x \geq 0, y \geq 0$ **(5)** yes

(6)
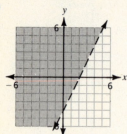

(7)
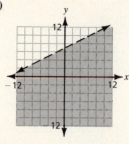

(8)
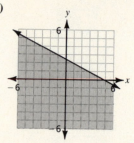

(9)
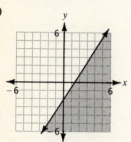

(10)
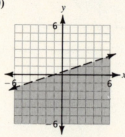

(11) $x = 4, y = 2$ **(12)** $x = 6, y = 0$

(13) $x = 2, y = 0$ **(14)** $x = 6, y = 0$

(15) $x = 6, y = 3$ **(16)** 166 2/3 regular sandwiches, no jumbo sandwiches

(17) Minimize cost; let x = no. regular sandwiches, let y = no. diet sandwiches, let C = cost; $C = 0.80x + 0.90y$; $(1/2)x + (1/3)y \geq 30, (1/4)x + (1/8)y \geq 55/4, x \geq 0, y \geq 0$

(18)

(19) $x = 60, y = 0$ **(20)** $x = 8, y = 0, (x_1 = 0, x_2 = 24, P = 24)$ **(21)** $x = 6, y = 0, (x_1 = 0, x_2 = 8, P = 30)$
(22) $x = 0, y = 7/3, (x_1 = 0, x_2 = 2/3, P = 7)$ **(23)** $x = 1, y = 0, z = 2, (x_1 = 0, x_2 = 0, x_3 = 4, P = 8)$
(24) $x = 0, y = 0, z = 9/4, (x_1 = 7/4, x_2 = 53/4, x_3 = 0, C = 27/2)$

Chapter 10

Section 10.1   (page 385):   **(1)** $\{1, 2, 3, 4, 5, 6\}$ **(3)** $\{R, R, R, W, W, W, W, B, B, B, B, B, B\}$ **(5)** $\{KH, QH, JH, KD, QD, JD, KS, QS, JS, KC, QC, JC\}$
(7) $\{BBB, BBG, BGB, BGG, GBB, GBG, GGB, GGG\}$ **(9)** $\{PN, PD, NP, ND, DP, DN\}$ **(11)** 1/6   **(13)** 4/9
(15) 1/4   **(17)** 1/3   **(19)** 1/3   **(21)** 0   **(23)** 1   **(25)** 3/4   **(27)** 2/3   **(29)** 4/13   **(31)** 1/13

Section 10.2   (page 390):   **(1)** 216   **(3)** 900   **(5)** 450   **(7)** 676,000   **(9)** 60
(11) 24   **(13)** 6   **(15)** 720   **(17)** 720   **(19)** 24   **(21)** 24
(23) 3   **(25)** 120   **(27)** 120   **(29)** 12   **(31)** 15   **(33)** 28   **(35)** 56   **(37)** 45   **(39)** 6

Section 10.3   (page 394):   **(1)** 1/4   **(3)** 1/36   **(5)** 1/16   **(7)** 0.132651   **(9)** 0.2499
(11) 4/169   **(13)** 1/36   **(15)** 1/12   **(17)** 16/663   **(19)** 1/66
(21) 1/11   **(23)** 1/12497500   **(25)** 1/24995000   **(27)** 1/10   **(29)** 1/10

Section 10.4   (page 400):   **(1)** 12   **(3)** 41   **(5)** 13   **(7)** 6 3/4   **(9)** 12 2/5   **(11)** 12
(13) 41   **(15)** 10   **(17)** 4   **(19)** 8   **(21)** none   **(23)** 32
(25) 10   **(27)** 1   **(29)** none   **(31)** median   **(33)** mean   **(35)** mean   **(37)** mean   **(39)** mode

Section 10.5   (page 404):   **(1)** 12, 7   **(3)** 12, 24   **(5)** 43, 33   **(7)** 43, 28   **(9)** 55, 67
(11) 6   **(13)** 106   **(15)** 93 1/3   **(17)** 169 2/3   **(19)** 344 4/7
(21) $\sqrt{5}$ or about 2.24   **(23)** $\sqrt{30.5}$ or about 5.52   **(25)** 2   **(27)** $\sqrt{33\ 2/3}$ or about 5.80   **(29)** $\sqrt{18\ 1/3}$ or about 4.28

Section 10.6   (page 410):   **(1)** 1.50   **(3)** 2.50   **(5)** -2.75 **(7)** -1.43 **(9)** 2.43
(11) 4.7 percent   **(13)** 81.6 percent   **(15)** 6.7 percent
(17) 7.6 percent   **(19)** 33.2 percent   **(21)** 36.7 percent   **(23)** 92.7 percent   **(25)** 62.5 percent
(27) 33.7 percent

Section 10.7   (page 415):   **(1)** 10 min. 4.864 sec. – 10 min. 11.136 sec. and 10 min. 3.872 sec. – 10 min. 12.128 sec.   **(3)** 10 min. 31.12 sec. – 10 min. 42.88 sec. and 10 min. 29.26 sec. – 10 min. 44.74 sec.   **(5)** $842,752 – $877,248 and $837,296 – $882,704   **(7)** $22,916 – $24,484 and $22,668 – $24,732   **(9)** $14,206 – $14,794 and $14,113 – $14,887   **(11)** 0.99   **(13)** 0.95   **(15)** 0.99

Chapter Test 10   (page 417):   **(1)** $\{R, R, R, R, R, G, G, G, G\}$ **(2)** $\{1, 2, 3, 4\}$ **(3)** 5/9
(4) 0   **(5)** 1   **(6)** 65,000   **(7)** 6   **(8)** 12   **(9)** 10
(10) 21   **(11)** 1/4   **(12)** 1/216   **(13)** 9/169   **(14)** 11/221   **(15)** 4/17   **(16)** 7   **(17)** 7   **(18)** 4
(19) median   **(20)** mean   **(21)** 43,31   **(22)** 132   **(23)** 98   **(24)** $\sqrt{306}$ or about 17.49   **(25)** $\sqrt{15}$ or about 3.87   **(26)** -1.5 **(27)** 2.33   **(28)** 0.1 percent   **(29)** 2.3 percent   **(30)** 62.5 percent
(31) 108 – 112   **(32)** 77.02″ – 78.98″   **(33)** 161.08 lbs – 168.92 lbs   **(34)** 0.98   **(35)** 0.98

(36)

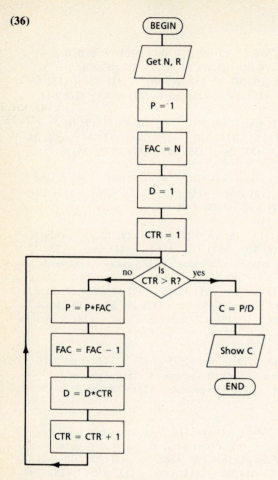

Appendix   (p. 424):

(1) 19   **(3)** −4   **(5)** −3   **(7)** −24   **(9)** −5   **(11)** 16   **(13)** −19   **(15)** 4   **(17)** 84
(19) −65   **(21)** −80   **(23)** −165   **(25)** −6   **(27)** 13   **(29)** 4   **(31)** −17

Index

Acknowledgments

All photos not credited are the property of Scott, Foresman and Company.

Chapter 1

22 The Granger Collection, New York

Chapter 3

68 Bell Laboratories
117 The Bettman Archive

Chapter 4

167 Culver Pictures

Chapter 5

168 Kathy Cunningham
207 The Bettman Archive

Chapter 6

208 Jean-Claude Lejeune
252 The Bettman Archive

Chapter 7

295 Culver Pictures

Chapter 8

343 The Bettman Archive

Chapter 9

344 Lynn M. Stone/ANIMALS ANIMALS
380 The Bettman Archive

Chapter 10

421 The Granger Collection

Front endsheets, left to right:

The Granger Collection, New York
The Bettman Archive
The Bettman Archive
The Granger Collection, New York

Back endsheets, left to right:

The Bettman Archive
Culver Pictures
The Bettman Archive
The Bettman Archive

MATHEMATICAL TIME LINE

Leonhard Euler
1707–1783

Carl Friedrich Gauss
1777–1855

Ada Byron
1815–1852

George Boole
1815–1864

Eighteenth Century

1703 Leibniz published a description of the binary system and generalized the idea for other bases. **(pp. 42, 71, 119, 318)**

1733 Abraham De Moivre wrote a pamphlet in which he formulated the concept of the normal curve. **(pp. 405, 421)**

1734 Leonhard Euler introduced the $f(x)$ notation for functions. **(p. 295)**

1750 Gabriel Cramer popularized the use of determinants (not the modern notation) for solving systems of linear equations. **(p. 318)**

1761 Johann Heinrich Lambert proved that π is irrational. **(p. 39)**

1796 Carl Friedrich Gauss discovered that a 17-sided polygon can be constructed using only compass and straight edge. **(pp. 308, 343)**

Nineteenth Century

1824 Neils Henrik Abel proved that the general quintic equation is not solvable algebraically. **(p. 235)**

1829 Peter Gustav Lejeune Dirichlet gave a modern definition of a function. **(p. 254)**

1831 Evariste Galois established the conditions under which an equation is solvable algebraically. **(p. 235)**

1834 Charles Babbage began work on his analytical engine. **(pp. 2, 167)**

1842 Augusta Ada Byron published her notes on Babbage's analytical engine. **(pp. 2, 71, 167)**

1854 George Boole published *The Laws of Thought*, developing logic and Boolean algebra. **(pp. 74, 101, 117)**

1857 Arthur Cayley originated the theory of matrices. **(p. 326)**

1874 Georg Cantor published his first paper leading to the founding of set theory. **(p. 24)**